"In taking us from 'castle' to 'condominium', Rob Harris takes us on a journey that might be seen as circular in nature. We start with private banks in grand residences and arrive in the era of home working assisted by Artificial Intelligence. In moving from 'market' to 'factory' to 'corporate' to 'digital' – the four office ages described in the book – we are given a rigorously researched expert view of not just how the office and office work has developed over time but how it might fare in the future."

Jeremy Myerson, *Professor Emeritus, Royal College of Art; Chairman, WORKTECH Academy*

A History of the Office and Office Work

This book brings together the office and office working as an economic construct, as a business function, as a building type and as a mode of work. It explores social and economic relations, and it traces the evolution of work and the environment and equipment needed to support it over the last 400 years. In so doing, it helps to fill a void in popular understanding of the office.

The role of the office has been examined over the past three years in ways that it has never been examined before. Whilst the existential crisis arising from COVID-19 has settled somewhat, there remain big questions over the future of office workers and office real estate. This book is a timely, well-researched and fascinating contribution to the debate.

Rob Harris explores how the nature of work has changed, and continues to change, placing the events surrounding COVID-19 into a longer-term perspective and asking what is the future of the office? Will it go through a radical reinvention? How will office work evolve in the future? Whither office buildings? The book emphasises the continuum of change and that today's events are simply the latest stage of change, rather than something entirely novel.

This book will be of interest to anyone with a stake in the built environment, whether as an investor, occupier, owner, manager or advisor. It is equally accessible to those with a non-technical background, including students on courses such as economic history, economics, geography, real estate and urban planning.

Rob Harris is a consultant and analyst in the commercial real estate sector, where he has spent over four decades advising developers, investors, occupiers and public sector bodies. He established Ramidus Consulting Limited in 2003 as a specialist, independent built environment research and advisory business. Rob presents widely on a range of property market issues.

A History of the Office and Office Work

From Castle to Condominium

Rob Harris

Routledge
Taylor & Francis Group
LONDON AND NEW YORK

Designed cover image: Wikimedia Commons

First published 2025
by Routledge
4 Park Square, Milton Park, Abingdon, Oxon OX14 4RN

and by Routledge
605 Third Avenue, New York, NY 10158

Routledge is an imprint of the Taylor & Francis Group, an informa business

British Library Cataloguing-in-Publication Data
A catalogue record for this book is available from the British Library

ISBN: 978-1-032-94891-1 (hbk)
ISBN: 978-1-032-91397-1 (pbk)
ISBN: 978-1-003-58220-5 (ebk)

DOI: 10.1201/9781003582205

Typeset in Times New Roman
by Newgen Publishing UK

This book is dedicated to DEGW whose founders and colleagues helped shape a new way of understanding offices and the relationship between people, work and place.

Contents

Figures

About the author

Rob Harris is a researcher and advisor in the commercial real estate sector, where he has four decades of experience. Rob started work at design practice DEGW in the early 1980s, and he was director of research at developer Stanhope Properties plc. in the 1990s. He established Ramidus Consulting Limited in 2003 as a specialist, independent built environment research and advisory business. Ramidus provides evidence-based research and advice for developers and investors, for public sector urban planners and for industry bodies including the British Council for Offices, Institute of Workplace & Facilities Management, Investment Property Forum and the Royal Institution of Chartered Surveyors.

Rob's experience and interests extend from the workplace to the city region. He is particularly interested in the relationship between work and built space: how social, economic and technological change impact on how organisations utilise buildings and workspace. He has conducted many innovative studies anticipating trends in demand to help clients manage risk in a highly uncertain environment.

Rob has published numerous papers on real estate matters and has previously published two books: *London's Global Office Economy: From Clerical Factory to Digital Hub* (Routledge, 2021) and *Property and the Office Economy* (2005). He is currently Chair of the Property Research Trust; Chair of the British Council for Offices Research Committee; and a Visiting Fellow at the Henley Business School (University of Reading).

Foreword

I have always been puzzled as to why the office has never received the cultural and economic attention that historians routinely give to other types of building. It's such a large part of our lives – a place where we spend so much of our time – and yet it has not been subject to the kind of forensic historical analysis afforded to, say, airports or opera houses, factories or department stores.

Our city skylines are today dominated by office towers and our transport systems built around ferrying workers to and from them. But despite all this, the office appears to be hidden in plain sight from public and professional consciousness – and badly too, like a large elephant behind a small tree. Only when the global pandemic hit in 2020, and the office was declared out of bounds for a short and memorable period, did we really acknowledge its existence – and immediately question its future.

One might attribute some of this relative neglect to an unwillingness on the part of some writers to untangle the complex web of commercial, social, environmental and technical interactions that bind the office into its wider economic content – it's just too big and unwieldy a beast. Alternatively, the office has been viewed as too prosaic and banal a subject for historians despite its ubiquity – too 'thin' to be worthy of in-depth consideration.

Whatever the reasons might be, there has been for some time a large gap for a detailed history of the office, which does not start with modernism and the late-nineteenth century burst of modern technologies such as the typewriter, electric light bulb, adding machine and the elevator, but stretches back much further over more than three centuries to the coffee houses and financial exchanges of the late seventeenth century.

In this book, Rob Harris not only explores 'quill pen and coal fire' roots of the office but also places the development of the office building in relation to the evolution of the economy, business and work. This is one of the author's two main reasons for writing the book. The other is to provide a firm contextual base on which the current debate about the future of work in the post-pandemic era can be more effectively conducted. It becomes so much easier to evaluate emerging hybrid trends around 'the where of work' when you have a clearer picture of how the office first came to be so central to working life, the planning of cities, and the destiny of economies.

In taking us from 'castle' to 'condominium', Rob Harris takes us on a journey that might be seen as circular in nature. We start with private banks in grand residences and arrive in the era of home working assisted by Artificial Intelligence. In moving from 'market' to 'factory' to 'corporate' to 'digital' – the four office ages described in the book – we are given a rigorously researched expert view of not just how the office and office work has developed over time but how it might fare in the future.

Jeremy Myerson
Professor Emeritus – Royal College of Art
Chairman – WORKTECH Academy

Acknowledgements

Several people have advised me during the preparation of this book, providing insight, critical commentary and inspiration in equal measure. I would like to acknowledge the following people.

Erik Brown
Joanna Eley
Adrian Forty
Kerry Harris
Peter McLennan
Jeremy Myerson
Nigel Oseland
John Parman
Andrew Rabeneck
James Woudhuysen
John Worthington

1 Perspectives on offices

This book has been written for two reasons. Office activities have grown over time to the point where they account for 30–40% of all jobs, and yet despite this growth, their collective impact is poorly understood. Both academic writing and broader socio-economic histories have neglected offices compared, say, to the history of manufacturing. The first reason, therefore, is to address this neglect of a fundamental element of the economy and society with an account of the origin and development of offices and office work.

Secondly, it addresses the enormous uncertainty over the future of the office and office work brought about by a combination of the COVID-19 pandemic, and the prospect of artificial intelligence and other contemporary trends in office work. Much has been written in recent years about the future of offices (even suggesting an existential crisis), and yet most of the commentary is acontextual. Today's changes are treated in isolation rather than as part of the broader sweep in the evolution of office work. The book provides the context against which today's trends can be properly evaluated.

Existential crisis and more

In March 2020, workers in all corners of the economy were ordered to stay at home in the first of a series of 'lockdowns' aimed at minimising the spread of the COVID-19 pandemic. Around one-third of those workers were, up to that point, dependent upon an office for their work. Almost overnight, they found themselves working from home – interacting with colleagues digitally and experiencing the vagaries of the newly popular 'virtual meeting'. This was an unprecedented situation, which, in the minds of many observers, left offices in an existential crisis; the simple narrative being, if we can all work from home effectively, then why do we need offices at all?

At the time of writing, in mid-2024, employers are still struggling with the impact of what has become known as 'hybrid working'. Most people have not returned to the office full time; many are there no more than 3 days a week. A small number of companies have given up real estate altogether and gone entirely 'remote'. Meanwhile, the real estate industry – designers, developers, investors, managers – is busy trying to reinvent the 'office experience' to attract reluctant commuters back to the office workplace.

Despite the shock of the pandemic, it is a fact that the nature of office work and the office workplace were undergoing a period of rapid change before 2020. Strong forces were acting on widely held assumptions about what offices were for, how they should operate and how they should be designed and managed. At the centre of the debate sat the sometimes liberating impact of technology. Agile working – workers using laptops and smart phone technology to release them from the traditional physical ties to the office – was growing in popularity and

DOI: 10.1201/9781003582205-1

practice before the pandemic. But there were also strong technological, business and socio-cultural trends, which were changing work.

Technological. The 1990s and 2000s saw transformative change in office work caused by the adoption of office technology. Desktop and laptop computing, the internet, email, smart phones and all the connective paraphernalia have utterly changed how workers engage with their work, their employer and their workplace. What work is, where and when it is undertaken and the tools which we use would be unrecognisable to an office worker time traveller from even the 1970s. Without the technological changes, the drastic transition precipitated by lockdowns would not have been possible.

The digital revolution has been a positive force not only in enabling modern workstyles, but also in transforming work itself with the creation of new jobs. However, it is clear that potentially a large number of jobs exist in the long shadow of artificial intelligence (AI), and will disappear as suddenly as did the clerks, file managers and secretaries of yesteryear.

As recently as late-2022, OpenAI released ChatGPT, and suddenly 'generative AI' became a household term. AI could lay waste to many of the skilled jobs currently regarded the preserve of trained, relatively well-paid professionals, as well as spreading to enormous types of activities across all sectors of the economy. However broad and deep it travels, there are two generic impacts. The first is that AI will destroy jobs through 'automation'; many jobs undertaken by humans will simply disappear forever. The second impact – often presented on the back of earlier phases of innovation – is that alongside the destruction there will be job creation.

There can be no escaping that while the first impact will reduce the quantum of demand for workplaces, the second impact could lead to an entirely new workplace design agenda, undermining the current models of designing, delivering and occupying real estate. How much office space will we need? What will it look like? And how will we engage with it?

Business. The large, multilayered, bureaucratic behemoths that bestrode the twentieth century economy are being challenged by new economic models and new forms of work. The vertical integration of supply chains is evolving with more specialist inputs and at the same time work processes are being fundamentally reconfigured by technology. The result is that the 'shape' of organisations is changing. In the twenty-first century corporatism is ceding to a more variegated economic landscape of large and small businesses, engaging in complex business ecosystems: the knowledge economy (see below).

In the corporatist era, it was the role of management to instil a common corporate culture on workers, almost a sense of identity, as if being part of a sports team: all aligned to a shared set of goals. The problem is that in the knowledge economy the 'managerial demiurge' loses its meaning. Knowledge workers do not need to be corralled and closely supervised. Many can work independently, and 'self-manage'. Braverman's distinction between conception (thinking, planning) and execution (doing) of work[1] no longer holds sway because there is much less 'doing'. There is, in short, a growing mismatch between the traditional imperatives of management and the new aspirations of workers.

As businesses lose their internal bureaucracies and departmental divisions, their management function becomes less about control and more about support. Static office environments become dynamic meeting places in which workers gather not to fill the hours between 0900 and 1700, but to interact, to concentrate, to learn, to collaborate, to meet and to socialise – not unlike the markets and exchanges of the eighteenth century. What does this mean for real estate and the traditional ways of designing, delivering and managing office space? Should we be thinking of the office less as a castle and more as a condominium?[2]

There is then the nature of office work itself. While much workplace commentary highlights the merits of workplaces in terms of creativity, collaboration, innovation and serendipity at the water cooler, it remains the case that much office work is dull, repetitive, reactive and routine. Even for relatively well-paid jobs undertaken by qualified workers, office work is often lacking in meaning. This phenomenon was captured recently by David Graeber in his book *Bullshit Jobs*.[3] It is certainly the case that much office work is not a daily journey of joyous innovation and creativity, but rather the daily chore to be endured.

Similarly, if a job is reduced to a transaction between employer and employee, then it simply joins the other items on the corporate potential savings list. Allied to the potential impact of AI, questions must be raised about the future of office work and the nature of the businesses employing office workers.

Socio-cultural. The question 'what is work?' is one that grows in importance as emerging evidence suggests that attitudes to work are shifting; in particular, there seems to be a widespread questioning of what has traditionally been referred to as the Protestant work ethic: "*the determination to pursue a moral code, to find salvation through work*".[4] Under this system, all manner of privations could be tolerated on the grounds that the very effort of work itself would bring personal satisfaction as well as wider social benefit. There is now a widespread questioning of the value and purpose of work. The pandemic brought this general trend into sharp focus with a dramatic rise in workers choosing not to work. While the level of 'quiet quitting' has reduced somewhat, the long-term trend remains.

In March 2024, in the UK, it was announced that over one-fifth of adults, around 9.2m economically active people, are not actually in work, nor looking for work. Further, only around one-third of those not in work cite long-term illness as the main reason for not working.[5] Attitudes to work are shifting, among both older and younger workers, with a greater emphasis on personal well-being, freedom to choose and reduced dependency on a salary to avoid absolute poverty. Workers today are far more vocal about the ethics and practices of their employers; they are less prepared to accept working conditions that do not align with expectations and they are less prepared to commit to long commutes and a 5-day working week. Employers are being challenged to introduce much greater flexibility in where and when people work, the facilities they provide in the workplace; their attitudes to mental health and the nature of the employment contract.

Alongside a rise in those not working, there has been a steady rise in the self-employed. The rate of self-employment has doubled since the 1980s across the economy, and now accounts for 15% of workers. The vast majority of the self-employed (85% in 2019) are sole traders with no employees (44% of the self-employed had employees in 1975, compared with just 15% in 2019). By the end of 2019, there were nearly 4 million solo self-employed workers in the UK, up from 2.3 million in 2000. These are all workers, not just white-collar workers; nevertheless the numbers indicate a broad trend away from corporate jobs to self-management.

Perspectives on offices

Technological, business and socio-economic trends have been transforming offices in recent years, but as the following paragraphs demonstrate, offices remain largely neglected from a number of perspectives; as an economic driver, as a facet of the city and as a defining element of the knowledge society.

This book addresses this neglect by presenting a history of offices and office work from the perspective of the interaction between the economy, business, work and the workplace. It seeks to demonstrate how these elements have co-evolved, as changes in the economy have driven changes in work, which, in turn, have driven changes in the built environment, specifically offices.

Perspective #1: the office as an economic driver. Office buildings have become such a ubiquitous part of the urban landscape that they are largely taken for granted. Rarely do they draw the attention of city dwellers or tourists in the way that say, cathedrals, museums and palaces do. In the main, offices have a more utilitarian feel; apart from the odd architectural statement, they exude anonymity as one office block blends into the next, into the next. Offices have also become such a central part of our economy and culture that their role and functionality are rarely discussed.

Despite their scale, offices as we know them today have had a relatively short life. In their modern format, offices are no more than 150 years old. At the time of writing, the author has been working for 40 years, which means that modern offices have been around for just four such career spans! Seen this way, offices are a relatively modern phenomenon, despite the fact that they are so deeply embedded in our lives.

Figure 1.1 provides some statistical context to the scale of growth in white-collar work, or the office economy, in the UK not only in absolute terms, but also in relation to a selection of other economic sectors. Financial, professional, technical, health and education jobs grew comparatively slowly until after the Second World War, with growth picking up in the sixties and seventies, before accelerating rapidly with the advent of office technology. It should be noted here that the financial, professional, technical, health and education sector includes non-office jobs in health and education, but it excludes office jobs in information & communications, public administration and elsewhere.

A similar pattern in the shift in jobs also took place in the United States. Table 1.1 shows the shift from goods-producing jobs to service-producing jobs between 1850 and 2023.[6] While the latter includes more than white-collar work (such as retail and leisure), the overall pattern is clear. Service-producing jobs rise from less than one-fifth of the total in 1850 to four-fifths of all jobs in 2022. Interestingly, while the proportion of goods-producing jobs continued to grow until 1982, this has halved since the 1980s.

As we shall see below, this phenomenal transformation in the nature of the economy and work passed by almost unnoticed. While the decline of manufacturing became the subject of intense academic attention, public policy and public discourse, the rise of white-collar work was largely ignored in each of these respects.

Economist and statistician Colin Clark published his *Conditions of Economic Progress* in 1940, in which he examined how economies evolve over time, passing through phases when the economy is dominated in turn by primary, secondary and tertiary production, broadly representing agriculture, manufacturing and service activities, respectively.[7] Clark's broad notion of progression towards economic maturity, was reinforced a couple of decades later by American economist Walt Whitman Rostow, whose academic legacy included the classic *The Stages of Economic Growth: A Non-communist Manifesto* (1960), which set out five progressive steps in economic development, culminating in the *age of mass-consumption,* featuring growing disposable income, consumerism and dense urbanisation.[8]

Both Clark and Rostow convey a sense of 'progress' towards 'higher' levels of economic activity. William Goodwin gave this dynamic a business dimension in the mid-1960s by discussing office activity on terms of 'management centres' – cities in which "*there is a concentration of headquarters offices of nationally important companies. It is a place apart from the production centers*". He then isolated the workers as a key dynamic of management centres, in so far as they are managers "*who sell their managerial talents irrespective of the nature of the company that employs them. A management center is a reservoir of managerial talent available for hire*".[9] Offices are thus seen as a higher-order activity.

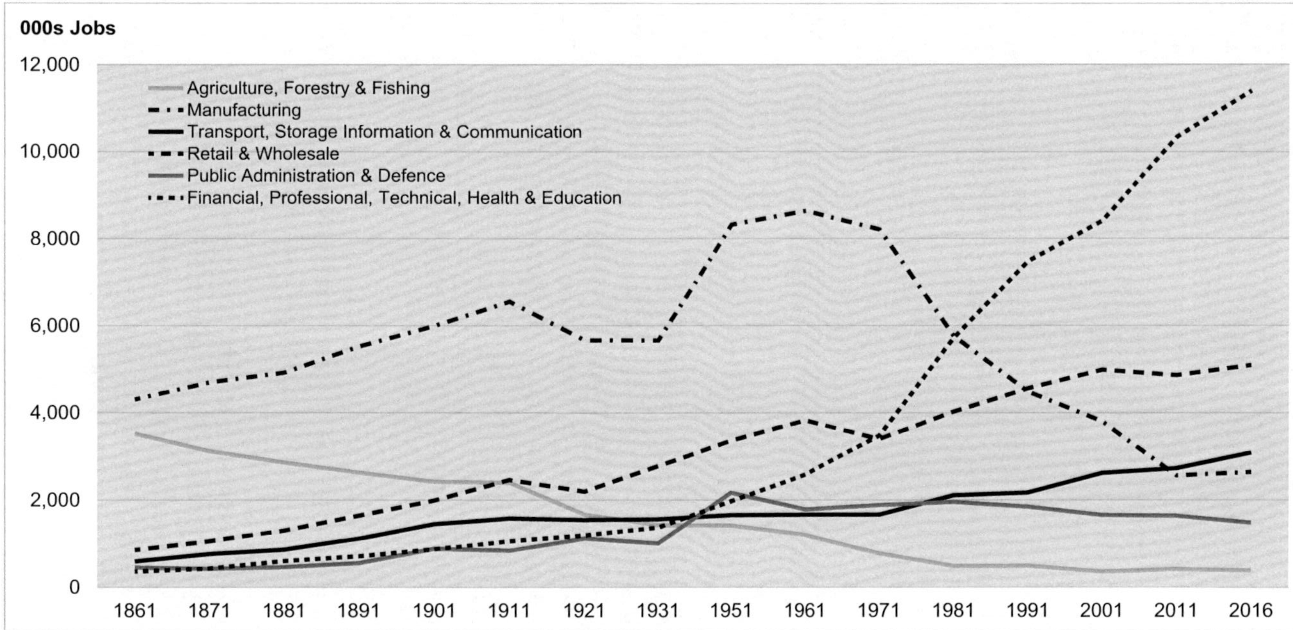

Figure 1.1 The rise of white-collar work in the UK, 1861–2016.

Sources: Thomas R & Dimsdale N (2017) A Millennium of UK Data, Bank of England OBRA dataset; Broadberry S; Campbell B; Klein A; Overton M & van Leeuwen B (2015) British Economic Growth, 1270–1870 Cambridge University Press.

Table 1.1 Distribution of employment by major sector, United States, 1850–1982

Year	*Percent of jobs*		
	Agriculture	*Goods-producing*	*Service-producing*
1850	64.5	17.7	17.8
1860	59.9	20.1	20.0
1870	50.8	25.0	24.2
1880	50.6	25.1	24.3
1890	43.1	28.3	28.6
1900	38.0	30.5	31.4
1910	32.1	32.1	35.9
1920	27.6	34.6	37.7
1930	21.8	31.7	46.6
1940	18.3	33.1	48.6
1952	11.3	35.5	53.3
1962	7.8	33.1	59.1
1972	4.4	31.4	64.2
1982	3.6	27.2	69.2
2012	1.4	12.6	80.0
2022	1.3	12.8	80.3

Sources: 1850–1982, Urquhart (1984); 2012 and 2022, US Bureau of Labor Statistics (2023).

And so it was, as the 1970s beckoned, and with the exception of a small number of key texts, there remained a primitive understanding of offices as a distinct facet of the economy as a whole. It had been largely ignored even though offices had over the previous three-quarters of a century come to dominate many city centres both in terms of land use and employment. The following two quotes from eminent geographers in the mid-1970s sum up the situation well.

> Peter Daniels ... *our understanding of the office system is in its infancy compared with progress in our knowledge of the role of retail or industrial land uses, for example, in the function and structure of cities.*[10]

> Harold Carter … *To a great degree our understanding of the operative processes and consequent usage patterns in the city centre is completely inadequate.*[11]

Perhaps the seminal work of the time was Daniel Bell's *The Coming of Post-Industrial Society.*[12] Bell clearly saw the socio-economic significance of the declining role of manufacturing and the rearrangement of the industries in which people work, the emergent pattern of occupations and the kind of work undertaken. He cited the rise of white-collar work as illustrative, suggesting that the concept of a post-industrial society could be understood in terms of "*five dimensions, or components, of the term*", as follows.

- economic sector: the change from a goods-producing to a service economy;
- occupational distribution: the pre-eminence of the professional and technical class;
- axial principle: the centrality of theoretical knowledge as the source of innovation and of policy formulation for the society;

- future orientation: the control of technology and technological assessment and
- decision-making: the creation of a new intellectual technology.[13]

Bell saw the combined impact of these changes as constituting the emergence of a new economic landscape, what he called the 'post-industrial' economy. But Daniels' focus was on mega trends in society rather than more specific trends within the economy; let alone within the office economy.

There also emerged a whole series of studies that had as their focus the changing nature of business, and particularly the growing role of the service and office activities, not least because of the policy consequences of the coincident decline in manufacturing activity. Gershuny (1978)[14], Gershuny and Miles (1983)[15] and Harris and Taylor (1978)[16] all produced excellent work. However, as late as 1988, Howells and Green were able to state that the

> *problem is that the emphasis on manufacturing as the main locomotive force in the economy, which has persisted until very recently, has left us with a very limited understanding of the service sector in relation to economic development.* [17]

They highlighted a research tradition in which service activity was painted as something which passively follows population and manufacturing growth; but they proposed that it was "*no longer an adequate conceptualisation of the place of the service sector in national economic development*", and that evidence was increasingly suggestive that services were a "*leading rather than lagging element in economic growth*".

The most startling aspect of Howells and Green's comments is that they were made in the 1980s. To this point there was no theoretical framework for the position of office activity in the wider economy; there was no economic policy to encourage its growth; there was no spatial planning to accommodate its physical needs.

John Marshall suggested that the growing role of information in the economy led to "*the development of an integrated office-based sector of employment*", in which "*the office block has replaced the factory as a symbol of modern urban development.*"[18] Really? This is a remarkably naïve comment given the scale of office activity in the late-1980s – written two years *after* Big Bang in the City – and reveals the paucity of academic understanding of the role of offices in the modern urban economy.

Increasing numbers of studies of metropolitan areas empirically evidenced the 'glue' that held the companies so tightly together, in terms of the location of markets (clients and suppliers), highly skilled labour markets and the agglomeration benefits associated with minimising transaction costs.[19] Other studies showed how office-based firms created competitive advantage in urban areas including creativity and innovation, spatial proximity, flexibility, timeliness of delivery, scope of services offered and adaptability to client needs – all a long way from the simpler explanations based on classical economics of earlier times.[20]

While our understanding of offices remained in its infancy, however, there was emerging the first tentative thinking about the potential role of offices in a post-industrial, or knowledge, society.

The office infrastructure of the nineteenth century included the typewriter, telephone and telegraph; to be replaced in the mid-twentieth century office by the calculator, copier, fax machine and word processor. Perhaps *the* defining feature of the final decade-and-a-half of the twentieth century was the arrival and then ubiquitous availability of digital technology to almost everyone. The unfolding technologies touched all aspects of society in a truly fundamental manner. The first computers in offices were large, centralised, clunky machines; they required dedicated space (due to the noise, heat and miles of cabling) and they were ruinously expensive. But by the 1970s large businesses were turning to the new machines to automate office processes such as customer billing, payroll, inventory and financial statements.

Perspective #2: the office as a facet of the city. Well into the second half of the twentieth century, there remained among economists, geographers and planners only the crudest understanding of the internal dynamics of the office economy: offices as a distinct facet of the city were simply not recognised.

A lone beacon of insight was John Rannells' 1956 study of central Philadelphia, *The Core of the City*.[21] This work is important because of its emphasis on the role of uses, or activities, in shaping the morphology of the city. For example, Rannels made the critical distinction between broad land use changes associated, in classical economics terms, with utility maximisation as a factor of access, and the influence of operational decisions on the level and structure of final demand for space:

> *It is necessary to take in somewhat more than relates directly to land use, since changes in demand for space or location may come about as secondary results of business decisions or consumer preferences which, in themselves, have no concern with questions of land use.*[22]

Rannells shifted the emphasis sharply away from broad economic factors to the internal dynamics of the office economy, specifically the operational requirements of the companies themselves, and their linkages to one another. As he argued, the locational characteristics of activities are "*visible manifestations of the many systems of activity into which the city's life is organised.*"[23]

The *Core of the City* was very quickly followed in 1959 by Edgar Hoover and Raymond Vernon's seminal work, *Anatomy of a Metropolis*, in which Chapter 4 was devoted to "*the white-collar corps*", one of the earliest attempts to separately articulate the *raison d'etre* of the office economy. Noting that nearly a fifth of the jobs "*are the accounting and legal services, the advertising and employment agencies, the architects and engineering services, and a variety of similar activities*" the authors noted that such firms "*look for much of their business to the office community and to one another*". They suggested that while there is "*a temptation to think of this clustering of office work as being based on some nebulous preference of executives for the fast tempo and bright lights of the big city*", the "*forces that have produced the clustering can be much more objectively defined*".[24]

As well as through economic and business linkages, offices were also defined in spatial terms. Edgar Horwood and Ronald Boyce,[25] separated the core of the central business district (or CBD, the area where the office economy is at its most intense) from the frame in which it was located. Similarly, French geographer Jean Gottmann's *Megalopolis* addressed *The White Collar Revolution*, and the process and the consequences of its rapid growth. He described the centripetal forces bringing about the clustering of large offices in downtowns, among which were "*the need for good and reliable information*" and "*the need for a pool of labor*". Gottmann refined his thinking on white-collar work in a later paper, *Why the Skyscraper?*, suggesting that those sectors of employment which were

> *regarded by the classical economists as external activities because they did not seem to be important elements in the economic process (production through processing and distribution to consumption) ... are now becoming major sectors of employment.*[26]

Finally he hinted at the nature of the matrix that binds the office economy together in the city centre.

> *The daily work of decision making requires consultations with many specialists; the specialists cannot be expected to gather where there are only one or two customers, but rather where there are several thousand. Also, the specialists prefer to be near one another.*[27]

This was written in 1966, just 15 years before the first IBM PC; ten years before financial deregulation in the US and just 20 years before Big Bang in London. Why were offices, or the 'office industry' as Gottman called it, still seen as so novel? Despite these few exceptions, throughout the post-war period, there remained a myopic focus in academia, planning and economic policy on the decline of manufacturing and a widespread neglect of the rise of the service industries, particularly the office economy.

Beyond academic work, the office as a distinct urban function has also been largely neglected in mainstream publications tracing urban history. In London, in many senses the birthplace of the modern office economy, offices have not featured as a distinctive part of its written history. Scanning the contents lists and indices of most histories of London reveals barely any particular treatment of office buildings or the activities therein. Kynaston's four volume history *The City of London* has no index reference to offices; neither does Ackroyd's *London: the Biography*; Gray's *History of London*, Jenkins' *Short History of London*, Porter's *London: a Social History* or Sheppard's *London: a History*.

The first comprehensive work to address the office economy as a distinct urban function was that by Peter Cowan and colleagues (of University College London and the London School of Economics), who traced the office economy back to the late-nineteenth century, describing how, as the industrial revolution reached its climax, "*the role of the city changed utterly*".[28] As the industrial processes evolved and changed, so too did their administration and governance. Office activities grew in importance and began to assume their own, distinct identity in the urban environment. Cowan and colleagues argued that until the industrial revolution,

> *the office function was generally an integral part of other facets of urban life. But it is a commonplace that, as an organisation or an organism grows in complexity, certain parts or functions become separate from others. The office function of communication and control is just such a case. When an organisation is fairly simple, as in a cottage industry or a small workshop, the office function is integral with the production process itself. But as the enterprise grows and increases in complexity ... so the office function begins to separate off.*[29]

The introduction of the telephone and other contemporary innovations led to new forms of organisational structure and, as Cowan and colleagues observed, helped create an office economy that was no longer simply part of other facets of urban life but a facet in its own right.

> *But if we look back to the end of the nineteenth century we can see that the office function had become well established as a separate entity in the city: new buildings were being constructed to house it; more and more of the labour force were working in offices, and women especially were beginning to go into office work in large numbers. The outlines of the office as we know it were laid down.*[30]

Cowan's work was followed by Peter Daniels' milestone book of 1975, *Office Location*,[31] which took an activity-based approach to untangle the web of relationships within the office economy in order to explain its physical expression. It helped describe the glue that held offices tightly together in the CBD, and it examined the nature of change that led to new demands being placed on buildings.

Perspective #3: the office in a knowledge society. It was at this point that offices began to lose their association with 'paper factories', or 'white-collar factories' and start a new association with information work or, more particularly, knowledge work and more widely the emergence of a knowledge-based society. One of the earliest examinations of the impact of knowledge in the economy was by Fritz Machlup in the US in the 1960s. Alongside Peter Drucker, Machlup described the emergence of innovative industries and their growing impact on economic performance.[32] Similarly, Giuliano described how advances in solid-state circuit components and microelectronics led to the computer becoming smaller and cheaper.

> *Remote terminals, consisting of either a teletypewriter or a keyboard and a video display, began to appear, generally tapping the central processing and storage facilities of a mainframe computer. There was steady improvement in the cost-effectiveness of data-processing equipment. All of this was reflected in a remarkable expansion of the computer industry.*[33]

As Giuliano said, the late-1960s and the 1970s also saw the advent of inexpensive copiers, minicomputers, small and affordable private automated branch exchanges (electronic switchboards), the word processor (the typewriter's successor) and then, toward the end of the 1970s, the microcomputer. Giuliano was describing the beginning of the digital revolution; but one of the earliest to recognise the implications was Peter Drucker who, in Chapter 12 of his *The Age of Discontinuity* (1969) referred to the 'knowledge economy', and of knowledge (systematic, purposeful, organised information) being the central economic resource of the key industries of tomorrow.[34]

During the 1970s, there emerged a growing interest in the knowledge economy as innovations in information technology opened up new markets in telecommunications, computing, media and entertainment. Indeed a fourth stage of 'progress' was identified – the quaternary economy or sometimes post-industrial economy. This new stage was seen to be defined by the prevalence of 'knowledge work', where the production, analysis and dissemination of knowledge and information supplant production as the driver of economic growth. The launch of the IBM personal computer (PC) in 1981 was a signature moment: the PC represented the democratisation of computing power – placing it in the hands of every worker rather than a few experts.

Among the earlier exponents of the influence of technology as a driver of structural change in urban morphology was Brian Berry who, in 1973, spelt out the concept of technology as a decentralising agent, suggesting that the new communications technologies would "*concoct a solvent*" to dissolve "*the core-oriented city in both time and space, creating what some refer to as urban civilisation without cities.*"[35] There emerged a whole series of studies that had as their focus the impact of technology on the economy and society.[36]

The role of information in the context of the post-industrial society was explored widely in the 1980s. Jean Gottmann's *The Coming of the Transactional City*[37] and Thomas Noyelle and Thierry Stanback's *The Economic Transformation of American Cities*[38] are two of the more widely read. Mark Hepworth's paper *The Information City*[39] and *Planning for the Information City*[40] examined the office economy in terms of the interrelated roles of information and the new information technology in the process of economic development. John Marshall's *Services and Uneven Development* suggested that the growing role of information in the economy led to "*the development of an integrated office-based sector of employment*", in which "*the office block has replaced the factory as a symbol of modern urban development*".[41]

William Coffey argued that "*research on services was conceptually stagnant for several decades*" and that it was only in the later 1980s and into the 1990s that producer services (aka the knowledge economy) "*were finally recognised as important economic phenomena in their own right – no longer simply 'the nonproductive remainder' of the economy*".[42]

> *The 'modern' era of research on services dates only from the early- to mid-1980s; it is only since this time [that researchers] have begun to analyse in a concerted and systematic way questions concerning the economic role and the locational dynamics of the full range of services activities, including producer services.*[43]

Coffey also made a direct link between the economic activities themselves and the buildings they occupied (offices), with the critical link being information technologies.

> *Existing structures are often technologically insufficient to support the new information-related functions of firms, producing a demand for new office buildings capable of meeting modern technological and spatial conditions.*[44]

In 1992, management guru Peter Drucker predicted that the traditional factors of production – land, labour and capital – would become secondary to knowledge.[45] He was describing the ascendancy of the knowledge economy, and the knowledge worker. Drucker was correctly interpreting the implications of the propensity within business, across all sectors, to collect, analyse, manipulate and extrapolate information and data.

Of course, the question that emerges in the context of this book is: what will work and the place of work look like in the knowledge economy of say the mid-twenty-first century? Crudely speaking, if farming, manufacturing and services were typified by farmland, factories and offices, what will be the physical manifestation of work in a knowledge economy? Will office work and the office as we now understand it be replaced?

Perspective #4: the office as a social and economic matrix. As described above, the role of the office as a fundamental element of society and the economy was neglected for decades; and this neglect was thrown into sharp relief in recent years by strong technological, business and social forces of change. The previous three perspectives highlighted the role of office work as a key driver of the economy, as a facet of the city and as a defining aspect of the knowledge society. A fourth perspective, and the subject of this book, is the historic relationship between the economy, business, work and the workplace. It seeks to demonstrate how these elements have co-evolved, as changes in the economy and society have driven changes in work which, in turn, have driven changes in the built environment, specifically the office workplace.

The book narrates the chronological history of the office and office work over three-and-a-half centuries. Along the way, the story takes in economic change and the nature of organisations; social relations and the rise of the middle class; the mechanisation of the office and the feminisation of the workforce; the rise of corporatism, managerialism and company man; the interaction of office work, work landscapes and changing working styles, and the arrival and transformation of the workplace by the digital revolution.

Previous work by the author proposed that the office had passed through four eras in its evolution – market, factory, corporate and digital, and was about to begin a fifth – the network

Office Era	Time period	Work & Place Themes	Office Technology
Market	Part I: 1670-1770	**Trade**. Goldsmiths as bankers. Central banking. Coffee Houses. Joint Stock Companies. Exchanges. Insurance.	Coal fires. Candles. Copying. Ledgers. Paper. Bills of Exchange. Quill pen. Cash book. Shop counter.
Market / Factory	Part II: 1770-1870	**Organisation.** Stock Exchange. Clearing House. Private and merchant bankers. Joint Stock Banks. Office buildings. Speculative development.	Gas lighting. Copying. Ink Pen. Ledgers. Paper. Stenography. Polygraph. Correspondence book.
Factory / Corporate	Part III: 1870-1945	**Production**. Rise of corporation. Division of labour. Work: fixed, manual, repetitive and mechanical. Female workforce. Rise of professions.	Mechanisation. Adding machine. Electricity. Telegraph. Telephone. Typewriter. Copiers. Elevator. Air conditioning. Fluorescent light.
Corporate	Part IV: 1945-2000	**Process**. Globalisation. Scale. Bureaucracy. Command & Control. 'Corporate man'. Work landscapes. Birth of digital office.	Photocopier. Mainframe. System Furniture. Golf Ball Typewriter. PC. MS Office. Mobile phone. Internet. Email.
Digital	Part V: 2000 onward	**Knowledge**. Agile work. Data. Power.	Smart phone. The Cloud. Social Media. Augmented Reality. Automation. Artificial Intelligence. Robotics. Virtual Reality.

Figure 1.2 Four eras of the office economy.

office. These reflected stages in the changing nature of the organisations occupying offices, the role of the office and the equipment and technology used.[46] This book progresses through distinct time periods reflecting the broader socio-economic context and Figure 1.2 illustrates how these relate to the four eras of the office.

One of the major sub-themes of the book is the evolution of what is referred to as the 'cultural office' in Europe and the 'rational office' in North America. The former carried centuries-old socio-cultural baggage into the workplace (its design, operation and management) in ways that the latter avoided simply by starting later in a hugely different socio-economic context. The effects on office buildings and office work can still be felt today.

The final part of the book explores the changing office in the context of the contemporary economic and technological forces of change outlined earlier in this Introduction. During the past four years, the role, function and purpose of the office have been debated more extensively than at any time in the past. The existential crisis arising from COVID-19 has settled somewhat, but there remain big questions over the future of office work, workers and real estate. The book closes with some speculation on the future of the office.

It is important to note at this point that the final two chapters of this book move beyond a fully objective account of history, and into the realms of lived experience. The author has spent the past four decades undertaking applied research on offices so while the final chapters continue to draw upon secondary sources, the writing also draws directly upon experience and knowledge of actual events.

As the longer-term technological, business and socio-cultural changes collided with the COVID-19 pandemic in 2020, there followed a wave of speculation about the purpose and role of offices. What is the future for offices? Will they need to be reinvented for a new style of working? What will office work look like, say, 20 years from now? Do we need a new building form? Or a new workplace design? This book seeks to provide a detailed, long-term social and economic context against which to consider such questions.

Book outline. The book is broken down into five consecutive time periods. Within these five broad parts, individual chapters describe distinctive themes in the evolution of the office and office work in each period.

Part I **1670–1770: building blocks of the office economy.** The story begins in Chapter 2 with the aftermath of the Great Fire of 1666, when merchants gathered in coffee houses and when the concept of a 'company' was born. While everyone has heard of the Industrial Revolution, which was yet to get underway, what is less well known is that it could not have happened without a radical and rapid reshaping of the nation's financial markets. This period saw the financial revolution and the birth of the modern office economy.

Chapter 3 describes the emergence of insurance companies and how merchants morphed into early bankers and joint stock companies. Along with the new businesses, the role of the clerk (the first office worker) expanded. Offices were part of residential properties, with the private banker living above the 'shop' – a model that continued well into the mid-eighteenth century.

Part II **1770–1870: from bank house to office building.** This section traces the evolution of banking, as the core of the office economy. Chapter 4 examines working at the Bank of England, the spread of private bankers and the growth of joint stock banks, which revolutionised the design and occupation of offices.

Chapter 5 describes the role of the clerk in terms of the conditions of work, employer clerk relations, reward, class and behaviour. It traces their laborious and monotonous work and the primitive offices in which they laboured. The social relations between 'owner and clerk' are also explored in depth.

Chapter 6 describes the transition of offices from converted dual-use dwellings into purpose-built office buildings. It describes how the joint stock companies pushed the boundaries of grandiosity and ostentation, and it outlines the first speculative office buildings.

Part III **1870–1945: the white-collar factory.** Part III is set in the industrial revolution and the massive social and economic upheaval of the later-nineteenth century. Chapter 7 traces the rise of modern corporations with their ever-expanding clerical workforces; comparing and contrasting European (cultural) and the US (rational) approaches to designing offices, before describing the arrival of the twentieth century office and scientific management.

Chapter 8 is concerned with work and workers, outlining changes in working conditions, reward, discipline and the feminisation of the workforce, along with the extinction of clerking and the arrival of 'office work'. Chapter 9 examines the impact of mechanisation on the office and debates whether it led to a new proletariat or to the rise of the new middle class.

Chapter 10 describes the evolution of the buildings in which the white-collar factories evolved. It examines corporate office buildings, the skyscraper and the relationship between building form and work processes.

Part IV **1945–2000: the corporate office.** Part IV moves into the middle of the twentieth century and the maturation of the office economy. Chapter 11 traces the rise of corporatism: the enormous bureaucracies with armies of 'company men'. These were the anonymous, identikit office workers who made up every tiny part of the vast corporations, all overseen by the 'managerial demiurge'.

Chapter 12 sets out how work and workplace co-evolved during this time, going through a number of phases of development. The period traces offices and office work up to the point where computers and digital technology became commonplace, and finishes with the transformative impact of the financial revolution of 1986 ('Big Bang') and its equally transformative impact on the workplace and office buildings generally.

Chapter 13 attempts to summarise the history of the office up to around the millennium: three centuries of change, from clerk to nomad, before the book turns its attention first, to the present and then, to the future. The key purpose of this chapter is to allow the reader to pause and to reflect on some of the key lessons of the previous three centuries of change before turning to the period that lies ahead.

Part V **1990 onwards: the digital office and beyond.** The final part is set in the period from the around the millennium through to the COVID pandemic in 2020 and its aftermath. Chapter 14 refers to the 'death of management' and is concerned with the adoption of personal technology and the rise of the knowledge economy, knowledge business and knowledge work. It suggests that the period saw the switch from corporatism to individualism with major impacts for social relations in the office.

Chapter 15 is speculative: looking forward and suggesting how office work, workers and workplaces will evolve over the coming few decades. The chapter is organised around implications for cities, the office economy, work, workplace and the supply process. The last of these does not feature in the book generally because of the focus on work, but the supply process will have a distinct influence on the future of the office workplace.

A note on terminology

There are three terms used repeatedly in this book that merit some definition at this early stage: the office economy, corporate/corporatism and white-collar.

The office economy. Government statisticians and economists break the economy and labour markets down into a series of sectors, including agriculture, wholesaling, retailing, finance, professional services, health education and so on. In some of these sectors, firms occupy only offices; while in others there is either no meaningful office activity or the office activity is ancillary to the main activity. The first category includes accountants, advertising agencies, insurance agencies, investment banks, lawyers, public relations firms and real estate firms. These sectors are characterised by their predominantly business-to-business nature and by the fact that they occupy *only* office buildings.

The second group includes retail banks whose headquarters support their regional branch networks, and technology firms where hardware firms have manufacturing plants and software firms also occupy studios and light industrial space. In the media sector, publishers are part of the office economy alongside their printing plants. The public sector is also a very large consumer of office space: central and local government, health, higher education, police and transport all have large office portfolios. Taken together the sectors thus described form the core of what is referred to in this book as the 'office economy'.

In the UK, jobs data are collected according to the Standard Industrial Classification (2007), which includes five sectors that can be said to account for the great majority of office jobs: information & communications; financial & insurance; professional, scientific & technical and real estate; administrative & support services, and public administration. At September 2023, these five sectors accounted for 11.8m jobs out of a total workforce of 36.8 m, or 32%.[47] If office jobs in health, higher education, transport and utilities are added, then it can be estimated that offices account for c14m jobs, upwards of 40% of all jobs. It should also be noted that the five sectors highlighted above have grown from 7.2m in 1996, when they accounted for just 25% of all jobs. Thus just over the past 28 years, the office economy has grown by 60%, much faster than other parts of the economy.

As well as these official descriptors of office activity, it is also possible to identify a number of features which firms in the modern office economy share in various combinations, including

- knowledge-based work;
- technology-enabled, 'agile and connected', workstyles;
- networks of relationships replace the 'corporate island';
- large and small firms undertake fundamentally the same work;
- occupation of purpose-built office accommodation;
- office work is physically separated from any manufacturing or other activity;
- tendency to concentrate spatially, normally but not exclusively, in downtown, and
- mostly business-to-business activities.

In these terms, the office economy is more than a building type; it is more than an employment type, and it is more than a locational characteristic. It is a blend of activities, work and place.

Corporate and corporatism. A corporation generally refers to an organisation, a legal entity, recognised in law, incorporated to carry out the objectives of a business. Early incorporated entities were established by charters, granted by monarchs, or passed by parliament. Today, the creation of a corporation involves a legal process called incorporation where legal documents containing the primary purpose of the business, name and location, and the number of shares and types of stock issued, are drafted.

As we shall see, early corporations held certain advantages over sole traders and partnerships, with limited liability spread amongst their shareholders. Today, corporations (public companies) issue stock to shareholders and belong to the stock market where their shares are openly traded between investors.

Another form of the company that has the limited liability shield of a corporation, and the more favourable tax treatment of either a sole proprietorship or partnership, is known in the US and UK as a limited liability company (or *Gesellschaft mit beschränkter Haftung*, GmbH, in Germany). While not a corporation, this type of entity became very attractive as an alternative for corporations not needing to issue stock.

While this book respects these formal defining features of corporations, it also adopts the more informal use of *corporation* as it refers to generally large, complex, multilayered and departmentalised businesses, comprising extensive white-collar workforces. A corporate organisation in these terms will typically be accommodated in a multiple building estate and be active internationally. A corporation can be either an entirely office-based organisation (bank, lawyer, accountant) or an industrial concern, which has a large office presence in terms of headquarters or regional offices. 'Corporate' can also refer to public sector organisations, such as local government, transport companies, utilities and so on, in terms of their office occupation and having large-scale and complex structures.

While the main focus in this text is upon how the evolving economy and the corresponding changing business landscape has influenced the nature of the workplace, work and workers, all bound up in what is referred to as the office economy, it is recognised that public administration, through national and local government, has also been an important component of overall office demand. Governments throughout Europe were using palaces to house the growing apparatus of the state before commercial businesses were occupying purpose-built buildings. The British government was building and occupying offices alongside the earliest commercial offices – The Admiralty and the Navy Office are examples (Chapter 3). Similarly, the legal profession occupied the Inns of Court and various professional institutions occupied early examples of offices. But this study is confined to the parameters described above.

White-collar. The etymology of the term 'white-collar worker' is not entirely clear, but it is widely attributed to Upton Beall Sinclair Jr (1878–1968), who was an American writer, political activist and the 1934 Democratic Party nominee for Governor of California, and who wrote nearly 100 books and other works in various genres. Sinclair's work was well known and popular in the first half of the twentieth century. He used the term to refer to clerical, administrative and managerial functions during the 1930s, defining a white-collar worker as a salaried professional generally working in offices. During this time, male office workers in Europe and America universally wore white-collared dress shirts, distinct from the attire of manual workers of the time. It is of course true that white-collar professionals work in non-office environments, for example, medicine and teaching. This book is chiefly concerned with those white-collar workers who labour in the corporate office environments described above, whether undertaking managerial, technical or administrative roles.

Notes

1 Braverman H (1974) *Labour and Monopoly Capital: the Degradation of Work in the Twentieth Century* Twenty Fifth Anniversary Edition, Monthly Review Press, New York.
2 Harris R (1996) *Less a Castle, More a Condominium: Taking a Look at the Future of the Office* Gerald Eve Chartered Surveyors.
3 Graeber D (2018) *Bullshit Jobs: A Theory* Simon & Schuster.
4 Donkin R (2001) *Blood, Sweat and Tears: the Evolution of Work* Texere, London and New York p49.
5 BBC (2024) More the a Fifth of UK Adults Not Looking for Work www.bbc.co.uk/news/business-68534537
6 Urquhart M (1984) Employment Shift to Services *Monthly Labor Review* April 1984 pp15–22; US Bureau of Labor Statistics (Sept 2023) www.bls.gov/emp/tables/employment-by-major-industry-sector.htm
7 Clark C (1940) *The Conditions of Economic Progress* Macmillan and Co, London.
8 Rostow WW (1960) *The Stages of Economic Growth: A Non-Communist Manifesto* Cambridge University Press.
9 Goodwin W (1965) The Management Center in the United States *Geographical Review* Vol 55 No 1 pp1–16.
10 Daniels PW (1975) Office Location: An Urban and Regional Study G Bell & Sons, London p2.

11 Carter H (1976) *The Study of Urban Geography* (2nd Edition) Edward Arnold p245.
12 Bell D (1973) *The Coming of Post-industrial Society: A Venture in Social Forecasting* Basic Books, New York.
13 Bell D (1973) *The Coming of Post- industrial Society: A Venture in Social Forecasting* Basic Books, New York p134.
14 Gershuny J (1978) *After Industrial Society? The Emerging Self-service Economy* Macmillan, London.
15 Gershuny J & Miles I (1983) *The New Service Economy* Frances Pinter, London.
16 Harris DF & Taylor FJ (1978) *The Service Sector: Its Changing Role as a Source of Employment* Research Series 25, Centre for Environmental Studies Ltd, London.
17 Howells J & Green A (1988) *Technological Innovation, Structural Change and Location in UK Services* Gower Publishing Company Limited, Aldershot p2.
18 Marshall JN (1988) *Services and Uneven Development* Oxford University Press, Oxford p32.
19 Coffey WJ (2000) The Geographies of Producer Services *Urban Studies* Vol 21 No2 pp170–183.
20 Lindahl DP & Beyers WB (1999) The Creation of Competitive Advantage by Producer Services Establishments *Economic Geography* Vol 75 pp1–20.
21 Rannells J (1956) *The Core of the City* Institute of Urban Land Use and Housing Studies, Columbia University Press.
22 Rannells J (1956) *The Core of the City* Institute of Urban Land Use and Housing Studies, Columbia University Press p5.
23 Rannells J (1956) *The Core of the City* Institute of Urban Land Use and Housing Studies, Columbia University Press p35.
24 Hoover EM & Vernon R (1959) *Anatomy of a Metropolis: the Changing Distribution of People and Jobs within the New York Metropolitan Region* Harvard University Press, Mass p99.
25 Horwood EM & Boyce RR (1959) *Studies of the Central Business District and Urban Freeway Development* University of Washington Press, Seattle.
26 Gottmann J (1966) Why the Skyscraper? *Geographical Review* Vol 56 No 2 pp190–212.
27 Gottmann, Why the Skyscraper? *Geographical Review* Vol 56 No 2 pp190–212.
28 Cowan P; Fine D; Ireland J; Jordan C; Mercer D & Sears A (1969) *The Office: A Facet of Urban Growth* Heinemann Educational Books Ltd, London p26.
29 Cowan P; Fine D; Ireland J; Jordan C; Mercer D & Sears A (1969) *The Office: A Facet of Urban Growth* Heinemann Educational Books Ltd, London p28.
30 Cowan P; Fine D; Ireland J; Jordan C; Mercer D & Sears A (1969) *The Office: A Facet of Urban Growth* Heinemann Educational Books Ltd, London p34.
31 Daniels PW (1975) Office Location: An Urban and Regional Study G Bell & Sons, London p221.
32 Machlup F (1962) *The Production and Distribution of Knowledge in the United States* Princeton University Press.
33 Giuliano VE (1982) The Mechanisation of Office Work *Scientific American* Vol 247 No 3 September 1982 pp148–165.
34 Drucker PF (1969) *The Age of Discontinuity: Guidelines to Our Changing Society* Harper & Row, New York.
35 Berry BJL (1973) *The Human Consequences of Urbanisation* St Martins, New York p54.
36 See for example: Goddard JB & Pye R (1977) Telecommunications and Office Location *Regional Studies* Vol 11 pp19–30; Thorngren B (1977) Silent Actors: Communication Networks for Development In: Pool I de S (Editor) (1977) *Social Impact of the Telephone* MIT Press, Boston; Pye R (1979) Office Location: the Role of Communications and Technology In: Daniels PW (Editor) (1979) Spatial *Patterns of Office Growth and Location* Wiley, London.
37 Gottmann J (1983) *The Coming of the Transactional City* University of Maryland Institute for Urban Studies Monograph Series No2 College Park.
38 Noyelle TM & Stanback TJ (1984) *The Economic Transformation of American Cities* Allanheld Osmun, Totowa, New Jersey.
39 Hepworth M (1987) The Information City *Cities* Vol 4 pp253–262.

40 Hepworth M (1990) Planning for the Information City: The Challenge and Response *Urban Studies* Vol 27 No 4 pp537–558.

41 Marshall JN (1988) *Services and Uneven Development* Oxford University Press, Oxford p32.

42 Coffey WJ (2000) The Geographies of Producer Services *Urban Studies* Vol 21 No2 pp170–183.

43 Coffey WJ (2000) The Geographies of Producer Services *Urban Studies* Vol 21 No2 pp170–183.

44 Coffey WJ (2000) The Geographies of Producer Services *Urban Studies* Vol 21 No2 pp170–183.

45 Drucker PF (1992) The New Society Organisations *Harvard Business Review* September-October pp95–104.

46 Harris R (2021) *From Clerical Factory to Digital Hub: London's Global Office Economy* Routledge, London p338.

47 Office for National Statistics, Census (2021) Employment & Employee Types www.ons.gov.uk/employmentandlabourmarket/peopleinwork/employmentandemployeetypes/

Part I

1670–1770

Building blocks of the office economy

The Bookkeeper by Philip van Dijk (c1719)

Office Era	Time period	Work & Place Themes	Office Technology
Market	**Part I: 1670-1770**	**Trade**. Goldsmiths as bankers. Central banking. Coffee Houses. Joint Stock Companies. Exchanges. Insurance.	Coal fires. Candles. Copying. Ledgers. Paper. Bills of Exchange. Quill pen. Cash book. Shop counter.
	Part II: 1770-1870	**Organisation.** Stock Exchange. Clearing House. Private and merchant bankers. Joint Stock Banks. Office buildings. Speculative development.	Gas lighting. Copying. Ink Pen. Ledgers. Paper. Stenography. Polygraph. Correspondence book.
Factory	**Part III: 1870-1945**	**Production**. Rise of corporation. Division of labour. Work: fixed, manual, repetitive and mechanical. Female workforce. Rise of professions.	Mechanisation. Adding machine. Electricity. Telegraph. Telephone. Typewriter. Copiers. Elevator. Air conditioning. Fluorescent light.
Corporate	**Part IV: 1945-2000**	**Process**. Globalisation. Scale. Bureaucracy. Command & Control. 'Corporate man'. Work landscapes. Birth of digital office.	Photocopier. Mainframe. System Furniture. Golf Ball Typewriter. PC. MS Office. Mobile phone. Internet. Email.
Digital	**Part V: 2000 onward**	**Knowledge**. Agile work. Data. Power.	Smart phone. The Cloud. Social Media. Augmented Reality. Automation. Artificial Intelligence. Robotics. Virtual Reality.

2 The basis of trading

There were very few offices, as we understand the term today, in the seventeenth century, nor for much of the eighteenth. There were some: the East India Company moved into its London headquarters, Craven House in Leadenhall Street, in 1648; and Custom House (for the collection of customs duties), at Sugar Quay was a grand three-storey building, with octagonal staircase towers (destroyed in the Great Fire). 70 Lombard Street, one of the earliest purpose-built private bank buildings, was built in 1757 for Sir Charles Asgill (1714–1788); 3 years later Andrew Drummond's private bank opened its new building at 49 Charing Cross. But, by and large, office activity was contained within domestic premises where goldsmiths, merchants and brokers ran their family owned businesses, often with the assistance of one or two clerks.

There were, however, three major developments during this time, which were critical to the birth of the office economy. While largely taken for granted today, one of its longest-lasting and most profound impacts was in the development of corporate structures and joint stock companies. These had share capital to allow investment and expansion and led to stakeholders and managers challenging the singular authority of owners, and ultimately to corporate structures.

The second development was the rapid expansion of coffee houses and exchanges, which were essential to establish some of the social relations of the nascent office economy, forming the basis of the business ecosystem and creating the means of communication, information gathering and deal-making.

The third critical development was the financial revolution. While everyone has heard of the Industrial Revolution, which was yet to get underway, what is less well known is that it could not have happened without a radical and rapid evolution of the nation's financial markets beforehand. The massive expansion of industry and infrastructure, not to mention the rapid development of international trade and investment was dependent upon a mature and sophisticated financial services sector. Thus it was, in the crucial role of "*the circulation of capital*", that London was a fundamental feature of the industrialisation of the national economy. "*London was the sluice through which, by means of bills of exchange, savings – often originating as surpluses from agricultural areas – could be channelled to areas – often mainly industrial – where credit and capital were needed*".[1]

We see during this period the main elements of that financial revolution coming together, including the shift from a royal debt to a national debt and long-term borrowing, the founding of a national bank, the development of financial and commodity exchanges, and the emergence of banks, insurers and brokers, alongside a proliferation of financial products and services. It was this financial revolution that laid the foundations for the emergence of widespread office work and the office economy in subsequent periods.

DOI: 10.1201/9781003582205-3

Plague, conflagration and rebuilding

Restoration of the monarchy in England in 1660 saw the return of King Charles II (1660–1685), following Oliver Cromwell's 5-year rule as Lord Protector, and the reaffirmation of the Anglican orthodoxy. The ensuing Restoration period, including the short reign of James II (1685–1688) was a period of revival and expansion, albeit dogged by the Anglo-Dutch Wars, and was a notable period of cultural revival in the form of literature and drama.

By 1660, London was by far England's largest city: at around 350,000 souls. At this time the largest provincial town in England was Norwich, with a population of probably less than 25,000. Just as they do today, London's wealthy lived cheek-by-jowl with the very poor. Open sewers and overcrowded homes sat next to expensive homes, often occupied by merchants who enjoyed the exotic and the luxurious in the elaborately decorated shops of the City's main shopping street, Cheapside.

London had grown quickly, chaotically and dangerously, and alongside the success and wealth driving the city's growth came widespread poverty and squalor. Overcrowding was acute and for many, sanitation and housing conditions were appalling. This was a pattern of growth for which the twin catastrophes of 1665 and 1666 were to punish the city dreadfully.

Queen Elizabeth I had made an early attempt to alleviate the problems by giving Royal Assent to the Restrictions on Building Act in 1592, which decreed that "*No new buildings shall be erected within three miles of London or Westminster*". In so doing, she set up a Tudor Green Belt. The same Act sought to tackle overcrowding by preventing dwellings being sub-divided – four centuries before the Parker Morris Standards in housing. However, the actions were a case of 'too little, too late'. Living conditions barely improved through much of the seventeenth century; indeed they were worsened by the sheer density of humanity and the persistent poor building and unsanitary conditions. The ensuing twin catastrophes – of plague and conflagration – were well-nigh inevitable.

First there was the bubonic plague. Following relatively minor outbreaks in 1625 and 1636, the Black Plague of 1665 killed upwards of 100,000 people. Then there was conflagration. Although directly causing relatively few deaths, the Great Fire of 1666 razed about four-fifths of the medieval city within 3 days, along with around 87 parish churches, 44 livery company halls, 13,200 houses, the Royal Exchange and the Custom House, with extensive damage also being inflicted on St Paul's, Bridewell and the Fleet Prison. The devastated area extended over 150 hectares (373 acres) within the City itself, and 25 hectares (63 acres) extramurally.[2] Writer, landowner and founding Fellow of the Royal Society, John Evelyn (1620–1706), famously declared: "*London was, but is no more*".

Despite the scale of devastation, and the enormity of the rebuilding required, within one week of the fire being finally extinguished, Charles II issued a proclamation promising "*a much more beautiful city than is at this time consumed*". He also outlined his wish to impose "*order and direction*", and promised main thoroughfares like Cheapside and Cornhill would be "*of such breadth as may with God's blessing prevent the mischief that one side may suffer if the other be on fire*". And within weeks from this, multiple proposals for a phoenix-like rebirth of the city were being formulated. John Evelyn, Robert Hooke, Valentine Knight, Richard Newcourt and Sir Christopher Wren (Figure 2.1) all advanced rebuilding plans.

The various designs all envisaged a more rational plan based on wide streets to replace the chaotic medieval street pattern, and were influenced by Baroque European cities; Wren took inspiration from Versailles. However, and despite the pedigrees of the various architects, none of the plans was adopted.

King Charles II admired Wren's design and appointed him as one of six commissioners to oversee rebuilding work. However, landed interests rose to oppose the plans and began piecemeal redevelopment, and there was a lack of appetite in a cash-strapped Parliament engaged in the Anglo-Dutch war for a legal battle with London's wealthy merchants and aldermen. The result was a more pragmatic approach to rebuilding. For example, the Rebuilding Acts of 1667 and 1670 addressed fire risk and sanitation with measures to ensure that new buildings would only be built in brick or stone; the use of flammable materials was restricted; streets were to be made wider and straighter, while jettying upper storeys and protruding signs were banned. A maximum number of storeys per house was also imposed in a desperate attempt to reduce overcrowding.

The lack of a comprehensive rebuilding plan acted as a magnet to speculative builders – some good, some otherwise. One of the most interesting characters to emerge was Nicolas Barbon (1637–1698). A somewhat mercurial character, Barbon was a financier who was instrumental in establishing building and property insurance. But this was by no means his only interest. Barbon was also a physician (graduating as a Doctor of Medicine at Utrecht in 1661, and admitted as Honorary Fellow of the College of Physicians in 1664); Member of Parliament (elected 1690 and 1695 for Bramber, Sussex) and economic theorist (including authoring *A Discourse on Trade* in 1690). Barbon's *Discourse* attracted critical praise from contemporary economists, including John Maynard Keynes and Joseph Schumpeter.

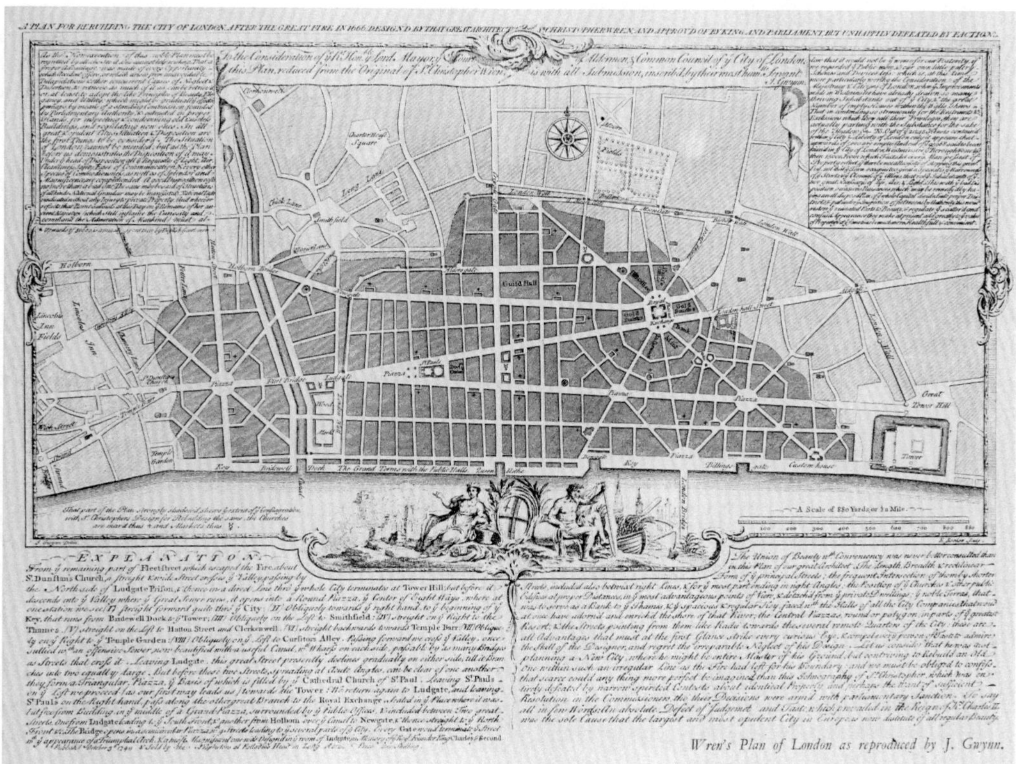

Figure 2.1 Christopher Wren's plan to rebuild London. Licenced under Creative Commons Attribution 4.0 International licence.

Figure 2.2 Nicholas Barbon: financier, economist, doctor and politician.

Despite these strengths, Barbon (Figure 2.2) also earned a reputation for being determined and ruthless: his methods for procuring and developing land have been more politely described as unscrupulous, bending people to his will, and often riding roughshod over royal declarations and Acts of Parliament to achieve his ambitions. Summerson suggested that Barbon's "*extraordinary capacity for finance, combined with a knack of profiting from the amiability as well as the stupidity of human nature marked him out as a formidable adventurer*." Further, that he lived in a "*lordly way, directing his mysterious, complex schemes, rarely without great profit and rarely with complete honesty*".[3]

It might seem odd in today's context that a man of such education and social standing could also be a ruthless speculator; but the seventeenth century was very different, with the building industry comprising a raft of specialist craftsmen and a number of organising speculators. Summerson described the speculative builder, the mainspring of London's expansion for 300 years, as having always been a person of the most various characteristics.

> *Sometimes he has been a lord, sometimes little more than a labourer; sometimes a substantial capitalist, sometimes a craftsman, with only his skill and times to adventure; sometimes an architect, sometimes a bricklayer or carpenter; sometimes a lawyer, a mechanic, a schoolmaster, a quack, an actor – indeed, almost any class, trade or profession.*[4]

Barbon was a prolific builder, exploiting standardisation and mass production to build hundreds of homes that were built "*all very much alike, economically planned to the point of meanness, with coarse ornaments which repeated themselves over and over again*".[5]

Before Barbon – and indeed for a long period after – most buildings were custom built for individuals on their own land, using their own labour and materials, whether residential or commercial. The buildings were relatively simple in form and straightforward to construct. Commercial buildings were typically general purpose in nature, occupied by small enterprises who did not make particular demands on building form.

Barbon was most active in rebuilding around the area now known as Midtown, and it was here that he started his largest project to date, Red Lion Square, in 1684. At this time, the land in the area was used as fields and paddocks, but Barbon had visions for a seven hectare (17 acre) housing project. Red Lion Fields lay to the north of High Holborn and to the west of the Inns of Court on the west side of Gray's Inn Road. However, as he started to divide up the fields to create a new London square, he was robustly challenged by a group of lawyers from the nearby Inns of Court, many of whom objected to losing their 'rural' vista.

The lawyers lost an expensive legal case against the developer, and relations between the two parties deteriorated. So much so, that on the 11 June 1684, a pitched battle broke out between roughly 100 lawyers and site workmen, with the main offensive weapons being bricks and other building materials. Barbon led from the front, and there were injuries sustained by both sides, but eventually Barbon and his workmen prevailed and work resumed. However, in acting as he did, Barbon helped to establish a reputation for developers as a rapacious and single-minded class, sometimes with questionable ethics. To some degree, this reputational profile survives in the popular psyche to the present day.

Barbon was at the vanguard of a new role: the speculative developer, who emerged to assemble land and capital, to organise the labour, to oversee the delivery of the physical building and to earn a financial gain upon completion. Initially, the developer acted as an intermediary between the landowner and the purchaser of the building: he assumed a level of risk and responsibility, and was rewarded accordingly. As the template became established, developers began to purchase land and develop in their own right. But whether as an intermediary or as a direct player, as projects became larger and more complex, so the need for working capital and access to credit became paramount. In the early days, wealthy individuals and building materials firms were active in this sense, but increasingly there emerged a critical role for financing development.[6]

London's population grew from c600,000 in 1700 to one million by the end of the century. In doing so, its physical extent grew markedly and its density increased. The eighteenth century witnessed a westward drift from the City of London to what we now know as Westminster, or the West End. New, often highly desirable districts sprung up, based on a patchwork of squares. Berkeley Square (1733), Bloomsbury Square (1660), Cavendish Square (1717), Golden Square (1673), Grosvenor Square (1731), Hanover Square (1714), Portman Square (1764), Red Lion Square (1684) and St James's Square (1684), among others, were all completed during this period. Indeed the City began to lose its lustre.

> *Beautiful City houses like those at 1-2 Laurence Poutney Hill, dating from 1702, were no longer sufficient lure. Even aldermen were choosing to live in nearby country houses or in the more fashionable suburbs to the west.*[7]

The westward drift was the bow wave of a wholesale shift of population out of the City, a process, which over the next two centuries would "*transform the City from a place to live into a conglomeration of offices deserted every night*".[8]

As well as new homes, shops and leisure expanded rapidly, in areas such as Mayfair, Piccadilly and the Strand. Cheapside in the City had been the hub of shopping activity, "*although from the Restoration onwards, the central business district for retailing shifted north and westwards, to Covent Garden, the Strand and (later) Oxford Street and Regent Street*".[9] William Hamley founded a toy shop in 1760 at 231 High Holborn; while James Christie founded an auction house in 1766 in Pall Mall. It was during this period that high street shopping matured into a mainstream leisure activity.

Shops not only grew in number rapidly, but many began to sell 'luxury' goods, reflecting the growing affluence of society at large. Booksellers, furniture-makers, glassmakers, haberdashers, jewellers, linen drapers, milliners, tailors, tea and coffee merchants and stationers were all growing in number in the changing retail scene. At the same time, shop frontages became more extravagant to attract the attention of shoppers and to entice them to examine the goods inside. For example, bow windows became very popular, where mirrors and lights would focus attention on glistening goods.

> *Watchmakers and glassmakers eclipsed even goldsmiths' and jewellers' shops with their displays of fanciful clocks set in alabaster ... gold and silver. And the riches cut glass lighted by patent lamps at night.*[10]

César de Saussure, a young French Protestant from Switzerland, described London's four main shopping streets in the 1720s – the Strand, Fleet Street, Cheapside and Cornhill – as "*the finest in Europe*".[11]

Many of the wealthy descended upon London for 'the season' (broadly winter through to spring, and coinciding with the sitting of Parliament). Landed gentry and merchants travelled there from their country homes and estates for a prolonged period of business and social activities. As well as balls, dinner parties and theatre, the more serious business of matchmaking young people, and engaging the services of architects, bankers, brokers, doctors, lawyers and scriveners (clerks for the wealthy) all took place in a frenzy of activity.

As the number of wealthy residents grew so too did the number of banks catering for private trade. The 'West End' banks, of which Hoare's in Fleet Street, Coutts' in the Strand and Drummonds at Charing Cross still exist (in name anyway), dealt mainly with the landed gentry and aristocracy, for whom they provided short-term loans or mortgages, transfers of cash between country and town, and management of investments in Government stocks.[12]

Birth of the company

In the sixteenth century the guild system formed the elementary basis upon which business was conducted. Some guilds can be traced back to the twelfth century, with the earliest charter still in existence being granted to the Weavers' Company in 1155. However, guilds grew strongly through the fifteenth century; and by 1515 there were 48 guilds in London.

One characteristic of guilds was that they were often place-specific, and members lived and worked as part of the local community. In London, this feature expressed itself in terms of what we now call business clusters. The coachmakers lived in Long Acre, the mercers and drapers in Cheapside, Ludgate Hill and Gracechurch Street. There were orange merchants and wet salters in Billingsgate, fishmongers on Fish Street Hill, seamstresses in Paternoster Row, chairmakers in St Paul's Churchyard and booksellers in Little Britain.[13]

Over time various guilds introduced distinctive clothing and regalia – or livery – to give their members an exclusive identity and, because of this, the guilds came to be known as livery companies. They were, in some respects, primitive examples of 'professional organisations'. They regulated competition within their trades; sought to maintain what we might call today 'trading standards', and oversaw the development of craftsmen with training. And just like the professions of later times, as they matured, they often established their own premises in large houses or halls to convey their important role. These premises served not only for meetings and social interaction, but also for settling trade disputes.

Although they protected local monopolies over trades and often enforced their rules, the guilds also liked to convey a sense of social responsibility. Thus the leaders of guilds argued that the private gains facilitated by a guild's monopoly were balanced by the public benefits that arose from their monopoly powers.[14] Nevertheless their limitations became evident as private business expanded and became more sophisticated: for example, their activities and influence were restricted to the Square Mile, and many could not compete with cheaper tradesfolk competing from further afield. The guilds were unprepared for change, they were protectionist and they lacked economic innovation.

As the guilds waned in influence governments had begun chartering companies to establish trade links around the world. A royal charter gave the companies monopoly rights to trade in a specified geography and in this sense they were a joint enterprise between government and the private sector. But beyond this privilege, the companies had two distinctly new features. First, they were 'joint stock companies' (JSCs), in which individuals exchanged cash (normally) for a share in the business. This then entitled the 'shareholder' to a slice of the profits, proportionate to the size of their holding. Secondly, and critically, individual shareholders were protected should business profit turn to loss: their liability being limited to the extent of their shareholding (limited liability).

Joint stock companies (JSCs). The financial structure of a JSC allowed corporations to outlive the interest or life of individual shareholders, and it also created a legal body that shared certain rights and characteristics with individuals. For example, a company or corporation could enter into contracts, own assets, pay taxes and take out loans.

In these important senses, the modern company was born.

While London had neither a centralised banking system nor a well-functioning market for long-term government debt before 1688, over the course of the century it did develop a coherent market in company shares via chartered companies. While at first JSCs financed single trips, their success led to more sophisticated and on-going arrangements. By issuing shares which could be bought and sold openly, the traders created a crude stock market.

One of the earliest known JSCs in England was the Company of Merchant Adventurers to New Lands. This was founded in 1551 with 240 shareholders, and later became the Muscovy Company. It was granted a royal charter in 1555, and endowed with exclusive rights on trade with Russia. It was followed in 1581 by the Levant Company (chartered by Elizabeth I in 1592). Both were soon followed by the infamous English East India Company, granted a royal charter by Queen Elizabeth I in 1600 with exclusivity over trade with the Indian subcontinent (soon to be competing with the Dutch East India Company, established on the Amsterdam Stock Exchange in 1602); while the first JSCs operating in America were the Virginia Company of London and the Virginia Company of Plymouth, both chartered by King James in 1606 to trade along the east coast of America. A later example is the Hudson's Bay Company, prominent in the economic and the political history of Canada, which was granted a royal charter in England in 1670, to seek a north-west passage to the Pacific.

These companies were granted monopoly rights to undertake long and often dangerous and risky trading voyages. Other JSCs were created for large, capital intensive projects such as canal building and, later, railways. Indeed, the transformation of the economy by the railways was only made possible by the development of JSCs, which enabled thousands of small investors to finance them.[15] The introduction of limited liability protected investors from bankruptcy, and effectively separated the managers from the owners and gave company men much greater independence.

During the latter part of the seventeenth century, the number of JSCs grew rapidly, and in 1692 John Houghton began to circulate a newspaper, including current share quotations, which in May 1694, showed share listings of some 52 companies. By 1695, share dealings were taking place in no less than 137 JSCs, engaged in activities including sawmills, stage-coach lines and industries manufacturing whalebone whips, leather goods, powder, glass, paper, broadcloth, wool-cards, sail cloth, ropes, cordage, sugar, rum and soap.[16]

Joslin describes the market for joint stock securities in the late seventeenth century as being a "*personal and somewhat haphazard affair*" which by 1720, following significant growth in the number of joint stock companies, in addition to government issues, was operating as "*a regular market for stocks and shares, which enabled the banker both to operate in it more efficiently and made his investments in securities more shuntable*".[17]

Despite their apparent benefits, the growth of JSCs, with their corporate structures, lines of management and shareholders was not popular with everyone. Indeed, no less a figure than Adam Smith (1723–1790) warned that, being the custodians of other people's money rather than of their own, the directors of such companies could not be expected to watch over it with the same vigilance as the partners in a private enterprise look after their own money.

In promoting the value of free trade, and in concert with his criticism of the guilds, Smith argued against any form of monopoly (as the early JSCs had) that curtailed commerce and led to market inefficiency. He also worried about the scale and power that the chartered companies had achieved, and the separation of ownership and management. However, given that company directors manage other people's money rather than their own,

> *it cannot well be expected that they should watch over it with the same anxious vigilance with which the partners in a private co-partnery frequently watch over their own. Like the stewards of a rich man, they are apt to consider attention to small matters as not for their master's honour, and very easily give themselves a dispensation from having it. Negligence and profusion, therefore, must always prevail, more or less, in the management of the affairs of such a company.*[18]

Smith argued against any form of monopoly on the grounds that it would diminish trade and lead to market inefficiency; but he was also concerned over the sheer scale and power vested in the chartered companies. In another example, English writer William Hazlitt (1778–1830) was caustic in his assessment:

> *Corporate bodies are more corrupt and profligate than individuals, because they have more power to do mischief, and are less amenable to disgrace or punishment. They feel neither shame, remorse, gratitude, nor goodwill.*[19]

Sir Edward Coke (1552–1634), jurist and politician, was concerned about their accountability:

> *a Corporation aggregate of many is invisible, immortal, & resteth only in intendment and consideration of the Law ... They may not commit treason, nor be outlawed, nor excommunicate, for they have no souls.*[20]

And Edward, Lord Thurlow (1731–1806), lawyer and politician, was likewise disturbed, stating that corporations have neither bodies to be punished, nor souls to be condemned, they therefore do as they like.

The depth of these concerns demonstrates to us today how radical a step was taken at this time in terms of creating corporate organisations that would eventually grow and evolve into the behemoths of the twentieth century. And as well as changing the relationship between customer and seller, there can also be little doubt that the establishment of corporate bodies changed the relationship between business and Government, reducing the power of the latter over the former.

> *This new 'corporation', this new 'Société Anonyme', this new 'Aktiengesellschaft', could not be explained away as a reform, which is how the new army, the new university, the new hospital presented themselves. It clearly was a genuine innovation ... It was the first autonomous institution in hundreds of years, the first to create a power centre that was within society yet independent of the central government of the national state.*[21]

It was not until the nineteenth century that JSCs really came into their own, because it was at this time that the Industrial Revolution led to fundamental economic change, and when many more companies grew very significantly in scale, risk and in their appetite for speculation, or borrowing.

In the eighteenth century, small, family owned businesses remained the norm, and the legal process through which to form a corporate entity or company could only be achieved by parliamentary statute or royal charter; and the earliest charters were generally granted by states. This situation was to be relaxed fully in the nineteenth century, and is returned to in Chapter 4. At this earlier time, the focus of much commercial activity was on the burgeoning coffee houses, and it would be a while before the full potential of the corporate economy was to be fulfilled. However, the birth of the corporation was a pivotal factor in the history of office work, allowing as it did for stakeholders and managers to replace owners, and for companies to use external capital scale-up with new hierarchies and more complex structures.

Coffee houses and exchanges

Until the mid-1600s, people met in ale houses and taverns to socialise as well as discuss business. But obviously such venues could become rowdy and inappropriate for business discussions. What was needed was communal space where news could be accessed, trades made and social relationships developed. The answer was the coffee house – a novelty at the time and in the spirit of such times, an idea imported from overseas – as seen in the origin of the first coffee house in London.

In 1922, the editor of *The Tea and Coffee Journal*, William H Ukers, published a long book with a short title: at around 850 pages, *All About Coffee*, did exactly what its title promised. It was, in many ways the definitive history of coffee up to that point, tracing the cultivation and trade of coffee, its spread around the world, the history of coffee drinking and customs, the spread of coffee houses and coffee in literature and the fine arts![22]

Although coffee had been known in Constantinople since 1517, "*it was not until 1554 that the inhabitants became acquainted with that great institution of early eastern democracy — the coffee house*"[23]. The first coffee house in western Europe opened in Italy. While some evidence suggests coffee was in general use throughout Italy in 1645, the first coffee shop opened in Venice in 1683,[24] and the first coffee shop in Britain was not in London, but in Oxford, in 1650, "*at the Angel in the parish of St. Peter in the East*", Oxford, where coffee "*was by some who delighted in noveltie, drank*".[25]

The first coffee shop in London was opened by the exotically named Pasqua Rosée, a native of Smyrna, who was brought to London by a dealer in Turkish merchandise, Mr Edwards, to whom he acted as a personal servant, with duties including the preparation of Mr Edwards' morning drink of Turkish coffee.[26] Mr Edwards allowed Rosée to set up the first coffee house in London, in 1652. Not so much a house in reality as a shack (what we might call today a 'pop-up shop') against a wall of St Michael's Churchyard in the labyrinthine alleys around Cornhill. Rosée imported coffee and was soon selling over 600 dishes of coffee a day. By the early-eighteenth century, London boasted more coffee houses than any other city in the western world, save Constantinople. This rose to 8,000 by 1800.[27] Just as many coffee shops and co-working centres are today used as venues for business so too was the case over three centuries past.

Coffee houses were generally spartan, wooden and no-nonsense and patrons sat around long shared tables. In the seventeenth century, the coffee room usually opened off the street. At first only tables and chairs were spread about on a sanded floor. Later, this arrangement was succeeded by the boxes, or booths, such as appear in the Rowlandson caricatures, the picture of the interior of Lloyds, etc. The walls were decorated with handbills and posters advertising the quack medicines, pills, tinctures, salves, and electuaries of the period, all of which might be purchased at the bar near the entrance, presided over by a prototype of the modern English barmaid.[28]

Figures 2.3 and 2.4 show illustrations of seventeenth century coffee house. Coffee houses from this time ignored social boundaries. According to Matthew Green they were "*socially inclusive spaces where lords sat cheek-by-jowl with fishmongers and where butchers trumped baronets in philosophical debates*". Customers simply sat where there was space and joined the conversation. Green cites Swiss noble César de Saussure who, in 1726 observed the "*great number of coffeehouses*" in London where "*workmen habitually begin the day by going to coffee-rooms to read the latest news*". Green captures the busy nature: "*some going, some*

Figure 2.3 Interior of a London coffee house, c1674.

Source: Ukers (1922) p60.

Figure 2.4 Interior of a London coffee house in 1668.

Source: Unknown, 1668.

coming, some scribbling, some talking, some drinking, others jangling, and the whole room stinking of tobacco."[29]

Thus all manner of folk congregated at coffee houses to gossip; read the newspapers (in an era when newspapers were relatively expensive); to buy tobacco; to correspond and to socialise. They were also enormously popular as venues for discussion over matters ranging from politics to literature to religion. Indeed so popular were they in this respect they were at one point seen as a threat to the stability of society by providing an opportunity for sedition.

Charles II famously attempted to suppress coffee houses, issuing a proclamation in 1675 to end their legality, as well as banning the sale of coffee, chocolate, sherbet and tea from any shop or house, before widespread public outcry forced a retraction of the proclamation.

Green describes coffee houses as following a common formula, "*maximising the interaction between customers and forging a creative, convivial environment*".

> *On entering, patrons would be engulfed in smoke, steam, and sweat ... Rows of well-dressed men in periwigs would sit around rectangular wooden tables strewn with every type of media imaginable – newspapers, pamphlets, prints, manuscript newsletters, ballads, even party-political playing cards.*

When coffee was served, "*a little Cupid-like boy in a flowing periwig would bring a dish of coffee. It would cost a penny and come with unlimited refills*".[30] A central function of the coffee houses was that of information exchange. Before any electronic form of communication, the coffee house was the place to hear the latest news, gossip and information from markets, near and far. The Edward Lloyd coffee house on the corner of Abchurch Lane and Lombard Street in

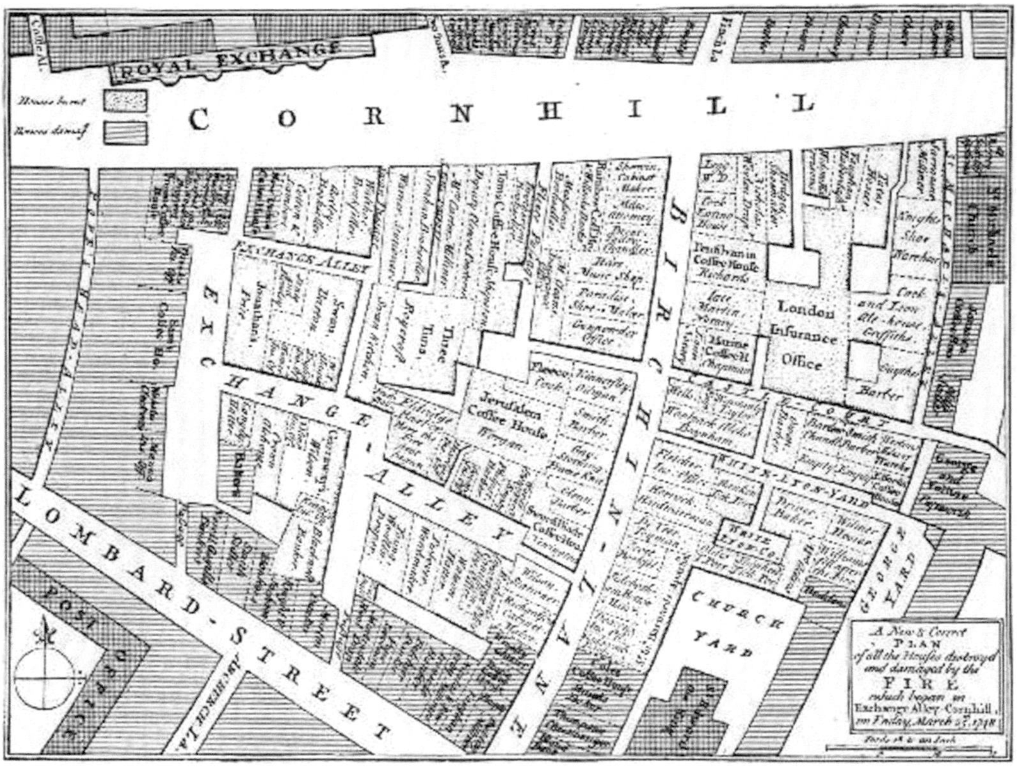

Figure 2.5 Coffee houses and other businesses in and around Exchange Alley, c1748.

Source: Ukers (1922) p76.

the 1680s grew in popularity with merchants and ship owners, who met there each day to gather intelligence of shipping, to auction cargoes and to report maritime disasters. Lloyd's eventually evolved into the marine insurance organisation we know today.

Figure 2.5 shows an early map of the area focused on Exchange Alley, sandwiched between Cornhill and Lombard Street. The Royal Exchange can be seen in the lower left corner of the image. Close inspection reveals many coffee houses in addition to Garraway's and Johnathan's; several early banks along Lombard Street, including Martin's; insurance offices, and a host of specialist shops including book sellers, milliners and stationers.

By the late-seventeenth century many London coffee houses catered specifically for different social groups (e.g. The Rainbow opened by James Farr in 1657 in Fleet Street provided a meeting place for freemasons and French refugee Huguenots); trades (e.g. book publishers gathered at the Latin coffee house near St Paul's Cathedral); professions (e.g. The Rainbow at Inner Temple Lane for lawyers), political groupings (e.g. Button's in Covent Garden for Whigs) and the arts and literature (e.g. Slaughter's in St Martin's Lane for artists).

But of course coffee houses were also venues for commerce. Jonathan's and Garraway's (Figure 2.6) were two of the most famous, both in Change Alley, Cornhill. They were renowned for the length of their hours and the variety of sales of commodities. Specific coffee houses also

Figure 2.6 Garraway's Coffee House, Exchange Alley, 1873.

sprung up to provide centres for trade with particular areas of the world: the Jamaica was largely frequented by those involved in the West Indies trade, the Jerusalem by those concerned with the East, the Virginia and Baltic (forerunner of the Baltic Exchange) by those whose merchandise came from either the American colonies or Baltic seaboard, and so on.[31]

Jonathan's coffee house, established by Jonathan Miles in 'Change Alley' in around 1680, which evolved into a sophisticated market. But being unlicenced and unregulated, dealings were subject to both deliberate frauds and unintended defaulting. Punishment for both was banishment from the coffee house, although this did not prevent the more determined. Curiously defaulters at this time were referred to as lame ducks – a phrase that survives to the present day with a wider application. It was also at this time the terms 'bulls' and 'bears' were coined.

It was from Jonathan's that in January in 1698 a broker by the name of John Castaing started to issue a twice weekly newspaper, *The Course of the Exchange, and Other Things*. The paper gave prices of foreign exchange in 14 cities, including Amsterdam, Antwerp, Hamburgh, Paris and Rotterdam. The paper also gave prices for a wide range of investments including shares, bonds and securities, as well as various commodities including paper, tobacco and wine. This was one of the earliest steps towards organised trading in securities and commodities in London.

Coffee houses had been critical in establishing some of the social relations of the nascent office economy, forming the basis of the business ecosystem and creating the means of communication, information gathering and deal-making. More importantly, they provided fore-runners for the more formal stock and commodity exchanges of the nineteenth century, whereupon coffee houses began to fall out of favour and tea became the drink of choice, consumed anywhere and by women as well as by men.

In some senses, Italy stole the march on the development of the office economy. Bills of exchange, accounting, legal structures and banking, even an office building, were all in place by the sixteenth century. Secular public banks also emerged, including Banco della Piazza di Rialto (Venice, 1587), Banco di Santo Spirito (Rome, 1591) and Banco di Santo Ambrogio (Milan, 1593).[32] But at this time, of course, what we know today as the nation state of Italy was in fact a collection of city states. The unification of modern Italy under one flag and one government, in a social and economic reform known as *il Risorgimento*, was not to occur until 1861. Before this, Italy lacked the national ambition for global expansion and trade that was growing elsewhere in Europe, notably in Great Britain, Holland and Spain. It was almost inevitable that the locus of financial power should shift from southern to northern Europe.

The Italian influence was not lost altogether. In the fourteenth century, Bruges became a centre of commerce and trade, strongly influenced by Italian traders. And by the beginning of the fifteenth century, Bruges had a fully functioning money market, reflecting the exchange rates of the leading European commercial centres, including Barcelona, London, Paris and Venice. The merchants and bankers met in a square known as the Place de la Bourse – hence the origin of the modern French bourse and the German borse.

While Bruges played an instrumental role in the development of stock exchanges, its influence had waned by the end of the fifteenth century. In its place, Antwerp assumed a growing role in European trade, providing accommodation for merchants and bankers. Indeed, the Antwerp beurs, or bourse, was the world's first purpose-built financial and commodity exchange (the building pre-dated stock trading). Built in 1531, the new Antwerp bourse was a voluminous, rectangular space with covered galleries on all four sides. It became the focal point for European trade as "*novel financial instruments evolved in Antwerp to finance the growing volume of international trade in that city over the course of the sixteenth century*".[33]

Antwerp was unable to consolidate its position as a leading commercial centre, but it had set a template for other exchanges, including London, which opened its version in 1565. The building's striking resemblance to the Antwerp exchange was no coincidence given that it was founded by financier and merchant Sir Thomas Gresham (1519–1579), who had been a royal agent in Antwerp. He described it as a "*comely bourse for merchants to assemble upon*"[34] and it was renamed the Royal Exchange following a visit by Queen Elizabeth I in 1571. Further European stock exchanges were opened in Rotterdam (1599), Amsterdam (1609), Middelburg (1616), Hamburg (1616) Delft (1621) and Nuremburg (1621).[35]

These buildings were markets, or exchanges, more than bank offices as we understand the term today. But they were forebears of the modern office, early prototypes where merchants and bankers congregated for trade, centralised away from the land and the factories but, significantly, adjacent to the key ports. The original Royal Exchange was destroyed in the Great Fire and rebuilt by 1669, re-establishing its role as the single most important meeting place for merchants and brokers.

As the number of JSCs grew through the seventeenth and eighteenth centuries, the number of tradeable stocks grew, eventually leading to the establishment of specialist stockjobbers and stockbrokers. An increasingly mature banking sector gave more companies the ability to borrow funds, which in turn led to a rapid growth in both the number and scale of JSCs in just a few years. Many brokers traded at the Royal Exchange alongside other merchants, including clothiers, druggists and grocers whose presence at the Exchange was increasingly unwelcome;

at the same time the Government sought in 1696 to constrain the number of jobbers and brokers by licencing and regulating them. The response of the brokers and jobbers was to up sticks and leave the Royal Exchange, most vacating by 1698, dispersing to various coffee houses between Lombard Street and Cornhill.[36]

The dispersal from the Royal Exchange was not a uniform success and it became clear that dealing in public places was not ideal, and continued to suffer from the intrusion of 'less respectable' gentlemen. There emerged a desire for a more exclusive setting and various options were tested. Then in 1762, 150 brokers formed a club and contracted with Jonathan's coffee house to use it exclusively. Each member would pay eight pounds per year membership, effectively transforming Jonathan's into a private club, from which they could exclude non-members and expel the unruly and unwanted, limiting the membership to 'more substantial' and the 'better sort' of brokers.[37] However, one ejected broker brought a suit against the newly formed club and the government found that Jonathan's coffee house did not have the right to exclude outsiders. Coffee houses became less popular, and brokers organized to purchase a building for their own.

In 1773, the new club opened a new and more formal exchange venue in Sweeting's Alley, just behind the Royal Exchange – New Jonathan's. This now had a set entrance fee, by which traders could enter the stock room and trade securities. Jonathan's was open to anyone so long as they paid the daily admission fee, which covered expenses such as rent. The building had a frontage of about 35 ft (10.6 m) with a dealing room on the ground floor and a coffee room on the first. In 1773, the *Gentlemen's Magazine* reported, "*New Jonathan's came to the resolution that instead of its being called New Jonathan's, it should be called The Stock Exchange, which is to be wrote over the door.*"[38]

It was, however, not an exclusive location for trading. In 1765, the Bank of England built a Rotunda where trading took place but this did not prove to be successful. Fraud was also rife during these times and in order to deter such dealings, it was suggested that users of the stock room pay an increased fee. This was not met well and ultimately, the solution came in the form of annual fees and turning the Exchange into a Subscription room.

The Subscription room created in 1801 was the first regulated exchange in London, but the transformation was not welcomed by all parties. On the first day of trading, non-members had to be expelled by a constable. In spite of the disorder, a new and bigger building was planned, at Capel Court. William Hammond laid the first foundation stone for the new building on 18 May. It was finished on 30 December when "The Stock Exchange" was inscribed on the entrance.[39]

The financial revolution

Modern finance evolved in southern Europe and particularly in what is now Italy, in cities such as Florence, Genoa and Venice. Over time and as methods of financing diffused northwards, to Amsterdam and Bruges, for example, techniques became more sophisticated. For example, the Amsterdam Exchange Bank (Wisselbank) was established in 1609 to resolve the practical problems created for merchants by the "*circulation of multiple currencies in the United Provinces, where there were no fewer than 14 different mints and copious quantities of foreign coins*". For the first time, one merchant could make a payment to another by arranging for his account to be debited and the payee's account to be credited.

> *By allowing merchants to set up accounts denominated in a standardised currency, the Exchange Bank pioneered to system of cheques and direct debits or transfers that we take for granted today.*[40]

Before very long, paper was beginning to replace coins and bullion for purposes of exchange. Through such innovation, Amsterdam quickly became the world's pre-eminent financial centre while Britain had no public bank "*or sophisticated bankers, much less a stock of metallic moneys stored in a central secure place*".[41]

The City was already an international centre of trade in Tudor times, a fact reinforced by Sir Thomas Gresham's establishment of The Royal Exchange in 1570, to compete with (and soon overshadow) continental European exchanges. But its global role was to become more dominant as its volume of trade outstripped that of Amsterdam. Thus while the population of Amsterdam stagnated, barely rising from 200,000 in 1700 to 217,000 in 1800, that of London rose rapidly from 575,000 in 1700 to 865,000 in 1800.[42]

Indeed, it is an enduring mystery as to why Paris fell so far behind London at such an early stage of modern trading. One theory lays the blame at the door of a little-known Scottish economist, John Law (1671–1729). Quite a colourful character, Law was sentenced to death in 1694 for killing a love rival with a sword in a dual in Bloomsbury Square. But he absconded and escaped to Amsterdam, before going to France. Eventually, Law persuaded the Banque Générale of the merits of establishing a centralised bank and in 1716 Law set up the Banque Générale Privée, which developed the use of paper money, as pioneered in England, Genoa, Holland, Sweden and Venice.

Law invested heavily in the Mississippi Company, causing wild speculation: shares rose to sixty times their original value. The bubble burst in 1720, and Law closed the bank and fled the country, leaving behind hundreds of ruined Parisian investors. The experience left the French with a deep suspicion of banking: "*The shock kept France from new attempts until in 1776 the Caisse d'Escompte was established.*"[43] The Banque de France was not founded until 1800, long after the Bank of Amsterdam (1609), and the Bank of England (1694).

And it was not just trade that underpinned growth in London and Great Britain more generally. Industry was expanding: rail and canal companies, ship builders, the extractive industries, the manufacturers, the textile makers: all had voracious appetites for capital and finance. This was the context within which a financial revolution got underway. The emergence of joint stock companies; the establishment of the Bank of England; the introduction of a widely circulating credit currency, the creation of a long-term funded national debt, and the emergence of an active market in securities all contributed to the revolution, which in turn, acted as a further spur to trade and industry.

The interdependency between industry and finance is a complex one: neither was the dominant driver of change. Ferguson poses the question of whether more sophisticated financial institutions encouraged industrial growth, or whether industrial growth spurred financial development, concluding that the two processes were interdependent, in a distinctly evolutionary fashion.

> *Recurrent mutation (technical innovation), speciation (the creation of new kinds of firm) and punctuated equilibrium (crises that would determine which firms would service and which would die out).*[44]

The economy was evolving from one based on credit to one based on a modern, financial system. Personal credit had been the order of the day and this was of course based upon social relationships and structures. They were personal credit relationships because, in the seventeenth century, enterprises were small, and typically family owned and funded. One's ability to secure credit rested more upon social standing and business reputation than on an audited balance sheet showing creditworthiness.

According to Muldrew, from the 1690s, a number of developments, including the formation of the Bank of England, and the broader investment opportunities that this facilitated, changes in the patterns of legislation and the development of small banks and corporate institutions led the economy to shift inexorably, changing the nature of credit.[45]

By the final decade of the century, a dynamic, integrated banking sector had emerged in London.[46] Formal financial intermediaries emerged, including the first English deposit banks; while some goldsmiths, merchants and scriveners were carrying out certain banking functions, including the acceptance of deposits, bill discounting and acting as factors for provincial merchants. Booker describes how, in the early-eighteenth century,

> *promisory notes became transferable to a third party, as the legal rate for interest fell to 5 per cent, and as paper money became widespread with the availability of Bank of England notes, so the private bankers created a framework of administration and procedure.*[47]

Banking as a specialist trade was emerging.[48] And one of the earliest was the bill discounting banks, which helped finance domestic and international trade by discounting the bills of exchange drawn by one merchant on another – merchant banking.

A bill of exchange is an unconditional written order, addressed by one person to another, signed by the person giving it, requiring the person to whom it is addressed to pay on demand or at a fixed or determinable future time a sum certain in money to or to the order of a specified person, or to bearer. Turner describes the bill as being originally developed to finance trade because of the inconvenience and risk of loss of shipping bullion over longer distances, and ubiquitous by the close of the century. Being transferable through time and distance the bill simplified the conduct of long-distance trade, and led to the concept that written acknowledgements of debts might be both transferred and sold.[49]

Bill discounting is the process that allows a business to get quick payment for their work and meet their operating expenses. Where a seller sells some goods or services to a buyer, the buyer has to make the payment as per the agreed credit period. However, if the seller needs money before that time, he can approach a bank and 'sell' the invoice to them. The bank then verifies the invoice with the buyer and makes the payment to the seller on their behalf. The payment is discounted by the bank's commission, hence, bill discounting. The process assists with the sellers' cash flow, while buyers gain an extended credit period.

The role of Government – as borrower rather than regulator – was also key to the changes taking place. For example, Murphy argued that the changes in state finance drew upon a wealth of knowledge and technical expertise that had already been established in London in the period 1685–1695, creating a network of jobbers and brokers that could trade securities and also allowed investors to develop a knowledge base of risk management.[50] Even by 1700, stock jobbing was cited as one of the principals of the prodigious conflux of the nobility and gentry from all parts of England to London, more than ever in former years.[51]

The scale of the changes to finance in the late-seventeenth century and very early-eighteenth century have often been overlooked. However, as Roseveare has argued, the changes in their totality, including a shift from a royal debt to a national debt amounted to a Financial Revolution and that, with respect to the adoption of long-term borrowing, the foundation of a great bank, the recoinage of the metallic currency and the circulation of a paper one, "*any one of these would make their decade [the 1690s] significant in financial history: together they unquestionably add up to a revolution*".[52]

The Bank of England. During the revolutionary times of the seventeenth century, Charles II took the extraordinary step of suspending the Exchequer – effectively suspending the payment of crown debts – and in the process ruining many goldsmiths. In the words of journalist and businessman Walter Bagehot (1826–1877), the credit of the Stuart Government "*never recovered from this monstrous robbery*", and by 1694 the credit of "*William III's Government was so low in London that it was impossible for it to borrow any large sum*".[53] The situation was made more urgent by the need for the Government to finance the war being fought by William III against Louis XIV of France.

In the face of such financial chaos, an audacious plan was hatched to raise £1.2m through public subscription. In return for their cash, the subscribers were granted a royal charter on 27 July 1694, under the title 'The Governor and Company of the Bank of England'. The Bank was set up as a private institution to provide the state with financial services in return for a profit.

The Bank of England was incorporated by Act of Parliament and charter in 1694 (Figure 2.7). The Act pledged taxes on shipping and liquor to a total of £140,000 pa, of which £100,000 would reward the subscribers of a £1.2m investment with 8%, and the management with £4,000 pa. The remainder would service the purchasers of £300,000-worth of annuities. In return the Corporation was allowed to deal in bills of exchange or bullion and to issue notes equal to its capital.[54] In the event the subscription was filled rapidly; in fact, between 21 June and 2 July.

Figure 2.7 Sealing the Bank of England Charter, 1694.

The promised charter was sealed on 27 July, "*and the Bank of England was immediately in business*".[55]

The Bank of England lent money to Government and developed a wider banking practice based on the elementary systems of credit exchange developed by goldsmiths. According to Booker, the early years of the bank were difficult and the ban on joint stock banking partnerships, introduced in 1708, was an attempt to increase its stability.[56]

Over the ensuing century, "*there were few pauses for peace, the national debt grew inexorably, and the Bank of England not only managed that debt but became a government bank, looking after the accounts of most of the departments of state*".[57] The Bank was the first English joint stock bank and by its charter of incorporation the Bank was allowed to discount bills of exchange, issue large volumes of notes and do other kinds of commercial banking. It thus combined under one roof elements of the commercial and the public worlds of finance. The "*Government had managed to construct a financial mechanism which in combination with the growing wealth of the nation gave it seemingly unlimited means of raising money from the population at large*".[58]

It is well known that the Bank supplied successive governments with funds to underwrite military campaigns, but it also built fiscal capacity more generally to underpin national economic progress, and by "*fostering the development of a more monetised economy, the Bank also made it easier for the state to collect taxes*".[59] The positive impact on fiscal policy meant that

> *it was possible for the state to increasingly demand funds without credibility problems affecting the ability to repay at equilibrium interest rates. A positive feedback loop ensued. The transformation of sovereign debt into money was, in turn, transmitted to the private economy, and this elastic supply of liquidity supported long-run economic growth and financial development.*[60]

The Bank opened for business on 1 August 1694 with a staff of seventeen clerks and two doorkeepers. Its first home was the Hall of the Mercers' Company in Cheapside, with tables "*furnished with pens, ink, paper and sand*". The Bank grew to 100 members of staff 25 years later; to over 300 in the 1780s and to over 900 by 1815. Murphy paints a graphic picture of its growth and development thereafter. For example, by the end of the eighteenth century, the Bank

> *was the largest private employer of white-collar workers in Britain. In 1783 there were 321 permanent clerks working at the Bank, more than double the white-collar staff of the East India Company and ten times the numbers employed by the large insurance companies, the Royal Exchange and Sun Assurance companies.*

And by this time, the Bank had 'divisionalised', with the workers

> *organised into two large departments: the Cashier's Office and the Accountant's Office and each of those departments was subdivided into more than a dozen offices, some employing just a few men, while managing the Three Per Cent Consols Office made work for more than fifty.*[61]

Consol. A consol was a British government security without a maturity date. The name is a contraction for Consolidated Annuities, a form of British government stock that originated in 1751. The first issue of consols carried an interest rate of 3% (reduced to 2.75% in 1888 and to 2.5% in 1902).

The work of those clerks partly concerned the provision of banking facilities to the London business community but the Bank's primary role by the late-eighteenth century was as banker to the state and manager of the national debt.

Importantly in terms of the wider growth of the banking sector, in 1708 and as part of the Bank's recharter, the Government granted the Bank a monopoly over joint stock banking in England and Wales, which was not removed until 1826 in the provinces and 1833 in London, and even then only for joint stock banks of deposit only. During this restive time, all other banks were limited to private partnerships with a maximum of six partners. Obviously this constrained their growth potential.

London's increasingly sophisticated financial structures, and its growing dominance over European centres, meant that by the turn of the eighteenth century it was established as *the* centre of world commerce.

> *In the eighteenth century came the great surge in British commerce, which took three main forms: a growing domestic demand for American and Asian consumer goods (above all sugar, tea, tobacco and coffee) and North European raw materials (such as forest products like timber); a growing European market for re-exports of American and Asian consumables; and a growing protected market for British manufacturers in the American colonies and Africa.*[62]

And the Bank of England was instrumental in this expansion.

> *It expedited the adventures and trades of London businessmen – from linen and diamond merchants to small-coal men, from the exporters of hats to the importers of sugar.*[63]

Summary. From the tragedy and devastation of the plague and conflagration the City of London was reborn. A new building agenda emerged with safety to the fore; the street pattern was changed and speculative building and new construction techniques all evolved. Alongside this development of a new physical environment, business relations also changed quickly, and the company in its modern share-based format evolved. Coffee houses emerged as venues for trade, gossip, information gathering and socialising, and these were quickly followed by formal exchanges with their commodity specialisms. The single most important meeting place in the City of London in the early-eighteenth century was the Royal Exchange.

The establishment of the Bank of England in 1694, the evolution of goldsmiths into primitive bankers and the emergence of an integrated banking sector based on deposit banks and bill discounting were all key events during this time. As well as trade in goods there was also trade in shares, money and credit. The markets were growing and becoming more complex with a proliferating array of instruments and products. Into this miasma grew a complex web of intermediaries who met, traded, gossiped and lunched amidst merchandise. Taken together the developments in the City of London at this time presaged and provided the necessary conditions for the development of the early office economy.

It was in this nascent centre of world commerce that the modern office was born. Goldsmiths became bankers and merchants did likewise, and both began to distinguish their activities with buildings that were dedicated to the new forms of activities. While continuing to look externally like domestic premises (underlined by clerks living on the upper floors), the buildings were arranged on the ground floor for commerce. However, it was not only the financiers that were beginning to organise around a new form of work: in the long shadow of the Great Fire, the insurers were growing rapidly and requiring space in which to create, manage and store growing volumes of policies.

Notes

1 Sheppard F (1999) *London: A History* BCA Books, London p234.
2 Porter R (1994) *London: A Social History* Penguin Books, London p109.
3 Summerson J (1953) *Architecture in Britain, 1530–1830* Penguin Books, London p44.
4 Summerson J (1953) *Architecture in Britain, 1530–1830* Penguin Books, London p22.
5 Summerson J (1953) *Architecture in Britain, 1530–1830* Penguin Books, London p45.
6 Summerson J (1953) *Architecture in Britain, 1530–1830* Penguin Books, London p45.
7 Gray R (1978) *A History of London* Hutchinson & Co, London p199.
8 Gray R (1978) *A History of London* Hutchinson & Co, London p200.
9 Berry H (2002) Polite Consumption: Shopping in Eighteenth Century England *Transactions of the Royal Historical Society* Vol 12 pp375–394.
10 Berry H (2002) Polite Consumption: Shopping in Eighteenth Century England *Transactions of the Royal Historical Society* Vol 12 pp375–394.
11 de Saussure C (1995) *A Foreign View of England in 1725–29* Cited in: Berry, Polite Consumption, p50.
12 Sheppard F (1999) *London: A History* BCA Books, London p228.
13 Bayne-Powell R (1937) *Eighteenth Century London Life* John Murray, London p114.
14 Davoudi L; McKenna C & Olegario R (2018) The Historical Role of the Corporation in Society *Journal of the British Academy* Vol 6 pp17–47.
15 Sampson A (1995) *Company Man: the Rise and Fall of Corporate Life* Harper Collins, London p25.
16 Walker CE (1931) The History of the Joint Stock Company *The Accounting Review* Vol 6 No 2 American Accounting Association pp97–105.
17 Joslin DM (1954) London Private Bankers, 1720–1785 *The Economic History Review* New Series Vol 7 No 2 pp167–186.
18 Smith A (1776) *An Inquiry into the Nature and Causes of the Wealth of Nations* Reprint edited by Campbell RH and Skinner AS (1976) Two volumes Clarendon, Oxford p741.
19 Hazlitt W (1822) *On Corporate Bodies* in *Table Talk, Essays on Men and Manners* Grant Richards, London 1901 p359.
20 Coke, Sir Edward (1826) The Reports Vol 5 Part 10 *The Case of Sutton's Hospital Joseph Butterworth, London* pp258–307.
21 Drucker P (1986) *Frontiers of Management* Dutton, New York p170.
22 Ukers WH (1922) *All About Coffee* The Tea & Coffee Trade Journal Company, New York.
23 Ukers WH (1922) *All About Coffee* The Tea & Coffee Trade Journal Company, New York p19.
24 Ukers WH (1922) *All About Coffee* The Tea & Coffee Trade Journal Company, New York p27.
25 Ukers WH (1922) *All About Coffee* The Tea & Coffee Trade Journal Company, New York p41.
26 Ukers WH (1922) *All About Coffee* The Tea & Coffee Trade Journal Company, New York p459.
27 Green M (2013) The Lost World of the London Coffee House *The Public Domain Review* https://publicdomainreview.org/essay/the-lost-world-of-the-london-coffeehouse/
28 Ukers WH (1922) *All About Coffee* The Tea & Coffee Trade Journal Company, New York p76.
29 Green M (2013) The Lost World of the London Coffee House *The Public Domain Review* https://publicdomainreview.org/essay/the-lost-world-of-the-london-coffeehouse/
30 Green M (2013) The Lost World of the London Coffee House *The Public Domain Review* https://publicdomainreview.org/essay/the-lost-world-of-the-london-coffeehouse/
31 Kynaston D (1994) *The City of London Volume One: A World of its Own 1815–1890* Chatto & Windus, London p10.
32 Pevsner NA (1976) *A History of Building Types* Thames & Hudson, London p199.
33 Carlos AM & Neal L (2011) Amsterdam and London as Financial Centers in the Eighteenth Century *Financial History Review* Vol 18 No 1 pp21–46.
34 The Royal Exchange: www.theroyalexchange.co.uk/christmas/changing-faces/ Accessed 12 July 2020.
35 Pevsner NA (1976) *A History of Building Types* Thames & Hudson, London p199.
36 Stringham E (2002) The Emergence of the London Stock Exchange as a Self-policing Club *Journal of Private Enterprise* Vol 17 No 2 pp1–19.

37 Stringham E (2002) The Emergence of the London Stock Exchange as a Self-policing Club *Journal of Private Enterprise* Vol 17 No 2 pp1–19.
38 Reprinted in: Jenkins A (1973) *The Stock Exchange Story* Heinemann, London p45.
39 Sheppard F (1999) *London: A History* BCA Books, London p227.
40 Ferguson N (2009) *The Ascent of Money: A Financial History of the World* Penguin Books p49.
41 Carlos AM & Neal L (2011) Amsterdam and London as Financial Centers in the Eighteenth Century *Financial History Review* Vol 18 No 1 pp21–46.
42 Carlos AM & Neal L (2011) Amsterdam and London as Financial Centers in the Eighteenth Century *Financial History Review* Vol 18 No 1 pp21–46.
43 Pevsner NA (1976) *A History of Building Types* Thames & Hudson, London p202.
44 Ferguson N (2009) *The Ascent of Money: A Financial History of the World* Penguin Books p54.
45 Muldrew C (1998) *The Economy of Obligation, the Culture of Credit and Social Relations in Early Modern England* Basingstoke pp328–329.
46 Turner GD (2015) *English Banking in the Eighteenth Century: Bankers, Merchants and the Creation of the English Financial System* Durham theses, Durham University Available at Durham E-Theses Online: http://etheses.dur.ac.uk/11297/ p40.
47 Booker J (1990) *Temples of Mammon: the Architecture of Banking* Edinburgh University Press p5.
48 Turner GD (2015) *English Banking in the Eighteenth Century: Bankers, Merchants and the Creation of the English Financial System* Durham theses, Durham University Available at Durham E-Theses Online: http://etheses.dur.ac.uk/11297/ p31.
49 Turner GD (2015) *English Banking in the Eighteenth Century: Bankers, Merchants and the Creation of the English Financial System* Durham theses, Durham University Available at Durham E-Theses Online: http://etheses.dur.ac.uk/11297/ p33.
50 Murphy AL (2009) *The Origins of English Financial Markets: Investment and Speculation before the South Sea Bubble* Cambridge University Press pp220–24.
51 Porter R (1994) *London: A Social History* Penguin Books, London p173.
52 Roseveare HG (1991) *The Financial Revolution 1660–1750* Routledge, London p47.
53 Bagehot W (2009) *Lombard Street: A Description of the Money Market* Seven Treasures Publications (Original work published 1873) p44.
54 Bayne-Powell R (1937) *Eighteenth Century London Life* John Murray, London pp118–119.
55 Roseveare HG (1991) *The Financial Revolution 1660–1750* Routledge, London pp36–37.
56 Booker J (1990) *Temples of Mammon: the Architecture of Banking* Edinburgh University Press pp2–3.
57 Kynaston D (1994) *The City of London Volume One: A World of its Own 1815–1890* Chatto & Windus, London p13.
58 Sheppard F (1999) *London: A History* BCA Books, London p139.
59 O'Brien PK & Palma N (2022) Not an ordinary Bank But a Great Engine of State: the Bank of England and the British Economy, 1694–1844 *Economic History Review* pp1–25.
60 O'Brien & Palma, Not an Ordinary Bank But a Great Engine of State.
61 Murphy AL (2017) Clock-watching: Work and Working Time at the Late Eighteenth Century Bank of England *Past & Present* Vol 236 Issue 1 pp99–132 (August 2017).
62 Kynaston D (1994) *The City of London Volume One: A World of its Own 1815–1890* Chatto & Windus, London p9.
63 Kynaston D (1994) *The City of London Volume One: A World of its Own 1815–1890* Chatto & Windus, London p307.

3 A new economy of mercantilist capitalism

London had a thriving insurance industry before the Great Fire, the specialist marine insurance industry, which had its origins in the early-seventeenth century, when Britain's role in international trading flourished. From early on, ships and their cargoes were insured with private capital, using an actuarial approach to underwriting: "*No merchant could be without insurance*".[1] The industry continued to grow in the latter half of the eighteenth century.

The Lloyd's coffee house was opened in 1686 by Edward Lloyd (c1648–1713) on Tower Street, before relocating to Lombard Street in 1691. The venue was popular with merchants, ship owners and sailors for dealing in insurance and ship broking as well as a source of shipping news. Lloyd's List began publication in 1734. By the middle of the eighteenth century London had become the most important centre for marine insurance in Europe.

Two examples of successful marine insurers were the Royal Exchange Assurance and the London Assurance. Both companies were formed in 1720 as marine insurance companies. They each received a royal charter under the *Royal Exchange and London Assurance Corporation Act 1719*, popularly known as the Bubble Act. While the charters gave the two joint stock companies exclusive corporate rights to write marine insurance, it did not prevent private underwriters, including those at Lloyd's of London, from continuing with their business.

The Royal Exchange Assurance survived as an independent company for over two centuries until merging with the Guardian Assurance Company in 1968, while London Assurance was absorbed by Sun Alliance in 1965. The first Governor of London Assurance was Walter Chetwynd, a member of the Staffordshire aristocracy, Member of Parliament and in 1717, created Viscount Chetwynd of Bearhaven, County Cork. Following the royal charter for Marine Insurance in 1720, a further charter was granted for Life business in 1721. The capital then authorised was £1,500,000, of which £896,550 was subscribed. Through a formal history of the first 200 years of the company, we also gain a glimpse into the working practices of an insurance company at this time.

> *If anyone conversant with the scientific and elaborate methods of today could transplant himself two hundred years back and approach the London Assurance as it then was, he would be truly astonished. The Directors sat personally to examine intending insurers, and their staff consisted of a secretary and two or three clerks.*

The Directors themselves examined applications for policies, dealing with both marine insurance and fire and life.

DOI: 10.1201/9781003582205-4

> *So a man proposing to insure his life would confront three Directors, and if they believed his favourable account of himself, and judged from his appearance that he was pretty certain to last a year they accepted his premium for that period. If not, they turned him down.*[2]

The same source illustrated the company's growth. Income from marine insurance grew from £9,740 in 1720 to £1,743,928 in 1917; and over the same period income from fire insurance rose from £570 to £1,231,617 and life insurance income grew from £170 to £247,386.

The London Assurance started corporate life in 1720 at The Rising Sun in Broad Street before moving in 1721 to Cornhill, near the end of Birchin Lane. The hours of business were 10.00 to 14.00 and 17.00 to 19.00 "*seeming to leave a liberal interval in the middle to prevent over-strain, and to refresh exhausted nature*"![3] By 1748, the corporation had moved to White Lion Yard Birchin Lane, before a fire on Cornhill: "*the Corporation was burnt out, but preserved its books and papers*". The Buffalo's Head in Threadneedle Street provided temporary premises. In 1752, the corporation was on the east side of Birchin Lane, and in 1845 moved to an office in the Royal Exchange, where it remained until 1922 at which point it bought a freehold site in King William Street to construct a building to its own design.

In 1773, the underwriters formed The Society of Lloyd's and took rooms of their own in the Royal Exchange, the birthplace of Lloyd's of London insurance market. By 1810, private insurers at Lloyd's were underwriting risks to the value of around £140m, with a total premium income of over £10m.[4] In 1811, the underwriters established a committee to manage their affairs.

The original Royal Exchange was destroyed by fire in 1838 and Lloyd's moved to South Sea House before returning to the rebuilt Royal Exchange in 1844, where it remained until 1928 when it moved into its own building at Leadenhall Street. By the end of 1950s, the building at Leadenhall Street became too small and Lloyd's moved to Lime Street, which has been home to the marketplace to this day. However, since 1986 it has been housed in its eponymous building, which was designed by the renowned British architect Richard Rogers.

Despite the success of marine insurance, it was general insurance that was to have the greatest impact on the office economy. The Great Fire famously started at a bakery in Pudding Lane and from there it spread rapidly, quickly overwhelming the pitiful attempts to control it; timber buildings were incinerated like matchwood and the fire raged for 5 days. In the process, around one-third of London was destroyed. But from this utter devastation rose a lasting benefit – building insurance and then the wider accident and life insurance industry. The devastation left by the conflagration changed the question of insurance "*from a matter of convenience into one of urgency*".[5]

The fire office

Following the fire, King Charles II declared there would be collections at churches across the country for the Lord Mayor of London to distribute amongst homeless and destitute Londoners. But the receipts proved woefully inadequate with the call coming so soon after a similar call following the plague a year earlier. The situation in London was worsened by the fact that, at the time of the Great Fire, tenants were liable for repairs to their houses; and they were also expected to pay rent while their destroyed houses were being rebuilt. Being entirely impractical in the circumstances, an emergency 'Fire Court' was set up to resolve disputes, with the judges empowered to decide who should rebuild, based on ability to pay. The Court ran from February 1667 to September 1672.

In 1680, the first joint stock company for fire insurance in London, the Fire Office, was set up by Nicholas Barbon. His role as a speculative developer was described in Chapter 1, but Barbon also distinguished himself as a founder of the property insurance industry. The Fire Office was renamed the Phoenix Office in 1705. In 1681, Barbon and a number of associates, founded the 'Insurance Office for Houses' next to the Royal Exchange, and in 1683 William Hale and Henry Spelman published proposals for 'The Friendly Society for Securing Houses from Loss by Fire', and in the following year set up the Friendly Society near Temple Bar. These were followed by the General Insurance Office (1685) and Lombard House (1694) and the Amicable Society (1706).

Others would follow, including the Sun Fire Office (1710), the London Assurance (1720), the Royal Exchange Assurance (1720), the Phoenix (1782), the County (1807), the Guardian (1821) and the Alliance (1824), whose backers included Alexander Baring, Samuel Gurney and Nathan Rothschild. A number of provincial companies were also formed, notably the Norwich Union (1797).

From the beginning insurance companies were well funded by private capital, often in the form of mutual and joint stock insurers, and often privately funded by local merchants and manufacturers. They were technically advanced and used an actuarial approach to underwriting. By the early-1720s, they had collectively underwritten 17,000 policies totalling £10 million – enough to cover the estimated cost of property that the Great Fire had destroyed. By the mid-eighteenth century most properties in London were insured.

In the early years, insurance companies employed their own fire crews and equipment, and the buildings they insured were identified by affixing a distinctive, lead-made (later tin-made) 'fire marks', stamped with the policy number, which made their identification easier in times of stress, especially before the introduction of street numbering in the 1760s. Insurance companies often had reciprocal arrangements with each other, such that one brigade could be reimbursed for dampening a fire at a property insured by another company. A single unified fire fighting force would not emerge until 1833, with the London Fire Engine Establishment.

Further, government legislation in areas such as employers' liability led to new lines in accident insurance. New developments in railways, steam and electricity all brought with them new forms of accident from which employers needed to protect themselves. During the late-1800s and early-1900s, insurers introduced cover for new forms of transport such as aeroplanes and motor cars. The first aviation liability policies in the UK were written before the First World War, and from 1919 the market expanded rapidly.

Life insurance companies were also established in the UK during the 1700s, but this market matured much less quickly than non-life. The Society for Equitable Assurances on Lives and Survivorships, founded in 1762, is generally considered to be the first dedicated life insurance company to appoint an "actuary" and conduct life insurance in much the same way as is done today. In the early-1800s, there were only a handful of British life insurers and customers – as might be expected for this time – were limited to the wealthy. However, by mid-century, the rapid expansion of the Victorian period was underway and society became infinitely more complex and affluent, allowing the life insurance sector to fulfil its potential.

Even more so than the banks, it was the rise of the insurers that generated the demand for platoons of black-coated workers – the clerks, and a new building type. The rapid growth in policies led to an enormous demand for writing, copying, bookkeeping and corresponding with an ever-growing customer base. There is scarce written evidence of how this played out in individual companies, but the following presents some such evidence from historical records of the Sun Insurance Office.

Sun Insurance. The Sun Fire Office, as it was originally known, was founded in 1710 by the eccentric Charles Povey. In 1960, University of Oxford Professor Peter Dickson published a meticulously detailed account of the company's 250 year history[6] and the following paragraphs pull out some of the more relevant elements for our story here.

Charles Povey was an eccentric, sometime entrepreneur, pamphleteer and author, whose first business was as a wholesaler of coals at Execution Dock in Wapping, with a strange and unbalanced personality whose interest in astronomy may have influenced his choice of name for his nascent insurance business. Povey's initial attempts to establish an insurance business were not a roaring success and he found himself having to reach out to a funding syndicate, leading to the creation of the Company of London Insurers, which then became the Sun Fire Office.

However, Povey caused disagreement among his syndicate members who "*quickly and painfully became aware that Povey was an irascible colleague whose presence at Board meetings made the formulation of policy extremely difficult*".[7] In short order Povey was encouraged to sell his interests in the concern to the 24 members of the Company of London Insurers in 1710. Thereafter Povey exercised no official control over the Sun Fire Office.

Like many businesses, the Sun Fire Office began life in a coffee house; more specifically Causey's Coffee House near St Paul's Cathedral. But, in the following year, evidently finding these first quarters too cramped, the Managers "*removed to two rooms in a house belonging to Mrs Alice Garway next door to the Amsterdam Tavern in Sweeting's Rents at the side of the old Royal Exchange*".[8]

> *Although the accommodation provided was not sumptuous, some care was taken in installing proper fittings. A handsome Lanthorne was bought for the staircase, genteel thick blue curtains and umbrellos (blinds) for the windows, carpets for the floors, a handsome press for books, and a scrutoir for the committee room.*[9]

In these early stages there were around 24 managers in the office and two clerks.

> *the Managers met quarterly to dine together at one of the London taverns, spending on the meal £6 to £8, a sum equal to a year's wages for many contemporaries, and conscientiously sampling the gastronomic resources of as many inns as possible.*

In 1717, the number of clerks grew to three, and by 1720 to five. Their duties included receiving premiums and instructions for insurances, paying claims, and filling and issuing policies. One of the clerks also supervised "*The more active tasks of soliciting new business and affixing the Office's fire marks to insured houses [which] were entrusted to a body of Walkers*".[10] Making up the remainder of the Office's payroll was the corps of thirty firemen: "*evidently a hardy and intractable group whose foreman had some difficulty in keeping them in order*". Their rudimentary equipment included helmets, leather buckets and 'preventers' – hooks on long staves for moving obstacles. The firemen could identify a building covered by a Sun policy because it displayed a fire mark, as shown in Figure 3.1.

The Sun's first decade gave little intimation of its future size and significance. A disparate and shifting body of managers, in the main lacking significant City connections, coupled with a limited number of staff and types of transaction, held back development. From about 1720, however, there occurred a series of events that were to help set the Sun firmly in the forefront of London's fire offices: a complete reorganisation of the firm's capital structure; the appointment of the first of a series of able and honest men to the two principal positions of treasurer and secretary – Colonel Robert Dalzell and Thomas Watts; and a restructuring of the Sun's management.

Control of the Sun's affairs had originally been confined to two bodies – the general meeting of managers and, of more practical importance, the Committee of Management appointed

Figure 3.1 The Sun Fire Office fire mark, issued between 1727 and 1732. Licenced under the Creative Commons Attribution 2.0 Generic license.

from among them. From about 1720, the latter body took the important step of appointing sub-committees to conduct and report on particular aspects of the office's business. By about 1730 there were four of these, each controlled by a manager and staffed by clerks. These sub-committees were the ancestors of the Sun's modern departmental system.

The Sun's managers were powerful, as shareholders exercised no control over their activities and were not permitted to see accounts, a state of affairs then typical of most British insurance companies and not remedied until the end of the nineteenth century.

In 1722, the Managers secured a lease on the whole of their building "*and busied themselves with making it comfortable*", but 5 years later another move was instigated. In 1727, one of the Sun's managers, Mr Lloyd, negotiated a new lease with a Mr Jackson on a building recently vacated by the South Sea Company's Annuity Office, next to the Crown Tavern in Threadneedle Street (Figure 3.2). It was agreed that the Sun should take a 21 year lease on the premises for a fine of £250 and £75 per annum rent.[11]

Figure 3.2 The Sun Insurance Office, Threadneedle Street, 1761.

After this move the business was evidently directed from the office in Threadneedle Street. Figure 3.2 shows the modest scale of the office with the company name emblazoned on the side of the building. Then, in October 1760, and due to the recent rapid increase in the scale of their operations, the Managers resolved

> *That it be referred to the Committee of Management to purchase a piece of ground in such a place as they shall think convenient and proper for the erecting a building for an office for this society or to take a lease of any ground or buildings as they shall think most for the convenience of the office.*[12]

In 1765, the Bank of England informed the Sun's General Meeting "*that it was prepared under the terms of a recent statute authorising the construction of a new thoroughfare from Cornhill to the Bank, Bank Street, to build a new office for them there and to lease it to them for ninety-nine years at a rent of £600*". The move to the new office took place in 1766, where it remained until the 1840s, supplemented by various extensions and the purchase in 1782 of a house next door, "*the present Office not affording Room for additional Clerks, which are, or maybe, necessary for carrying on the business*".[13]

Anticipating a growth in business in the residential areas west of the City, Sun established an office in Craig's Court in 1726, with one clerk and a messenger. By the end of the century this had expanded to a "*head clerk and four ordinary clerks handling a premium income roughly half as large as that received by the much bigger head office in Cornhill*".[14]

Both the City and West offices were growing the volume of business. In 1793, the Policy Department in the headquarters was divided into a Town Department and a Country Department, and the total number of clerks employed grew to 25.

In January 1838, the neighbouring Royal Exchange building was destroyed by fire, necessitating a further move for the business. A site was identified on the corner of Bartholomew Land and Threadneedle Street, almost opposite the old office. In March 1841, the Building Committee was in a position

> *to ask for tenders for the construction of the new office building to designs prepared by one of the leading architects of the day, Professor CR Cockerell, whose elder brother John was one of the Sun's Managers, Eight offers were received, of which Cubitt's (£21,980) was the highest and that of Webbs' (£18,740) the lowest, The latter was at once accepted and the work of building began.*[15]

The new building comprised ground, first and second floors; it "*was handsomely constructed and at first provided ample room for the clerks, for the Life Office and for the Personal chambers of Mr Richards, the Office Secretary*". However, as the volume of business grew

> *the formerly spacious quarters began to seem cramped. In 1865, the Fire Office was obliged to take over both the Office Secretary's chambers on the first floor and the private apartments rented by the Life Office for the use of the clerks and in 1881 to make extensive changes to the internal structure of the building.*[16]

All staff were based at Threadneedle Street except those in Craig's Court, which remained in occupation until 1865. At this time an opportunity arose for obtaining a long lease of numbers 60 and 61 Charing Cross on the south side of Trafalgar Square, as well as the freehold of a house at the back of the existing building.[17]

Table 3.1 shows how the number of workers at Sun Insurance grew during the nineteenth century. It was this increase in the size of the staff which compelled the unfortunate Office Secretary, Mr Fletcher, to give up his chambers for the benefit of the clerks in 1865. The number of letters written and received at head office, for example, nearly doubled between 1866 and 1880. Notwithstanding the increased number of clerks, reported a special committee of Managers in April 1881,

> *the Staff ... has been unable to cope in Office hours with the increased work and a considerable portion of such work has been done out of office hours. The Clerks have received extra payment for such work in accordance with the Regulations from time to time ...*[18]

Dickson provides some fascinating insight into the nature of clerks at this time. They included "*men of considerable administrative ability*" who served long tenures, often with successive members of the same family. Outside of office work, "*clerks were also active, in learned societies, in local government, in athletics, or in volunteer regiments*". In 1863, six clerks also

Table 3.1 Growth of Sun Insurance staff, 1806–1880

Department	*Number of staff*		
	1806	*1866*	*1880*
Accountants	9	15	15
Secretaries	2	9	10
Town	4	7	8
Country	4	11	21
Duty	1	1	–
Craig's Court	5	5	5
Messengers	1	8	10
Receipt & Issue	–	2	2
Surveyor	–	4	4
Guarantee	–	1	4
Foreign	–	–	2
Total	*26*	*63*	*81**

Source: Dickson (1960) p123.

Note: * Some data inferred from text

served in the London Rifle Brigade and nine in other regiments, "*partly supported by small grants from the Office*". Besides the clerks the other members of staff included seven Collectors and eight Messengers.

Sun Insurance retained its separate identity for almost 250 years, until 1959. In this year it merged with the Alliance Assurance Company, founded in 1824. The new organisation acquired the London Assurance Corporation in 1965, to become the Sun Alliance & London. It acquired the Phoenix Assurance in 1984 and Royal Insurance in 1996, whereupon it became the Royal & Sun Alliance Insurance Group (RSA).

This potted history of Sun Insurance demonstrates how at least one of the early office-based, corporate organisations evolved through various stages with successive offices to meet its changing requirements, each office larger and more sophisticated than the previous one. This same process was to be repeated across countless nascent corporate organisations over the coming decades.

The insurance sector was the first discrete element of the office economy to emerge, doing so at a time when most bankers still worked also as merchants or goldsmiths. As we shall see, banking quickly followed insurance as an identifiable sector, initially with small family owned businesses and then with joint stock banks and larger private companies employing managers to oversee expanding workforces. Legal and accounting firms were next on the scene and before long the 'financial and business services' sector emerged, with its complex system of supply chains and cross-business relationships. The stage was set for the emergence of office work, as we understand it today, and for the wider office economy.

Merchants become bankers

Bagehot reminds us of the peculiarly English origins of deposit banking, of which he can imagine

> *nothing better in theory or more successful in practice than private banks as they were in the beginning. A man of known wealth, known integrity, and known ability is largely entrusted*

with the money of his neighbours. The confidence is strictly personal. His neighbours know him, and trust him because they know him. They see daily his manner of life, and judge from it that their confidence is deserved.[19]

But he goes on to highlight the improbability of the central arrangement, the essence of which is that a very large number of people elect to trust a very few. For banking to profit, a small number of bankers deals with a much larger number of depositors.

But to get a great number of persons to do exactly the same things is always very difficult, and nothing but a very palpable necessity will make them on a sudden begin to do it. And there is not such palpable necessity in banking.

Indeed, Bagehot cites France where

you will not find any such system of banking as ours. Cheque books are unknown, and money kept on running account by bankers is rare". He goes on "If a branch, such as the National Provincial Bank opens in an English country town, were opened in a corresponding French one, it would not pay its expenses.[20]

In the seventeenth century, some London goldsmiths began exchanging foreign and domestic coins, exploiting the discrepancies in the weights and values of the coins.[21] They also began to accept deposits. The political and economic turmoil of the age, not least the unsettling backdrop of a civil war, encouraged many of the landed folk to take money and valuables out of their houses and deposit them in the safe vaults of the goldsmiths in London.[22]

Roseveare noted that goldsmiths had been quietly developing their financial activities for some time, and that the 'safe deposit' function was only the most basic of the services they performed.

They had long been dealing in gold and silver bullion and in the 1630s, when London merchants began to handle large quantities of Spanish silver, they were drawn more actively into the international exchange market which was shared with bills of exchange.[23]

Consequently, some goldsmiths, merchants and scriveners, such as Robert Vyner, were carrying out basic banking functions, including the acceptance of deposits, discounting bills and acting as factors for provincial merchants before the Civil War.[24] And Roseveare suggests that by the 1650s, "*several goldsmiths had conspicuously emerged as large-scale, full-time bankers serving extensive clienteles which included government departments and the state itself*".[25]

Paper money, or bank notes, evolved from the negotiable receipts that goldsmiths gave their customers for the money and valuables that they deposited for safekeeping. The goldsmith held the money as 'running cash', (similar to a modern current account), and paid interest to the customer on their deposited money. The goldsmith then earned interest on loans to other customers. This very basic banking activity, typical of many businesses operating in the eighteenth century, usually took place in converted residential premises. Many had a cellar where the safe was kept, a shop (what became in time the public banking hall, distinct from the parlour and meeting rooms) and parlour (what became back office) on the ground floor, and a sign outside.

Thus it was in the 1650s, with the development of paper credit money by the goldsmiths, that banking as we understand it today emerged. London's goldsmith-bankers were transformed from creditors into *the biggest debtors* in London as they began taking deposits on account from merchants and landowners and making loans and discounted bills. They learnt to issue promissory notes and made their deposits transferable by 'drawn note' or cheque. They became banks, in all but name.

Opportunities for other forms of investment were expanding, as economic activity increased and new assets became available during the Restoration period. Not least among these was growing activity in the land market during the latter half of the seventeenth century. And as trade expanded rapidly through the port of London so too did the demand for financing, insuring, broking, selling, negotiating and countless other specialist skills.

> *As an essayist noted in 1749, Britain's world role in overseas and entrepot trades demanded a rapid growth of Agents, Factors, Brokers, Insurers, Bankers, Negotiators, Discounters, Subscribers, Contractors, Remitters, Ticker-mongers, Stock-jobbers, and of a great variety of other dealers in money.*[26]

As business increased, bankers found a need to specialise, and a separation of activities occurred, which persists to the present day. First, banks in the West End developed their goldsmiths' origins by continuing to serve royalty, aristocracy and the landed classes, principally as banks for deposit and personal loans. Roseveare refers to West End banks "*along Fleet Street and the Strand, such as Hoare's, or Child's, or Gosling's, which served an aristocratic and landed clientele and tended to specialise in mortgages*".[27]

In contrast, those banks in the City increasingly served mercantile interests, providing discounting for merchants, shippers and other banks, and acting as the London correspondents for country colleagues. Roseveare argues that "*in Lombard Street there remained the City banks with a mercantile clientele whose needs were more varied and demanding. Loans on the credit of trade goods, foreign bills of exchange and investment in the entire range of stock market securities figure normally in their assets*".[28]

Merchants or wholesalers were primarily interested in their firms' extension of credit and in the management of their balance sheets, which were dominated by accounts receivable and debts owed. From this perspective they resembled financial firms rather than manufacturers. By the 1770s, one of the key similarities between the merchant and the banker was that they were both significant mobilisers of capital, and were among the most capital intensive businesses in Britain at the time.[29]

While Amsterdam remained the dominant European financial market for most of the eighteenth century, over the later decades of the century, this position began to change quickly "*and London assumed the position of 'first-amongst-equals' of European financial centres*". While other cities "*remained important in an era when markets remained regional or local given the limitations that technology placed on communications*"[30] the position shifted further in London's favour over time, ultimately leading to its dominant European role.

By the final decade of the eighteenth century, a dynamic, integrated banking sector had emerged in London, complete with the establishment of the Bank of England. While in 1720 London had a core of around 25 banks, this had nearly tripled to 70 by end of the century.

Even at this stage, business was conducted in what amounted to a domestic setting – a private house with ground floor 'shop' and residence above. General business was conducted in the shop, over a counter, while more discrete business was discussed in a parlour where the trappings of the social standing of the banker were evident. Furniture and fittings were domestic in scale and character (see Figure 3.3). Figures 3.3 and 3.4 show two renditions of 18th century offices which emphasise the small, domestic scale of enterprises of this time.

Figure 3.3 Office by Julius Bernhard von Rohr, c1719.

An example of the metamorphosis from metal to money is provided by one of the earliest banks, Child & Co. Goldsmith Robert Blanchard had been operating from premises in the Strand through the 1650s and 1660s, where his business was identified by the sign of the Marygold. Blanchard was joined in 1665 by Francis Child, and 8 years later they moved their trade and sign to a site at the west end of Fleet Street. By 1677, Child had become a partner and then married Blanchard's stepdaughter. In 1681 he inherited the entire business.

Figure 3.4 A merchant's office by Thomas Rowlandson, 1789.

In 1689, he was knighted and appointed 'jeweller in ordinary' to King William III, and he later served as a Member of Parliament. Supported by the patronage of the Earl of Dorset, who was Lord Chamberlain, he regularly advanced large sums of money to the Treasury. In 1698, he was elected Lord Mayor of London. Child died in 1713 and his three surviving sons ran the business in succession until the 1750s. Curiously, Francis Child's grandson Robert had no male heirs and so the Child fortune was eventually settled on his granddaughter, Sarah Sophia Fane, who married the 5th Earl of Jersey, and who acted as senior partner of the bank from 1806 to 1867 – an extraordinary feat for a female during this time.

The state bank. One of the oldest bank buildings in Great Britain and certainly the first purpose-built bank, was the extensive building, or collection of buildings, that is the Bank of England. The complex comprises a range of styles and forms, reflecting its varied architectural heritage, beginning with the oldest part, designed by George Sampson in 1733; then the lateral wings of the south front and the returns on the east and west sides, with several offices immediately attached, built by Sir Robert Taylor, between 1770 and 1786; and finally the many alterations and extensions made by Sir John Soane over his long tenure there (see 'The Old Lady').

Anne Murphy provided a valuable insight into eighteenth century office work with her analysis of working at the Bank of England.[31] She drew attention to the importance of time and time-keeping to the specialisation of work, co-ordination and scrutiny. For example, specific tasks were dictated by the clock as were breaks and co-ordination between offices. And so office hierarchy was thus important in terms of who was monitoring whose time keeping. Given the monotony and laborious nature of much work, this must have had a negative effect.

Murphy also confirms the length of the working day – long by modern office standards. She identifies the 09.00 to 15.00 slot, which focused on interaction with customers, and then the later shift "*from the late afternoon, sometimes into the very late evening, in order to update the ledgers and ensure that the records accurately reflected the various accounts at the start of the next working day*." Murphy cited one individual's account:

> *a basic working day that commenced at nine and commonly finished around 2 p.m. but he said that besides that job he was also employed, at a small additional salary, in posting bills and notes into the Discount Ledgers. That business could not start until late in the working day and had to be completed prior to the start of the next working day. Mr Southey estimated that on a moderate day he did not get away until six or seven in the evening and that frequently he stayed much later.*

This was not an uncommon situation in the later-eighteenth century when clerking remained an aspirational occupation for gentlemen. The overriding ethos was that the work had to be completed, and the day finished when the work was done. No overtime, flexi time or nine-day fortnight. Just do the work!

Even for those clerks who finished their shift by early or mid-afternoon, they rarely took a break, which meant that they typically worked between 6 and 7 hours. Similarly, those working in offices where business hours were 09.00–17.00 generally started a before 09.00 and usually took one and a half hours for lunch, again implying a working shift of around 7 hours. For such clerks this meant a working day that, although somewhat more onerous than other white-collar workers, was less hard than that of many of their fellow Londoners who worked close to a 12-hour day, starting work between 06.00 and 07.00 and finishing at around 19.00.[32]

And of course, like other workplaces, the Bank of England opened on Saturdays. The clerks were free on the Sabbath, but many probably thought, as Charles Lamb did, that Sunday was the 'very worst adapted for days of unbending and recreation'. It was a day when London was quiet, the shops shut and even the 'strollers in the fields on that day look anything but comfortable'.[33]

Murphy's description of the need to ensure accurate record-keeping and a consistent service to the public, with its timed routines and rhythms, shows how the clock became a measure of the performance of the clerks and the whole institution. Indeed, "*deadlines, especially as they related to the demands of public scrutiny, were in many ways tyrannical*".[34]

The overriding message in these passages of working life is one of relentless, mundane and repetitive work, often undertaken in poorly ventilated, ill-lit and cramped spaces. The work was often closely scrutinised, through fairly unforgiving management. All of these conditions were in place by the turn of the eighteenth century. And the same conditions simply magnified when the number of clerks grew from, say, 5, to 10, to 20, to 100. Indeed as the work scaled up, so clerking began to resemble factory production.

In order to cope with these changes in scale, it is also abundantly clear that the buildings within which the work took place would also have to evolve. What were in fact domestic buildings would have to give way to purpose-designed workplaces.

Banking houses. The banking houses themselves reflected the growing wealth and social stature of their owners, although they were *banking houses*, they were not bank buildings. While the owners had personal wealth, their businesses were not sufficiently capitalised to allocate funds to the construction of dedicated buildings. Instead, and perhaps even more importantly, their banking houses reflected their reputations for thrift and probity. It was not until the brash new joint stock banks emerged in the mid-nineteenth century that it became acceptable within banking to build grand and ostentatious premises, both to house their various operations and to serve as architectural expressions of their wealth, respectability and reliability.

As already stated, there were very few offices, as we now understand the term, in the eighteenth century; but the structure of the future office economy was taking shape. The building blocks of a sophisticated financial system were in place – joint stock companies, central banking, exchanges, banks and insurers – were all now evolving rapidly. Working conditions were better than factories, by and large, but they were a world away from today's office standards.

Virtually every bank before the mid-eighteenth century occupied a converted house; the buildings were adapted for a mix of business and residence – a mix which had long-lasting impacts on work culture and organisation. Following the Great Fire, the goldsmiths and early bankers had concentrated in and around Lombard Street

> *Private bankers had no need to move to another quarter. A house with a cellar below, a shop and a parlour on the ground floor, and a sign outside, was as suitable for a self-styled banker as it had been for a goldsmith, particularly if he was one and the same man.*[35]

The bank typically occupied the ground floor and basement, with the upper floors allocated to the owner and his family or a manager and family or clerks, or a combination of these.

A very large proportion of premises had a narrow frontage, with one room at the front and one room at the rear, and extensive depth: overall a narrow, oblong layout. Adding to this layout rudimentary lighting in many premises, together with the domestic furniture and accoutrements, and one can easily garner an image of a quite primitive business environment. Kynaston cites the example of one business owner, Bertram Currie, whose family bank in Cornhill was

> *'narrow, low, and not over well lighted', though 'of considerable depth'. Even the main banking hall of Glyns, the largest of the private banks, was a dim, shabby and fairly ramshackle affair.*[36]

Shops at this stage were a more established building use than offices and yet they served a similar function: small numbers of customers, often known to the owner personally, visiting at any one time, receiving the dedicated attention of the owner. The early office in this sense performed the same role as a shop; indeed private bankers uniformly referred to the ground floor of their premises, where the business was done, as 'the shop'. The building interiors were basic and functional.

> *To cross the threshold of one of these firms was usually to enter an austere world. Take Prescotts in Threadneedle Street: massive oak doors opened into a dimly lit banking hall, as elsewhere known as 'the shop', where clerks dealt with customers over an ancient oak counter and with quill pens and snuff boxes at hand to meet all needs; while beyond the oak-panelled partners' parlour, directly above the strong room in the basement.*[37]

Anderson describes the interior as comprising 'sombre little counting houses, with a table in the centre, a raised desk in a covered recess where a clerk may sit on a high stool, some cupboards and an iron safe in the wall' (see Figure 3.4). In such places Manchester merchants 'kept their books, counted their cash and laid the foundations of their fortunes'.[38] Goldsmiths and early bankers referred to the ground business space as 'the shop'. They often had basements where they kept cash and valuables in a secure environment.

Whether it was because they lacked sufficient capital to fund whole new purpose-designed bank buildings, or whether it was because of a desire to maintain certain social qualities such as discretion and trustworthiness, bankers chose to continue with the 'banking house' model. The compromise resulted in a template that resembled a large Georgian town house. Pevsner observed that

> *The headquarters of Georgian and even early Victorian banks were exactly like private houses, with the business rooms on the ground floor, left and right of the entrance, and the managers' living quarters above.*[39]

Black notes that the tradition of private banker living above the shop continued well into the mid-Victorian decades. He also describes the interiors of typical banking houses:

> *Everyday business, such as paying in bills or cashing, would be done at a long mahogany counter dividing the floor of the shop. More personal discussions and financial arrangements, especially those involving an aristocratic clientele, took place in the parlour, a private space usually modelled on a Georgian drawing room.*[40]

The tradition of the private banker living above the shop was as much a commitment to underlining the personal nature of his wealth as to any economic necessity. However, the shortcomings of adapted homes as offices did become apparent. The physical constraints and the domestic ambience combined to encourage growing banks to find alternative solutions. As we saw in the previous chapter, the first new purpose-built private banks in London were probably those erected in the 1750s and 1760s. Even with a purposeful design however, both of the examples cited here took the form of substantial town houses; and both properties maintained the tradition of domestic space over the business; while carpet, curtains, furniture and lighting all conveyed a sense of the domestic. A pure bank-only building (excepting the BoE), or indeed a pure office building of any kind, would not appear for at least another 5 decades.

Black observes how the early purpose-built banking houses showed striking similarities with the substantial Georgian town houses erected for the aristocracy and gentry.

> *The ground floors often possessed strong classical features, perhaps to hint at the strong rooms behind, but otherwise carried no overt advertising that these were, in fact, banking-houses. The contrast with the domestic simplicity of the upper floors had the useful effect of marking the division between place of business and place of residence within the same building, continuing a long City tradition of living above the shop.*[41]

The growth of purpose-designed banking houses can be illustrated here with two examples: Sir Charles Asgill's banking house at 70 Lombard Street (1757) and Andrew Drummond's private bank at 49 Charing Cross (1760).

Purpose-built banks. It is likely that the first purpose-built private banking house was built at 70 Lombard Street for Vere & Asgill. The bank started in goldsmithing in Lombard Street in c1730, with George Glegg and Joseph Vere at the helm. When Glegg died in 1740, Vere brought Charles Asgill (1714–1788) into the partnership. The firm was renamed Vere & Asgill in 1740, and Vere departed in 1753. Soon after, Asgill began to consider new premises. The result was 70 Lombard Street, designed by Sir Robert Taylor, and completed in 1757 (Figure 3.5). Asgill was knighted in 1761.

The exterior of the 1757 building was a classic Georgian design with plain but symmetrical lines on the upper levels and a more detailed ground floor, comprising a central main entrance flanked by two arched windows and four Doric columns. The building maintained many features of a large town house. Summerson describes Taylor's design as having "*just a hint of strong rooms behind its Doric order, but otherwise quite simply a gentleman's house*".[42] Black confirms the domestic character of 70 Lombard Street, with its similarities with substantial Georgian town houses:

> *The ground floor with strong classical features ... but no advertising that this was, in fact, a banking house. Nonetheless, the contrast with the domestic simplicity of the upper floors had the useful effect of marking the division between place of business and place of residence within the same building.*[43]

Asgill's banking house was part of what Booker calls "*the Lombard Street tradition*" and in general private banks built in the capital between the 1750s and the 1820s deviated little from such discreet, restrained facades.[44]

After several changes, the bank disappeared around 1796, and the following year the property was taken over by the Pelican And British Empire Life Insurance Office as their London office. The building was demolished in 1915.

Another purpose-built bank, 3 years after Asgill's, was that for Drummond's Bank in the West End. The private bank of Drummond & Co. can be traced back to the early-eighteenth century when Andrew Drummond (1688–1769) left a goldsmith business in Edinburgh to take up residence in London's Charing Cross in 1712. Black notes that while he was recorded as practising 'at the sign of the Golden Eagle', the bank's historians record that the recognised date for the establishment of the Bank was September 1717, when he opened his first ledger devoted entirely to banking.[45]

Drummond's was a highly successful bank and in 1758 commissioned a new building, which opened in 1760 at 49 Charing Cross. Black describes the building as having a Private Partners' Room, Managers' Room, strong rooms and an open yard with vaults underneath all lying behind the public banking hall. To the south of the yard there was an additional suite of offices, whilst to the north were various buildings for the maintenance of a coach and horses, in addition to strong rooms and what seems a private entrance to domestic quarters on the bank's upper floors.[46]

Andrew Drummond died in 1769, by which time the bank had amassed 1,500 customer accounts, many of which belonged to architects, artists, businesses and cultural institutions. Drummonds soon outgrew their new bank at 49, Charing Cross, taking 52 in 1766, another property in 1777, 50 in about 1825 and eventually 47 and 48. This process of building and then adapting and expanding was common at the time because the nascent banking sector was growing rapidly.

Figure 3.5 Vere and Asgill, 70 Lombard Street, 1757.

In administrative terms, the growth of business throughout London was overwhelming. Sites were outgrown and over-staffed. Banks which 50 years earlier had two or three principals and as many clerks, now had a payroll of perhaps 30.[47]

Booker also cites the case of private bank Coutts, which moved to 59 The Strand in 1739.

No sooner, it seems, had (architect) Paine remodelled the bank in 1770-1 than it outgrew itself, and the partners were forced to lease property around William Street and John Street. James Paine again made alterations in 1780-3, but within a few days another house had been added and two more were leased in 1799.[48]

Buildings of public office

The main focus in this text is upon how the evolving economy and the corresponding changing business landscape has influenced the nature of the workplace, work and workers. These are all bound up in what is referred to as the office economy. But of course, offices perform not only a commercial function but also one of civil governance, or public administration, through national and local government. Government's throughout Europe were using palaces to house the growing apparatus of state before 'business offices' were commonly understood and were creating national and local government buildings from early in the eighteenth century, contemporary with the rise of commercial offices.

In this important sense, early public offices contributed to the evolution of the office more generally, and helped to reinforce the cultural tradition of offices in Europe, distinct from the north American rational tradition (see 'Cultural and rational traditions' in Chapter 7).

Some of the earliest examples of public office buildings were in Italy: in the mid-sixteenth century the Uffizi (1560–1571, Giorgio Vasari) were built in Florence to house the administrative offices of Cosimo de Medici.

Middelburg Town Hall is situated in Zeeland, and is one of the finest Gothic buildings in the Netherlands. Construction began in 1452 and was supervised by several generations of the Flemish family of architects, the Keldermans. Construction was completed in 1520. Amsterdam Town Hall was built between 1648 and 1655 (Jacob Von Campen), facing the city docks, on 13,659 wooden piles.

St Petersburg saw its Admiralty building built as the HQ of the Admiralty Board and the Imperial Russian Navy. The edifice was rebuilt in the nineteenth century to support the tsar's maritime ambitions. The building was constructed to Adreyan Zakharov's design between 1806 and 1823.

Of course, Paris had a number of public office buildings, of which three classic examples are highlighted here. The Hôtel de Ville stands on the Place de l'Hotel de Ville in the 4th arrondissement. The south wing was built mid-sixteenth century by Francis I, and the north wing between 1605 and 1628 by Henry IV and Louis XIII. Having been destroyed by fire in May 1871, the building was rebuilt between 1874 and 1882.

Another key public building, for the national government is the Foreign Affairs building. The Ministère des Affaires Étrangères was built on Quai d'Orsay between 1855 and 1856 to a design by J Lacornée. And, of course, the Louvre. This iconic project of the Second French Empire was initially designed by Louis Visconti. Following Visconti's death in late-1853, the project was modified and completed by Hector-Martin Lefuel.

Outside of London, Britain's regional cities produced numerous examples of public offices. Major eighteenth century building projects included the Royal Exchange in Dublin following a competition (1769, Thomas Cooley) and the Register House in Edinburgh (1774–1788, Robert Adam) begun in 1774. Local government offices were also very common by the mid-nineteenth century, to name just three: Birmingham Town Hall (1832, JA Hansom), Liverpool St George's Hall (1840, HL Elmes) and Manchester Town Hall (1868–1877, Alfred Waterhouse).

One of the oldest examples of public offices in Britain is Somerset House, situated on the north bank of the Thames between Embankment and Strand. While often cited as the oldest office building, in fact the 'first' Somerset House was a Palace, occupied by Princess Elizabeth before she was crowned, and later by Anne of Denmark. The building then fell into disuse and it was demolished with the agreement of the King so that public buildings could be erected. In 1775, Parliament passed the Crown Lands Act, with the specific purpose of building public offices on the site.

The impetus for Somerset House was led in part by Edmund Burke, who felt that other European capitals had many palaces and grand public buildings where growing bureaucracies had been accommodated, and which engendered national pride. Newman reports that there "*had been a recurrent ambition to build a great new classical palace in London*", and that following a debate in Parliament, Secretary to the Board of Works, William Robinson, was instructed "*to design a building that would be an ornament to the Metropolis and a monument to the taste and elegance of His Majesty's Reign*". In October 1775, Robinson died suddenly, and within a month he was replaced by Sir William Chambers.[49]

Sir William Chambers (1723–1796) was born in Sweden and educated in England. Following a journey through India and China with the Swedish East India Company, he went to Paris in 1749 to study architecture at Jacques-François Blondel's École des Arts, Paris, and then to Italy for further study. Returning to England in 1755, he became the 'arch-professional and establishment architect in the England of George III'. A founding member of the Royal Academy, he was architectural tutor to the Prince of Wales, later George III, under whom he rose to become Surveyor-General.

Chambers was appointed to design the building in 1775, spending the final two decades of his life working on successive phases of building. His ambition was to reflect the favoured neo-Palladianism of the time and to create a building that reflected the munificence and grandeur of the monarchy. The site stretched from the Thames on the south side to the Strand on the north side. Chambers masterminded everything from the overall plan to design details such as cornices and mouldings.

His final designs included a single courtyard of 300 ft (91m) by 200 ft (61m), complete with terraces. Each block comprised six storeys: cellar, basement, ground, principal, attic and garret. The public offices and learned societies which were accommodated around the courtyard varied greatly in size, but each occupied all six floors of its allotted area, the upper floors also provided living space for officials. Large basement vaults for storing public documents were provided under the northern section of the courtyard.

Figure 3.6 shows the ground floor layout in 1801. The courtyard can be seen to be surrounded by many offices, generally accessed from a central corridor. The public offices are also clearly marked including: the Duchy of Cornwall and the Duchy of Lancaster, the Royal Academy, the Royal Society and the Society of Antiquarians, the Exchequer Office, the Lottery Office, the Victualling Office, the Navy Office and the Navy Pay Office, the Stamp Office and the Hawkers and Pedlars Office. There were several public offices on other floors, including the Hackney Coach Office, the Salt Office, the Surveyor General of the Crown Lands Office and the Tax Office.

The layout can be seen to reflect the tradition of corridors, interconnected rooms and small private spaces that one might expect to see in a palace and, albeit on a much grander scale, in

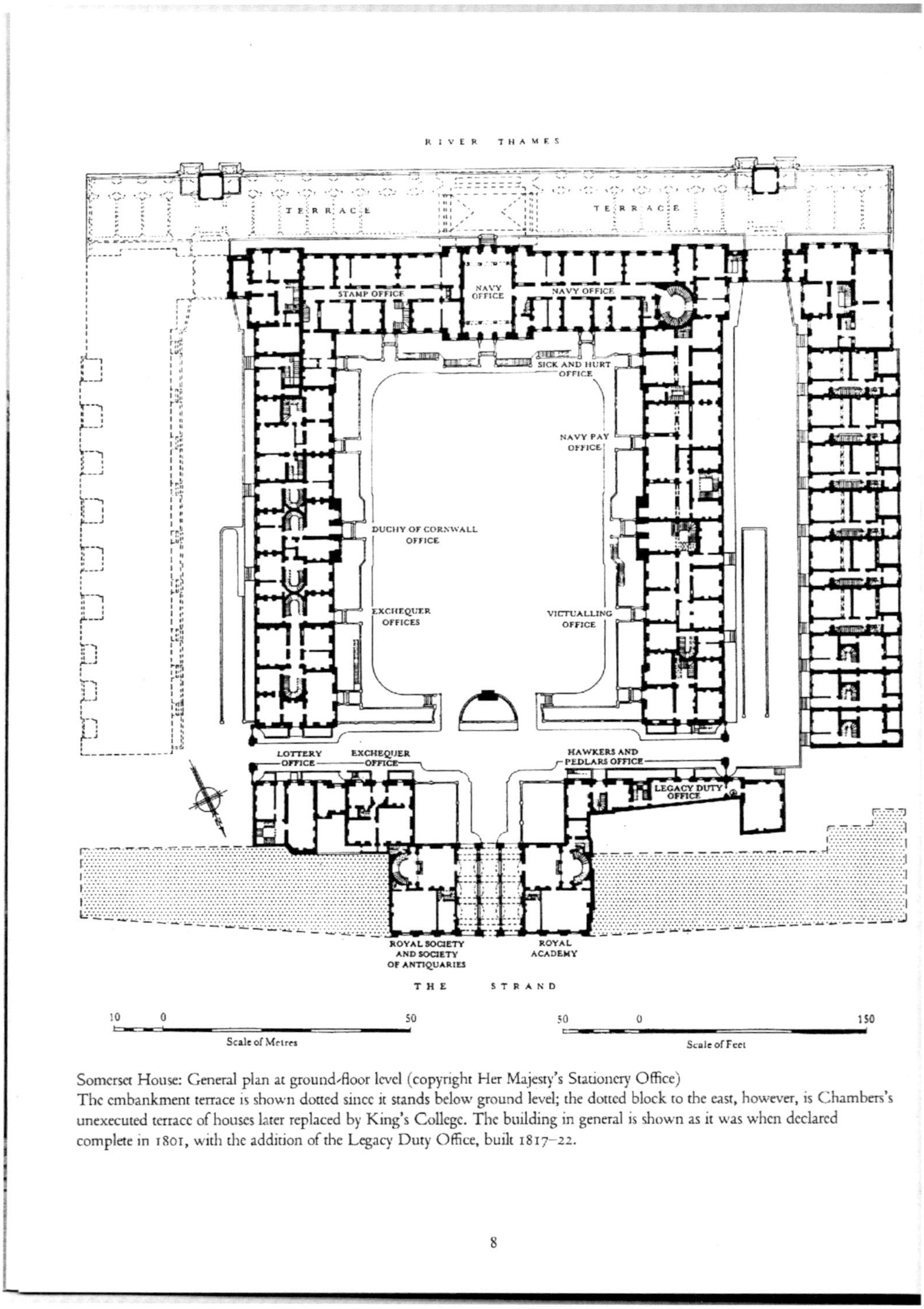

Figure 3.6 Somerset House, ground floor layout, 1801.

Source: His Majesty's Stationery Office.

the emerging bank premises of the eighteenth century. In this important way, the layout might be conceived of as part of the cultural tradition of European offices (see Chapter 10).

The North Wing, fronting Strand, was completed in 1779 and construction of the riverside wing was finished in 1786. The Royal Academy took up residence in the North Wing, followed by the Royal Academy and the Royal Society in 1780. Meanwhile work continued on the East and West Wings, which were occupied from 1788, and by 1790 the main quadrangle was complete.

While perhaps not the first office building, Somerset House demonstrated the key role of office work undertaken on behalf of the state as distinct, say, from insurers and bankers; and its sheer scale and internal complexity set it apart from most contemporary buildings. On a much smaller scale, another early public office building was the Navy Office on Crutched Friars, towards the east of the City of London (Figure 3.7). This building's most famous occupant was London diarist Samuel Pepys.

In 1656, Pepys became clerk to George Downing, teller of the receipt in the Exchequer, and in August 1658, he and his wife moved to Axe Yard, opposite Whitehall Palace, now on the site of the Treasury. He was still living there when he began his diary in January 1660 (writing every day until 1669), and in July of that year he secured a new job as Clerk of the Acts of the Navy

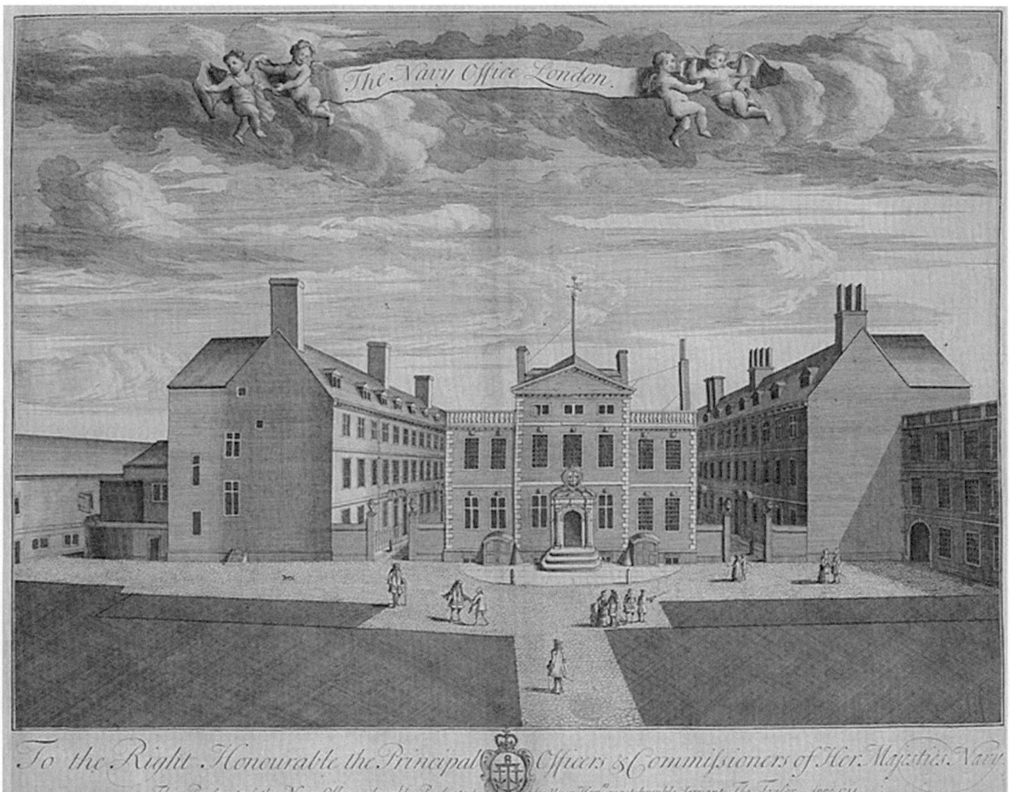

Figure 3.7 The Navy Office, Crutched Friars, 1714.

on Crutched Friars in the east of the City of London. He moved into official lodgings beside the office, opposite the church of St Olave, Hart Street.

The Navy Office was responsible for the day-to-day civil administration of the Royal Navy between 1576 and 1832. The original Navy Office was constructed in 1656 and, while it survived the Great Fire, it was also burnt down in 1673, only to be rebuilt to a design by Sir Christopher Wren in 1674–1675 and demolished in 1788 (to make way for warehouses for the East India Company), when the Navy Office moved to Somerset House. Most views of the second iteration of the building, such as that in Figure 3.7 from 1714, show a fairly modest building, mainly of three storeys, with a central door. The steps up to the door allow a raised basement space to be lit with natural light, possibly providing space for voluminous records.

The Navy Office was an example of office buildings performing state functions from an early stage. Another example was also a naval function, that of the Admiralty (the Navy Office was a subsidiary yet autonomous component of the Admiralty), which was housed in the building known officially today as the Ripley Building. This is the oldest of a complex of buildings between Whitehall, Horse Guards and The Mall, built to house the administration of the Royal Navy. The three-storey, U-shaped brick building was designed by Master Carpenter Thomas Ripley, and completed in 1726. The Ripley Building is often cited as the first purpose-built office building in Great Britain.

Summary. The successful eighteenth century banks were larger than their predecessors. Banks that previously had, say two partners and two clerks, suddenly found themselves with a payroll of 30 or 40. These are tiny numbers by modern standards, but they reflected a novel burgeoning international trading position in which the conditions were ripe for a rapid evolution of the financial services market. Financial products and company structures co-evolved in London, to enable the city to become the dominant economic force in Europe. What was also happening was the very beginning of the long close relationship between finance and government. The Government's involvement in various wars – Second Anglo-Dutch War, Third Anglo-Dutch War, Franco-Dutch War and the Nine Years' War among them – was expensive and required financing. The rise of the insurance sector – a key component of the emergent office economy – was a signature development.

Any vestiges of ambition held in Antwerp were extinguished as London marched forward. A whole new world of trading, dealing and exchanging emerged, initially in London's exotic coffee houses and later in its formal Exchanges.

Roseveare concludes his study with a quite profound observation. In first establishing that "*the development of banking borrowing, credit and investment in these 100 years [1660–1760] has every entitlement to be matched in importance with the Industrial Revolution yet to come*", he then argues that its significance was profound not just in the fiscal measures of a growing National Debt and lowering interest rates, but also

> *in more subtle but pervasive achievements – in the peaceful interchange of a commercial community, the secure prosperity of a middling class and, perhaps most importantly, in the political stability of a nation acquiescent in its fiscal system and trustful of its public credit.*[50]

We also see during this period the further development of corporate business structures and banks and insurance companies as we know them today; purpose-built office buildings, and the arrival of the office worker, in the shape of the clerk. Up to this point, the business landscape was dominated by family concerns, most of which operated from domestic premises, employing very small numbers of people. The emergence of corporate structures was regarded with suspicion for some time – for a perceived lack of accountability and prudence – while owners were reluctant to hand the keys to their businesses to non-family managers.

While the scale of buildings of public office in the eighteenth century reflected the growth in the administration of the state in their scale, the country's rapidly growing manufacturing base and increasingly global trading role was manifested during this time with burgeoning small office suites in private houses, suitably well-appointed to reflect the status and material success of their owners. By mid-century, the growing wealth of the London populace was reflected in the growth of private banks, and banking houses began to spring up, but as Cowan eloquently reminds us, the symbols of the city prior to the mid-eighteenth century remained the castle, the cathedral, the palace and the marketplace, the place of the office remained very small indeed.[51]

The combination of plague and conflagration caused utter devastation to 1660s London, but also precipitated the widespread rebuilding of the City. Indeed this first period of our story was one of very significant change, bringing to bear political and cultural context that would help shape the form of the office economy that was to emerge and mature during the nineteenth century. A fitting end to this chapter is a quote from Lewis Mumford, who brilliantly captured the wider social, economic and political implications of the economic changes that would play out over the nineteenth century, helping shape along the way the nature of office work.

> *Between the fifteenth and eighteenth century a new complex of cultural traits took place in Europe. Both the form and contents of urban life were in consequence transformed. The new pattern of existence sprang out of a new economy, that of mercantilist capitalism, a new political framework, that of a centralised despotism or oligarchy, usually focused in a rational state, and a new ideological form, that of mechanism, whose lines were laid down in the monastery and the army.*[52]

Notes

1 Kynaston D (1994) *The City of London Volume One: A World of its Own 1815–1890* Chatto & Windus, London p12.
2 Street GS (1920) *The London Assurance 1720–1920* Printed for private circulation p22.
3 Street GS (1920) *The London Assurance 1720–1920* Printed for private circulation p24.
4 Sheppard F (1999) *London: A History* BCA Books, London p228.
5 Dickson PGM (1960) *The Sun Insurance Office, 1710–1960: the History of Two and a Half Centuries of British Insurance* Oxford University Press p4.
6 Dickson PGM (1960) *The Sun Insurance Office, 1710–1960: the History of Two and a Half Centuries of British Insurance* Oxford University Press p18.
7 Dickson PGM (1960) *The Sun Insurance Office, 1710–1960: the History of Two and a Half Centuries of British Insurance* Oxford University Press p30.
8 Dickson PGM (1960) *The Sun Insurance Office, 1710–1960: the History of Two and a Half Centuries of British Insurance* Oxford University Press p33–34.
9 Dickson PGM (1960) *The Sun Insurance Office, 1710–1960: the History of Two and a Half Centuries of British Insurance* Oxford University Press p33–34.
10 Dickson PGM (1960) *The Sun Insurance Office, 1710–1960: the History of Two and a Half Centuries of British Insurance* Oxford University Press p36.
11 Dickson PGM (1960) *The Sun Insurance Office, 1710–1960: the History of Two and a Half Centuries of British Insurance* Oxford University Press p43.
12 Dickson PGM (1960) *The Sun Insurance Office, 1710–1960: the History of Two and a Half Centuries of British Insurance* Oxford University Press p44.
13 Dickson PGM (1960) *The Sun Insurance Office, 1710–1960: the History of Two and a Half Centuries of British Insurance* Oxford University Press p45.
14 Dickson PGM (1960) *The Sun Insurance Office, 1710–1960: the History of Two and a Half Centuries of British Insurance* Oxford University Press pp57–58.

15 Dickson PGM (1960) *The Sun Insurance Office, 1710–1960: the History of Two and a Half Centuries of British Insurance* Oxford University Press p114.
16 Dickson PGM (1960) *The Sun Insurance Office, 1710–1960: the History of Two and a Half Centuries of British Insurance* Oxford University Press p115.
17 Dickson PGM (1960) *The Sun Insurance Office, 1710–1960: the History of Two and a Half Centuries of British Insurance* Oxford University Press pp122.
18 Dickson PGM (1960) *The Sun Insurance Office, 1710–1960: the History of Two and a Half Centuries of British Insurance* Oxford University Press pp123–24.
19 Bagehot W (2009) *Lombard Street: A Description of the Money Market* Seven Treasures Publications (Original work published 1873) p118.
20 Bagehot W (2009) *Lombard Street: A Description of the Money Market* Seven Treasures Publications (Original work published 1873) p37.
21 Richards RD (1965) *The Early History of Banking in England* Augustus M Kelley, New York p36.
22 Clark DK (1941) A Restoration Goldsmith-Banking House: The Vine on Lombard Street In: Seymour C (Ed) *Essays in Modern English History: in Honor of Wilbur Cortez Abbott* pp3–47 Kennikat Press, Port Washington, New York.
23 Roseveare HG (1991) *The Financial Revolution 1660–1750* Routledge, London p12.
24 Turner GD (2015) *English Banking in the Eighteenth Century: Bankers, Merchants and the Creation of the English Financial System* Durham theses, Durham University Available at Durham E-Theses Online: http://etheses.dur.ac.uk/11297/ p31.
25 Roseveare HG (1991) *The Financial Revolution 1660–1750* Routledge, London p12.
26 Porter F (1994) *London: A Social History* Penguin Books, London p175.
27 Roseveare HG (1991) *The Financial Revolution 1660–1750* Routledge, London p72.
28 Roseveare HG (1991) *The Financial Revolution 1660–1750* Routledge, London p72.
29 Turner GD (2015) *English Banking in the Eighteenth Century: Bankers, Merchants and the Creation of the English Financial System* Durham theses, Durham University Available at Durham E-Theses Online: http://etheses.dur.ac.uk/11297/ pp80–81.
30 Michie RC (2000) (ed.) *The Development of London as a Financial Centre, Volume 1* Bloomsbury Publishing, London.
31 Murphy AL (2017) Clock-watching: Work and Working Time at the Late Eighteenth Century Bank of England *Past & Present* Vol 236 Issue 1 pp99–132.
32 Murphy AL (2017) Clock-watching: Work and Working Time at the Late Eighteenth Century Bank of England *Past & Present* Vol 236 Issue 1 pp99–132.
33 Murphy AL (2017) Clock-watching: Work and Working Time at the Late Eighteenth Century Bank of England *Past & Present* Vol 236 Issue 1 pp99–132.
34 Murphy AL (2017) Clock-watching: Work and Working Time at the Late Eighteenth Century Bank of England *Past & Present* Vol 236 Issue 1 pp99–132.
35 Booker J (1990) *Temples of Mammon: the Architecture of Banking* Edinburgh University Press p2.
36 Kynaston D (1995) *The City of London Volume Two: Golden Years 1890–1914* Chatto & Windus, London pp141–142.
37 Kynaston D (1995) *The City of London Volume Two: Golden Years 1890–1914* Chatto & Windus, London p30.
38 Anderson G (1976) *Victorian Clerks* Manchester University Press p9.
39 Pevsner NA (1976) *A History of Building Types* Thames & Hudson, London p200.
40 Black IS (2000) Spaces of Capital: Bank Office Building in the City of London, 1830–1870 *Journal of Historical Geography* Vol 26 No 3 pp351–375.
41 Black IS (2000) Spaces of Capital: Bank Office Building in the City of London, 1830–1870 *Journal of Historical Geography* Vol 26 No 3 pp351–375.
42 Summerson J (1977) The Victorian Rebuilding of the City of London *The London Journal* Vol 3 No 2 pp163–185.
43 Black IS (2000) Spaces of Capital: Bank Office Building in the City of London, 1830–1870 *Journal of Historical Geography* Vol 26 No 3 pp351–375.

44 Black IS (2000) Spaces of Capital: Bank Office Building in the City of London, 1830–1870 *Journal of Historical Geography* Vol 26 No 3 pp351–375.
45 Black IS (2003) Private Banking in London's West End, 1750–1830 *The London Journal* Vol 28 No 1 pp29–59.
46 Black IS (2003) Private Banking in London's West End, 1750–1830 *The London Journal* Vol 28 No 1 pp29–59.
47 Booker J (1990) *Temples of Mammon: the Architecture of Banking* Edinburgh University Press p9.
48 Booker J (1990) *Temples of Mammon: the Architecture of Banking* Edinburgh University Press p10.
49 Newman J (1990) *Somerset House: Splendour and Order* Scala Books, London pp4–5.
50 Roseveare HG (1991) *The Financial Revolution 1660–1750* Routledge, London p75.
51 Cowan P (1969) *The Office: A Facet of Urban Growth* Heinemann Educational Books, London p25.
52 Mumford L (1938) *The Culture of Cities* Secker & Warburg, London p75.

Part II

1770–1870

Bank house to office building

Jefferson's Polygraph (1804)

Office Era	Time period	Work & Place Themes	Office Technology
Market	**Part I: 1670-1770**	**Trade**. Goldsmiths as bankers. Central banking. Coffee Houses. Joint Stock Companies. Exchanges. Insurance.	Coal fires. Candles. Copying. Ledgers. Paper. Bills of Exchange. Quill pen. Cash book. Shop counter.
	Part II: 1770-1870	**Organisation.** Stock Exchange. Clearing House. Private and merchant bankers. Joint Stock Banks. Office buildings. Speculative development.	Gas lighting. Copying. Ink Pen. Ledgers. Paper. Stenography. Polygraph. Correspondence book.
Factory	**Part III: 1870-1945**	**Production**. Rise of corporation. Division of labour. Work: fixed, manual, repetitive and mechanical. Female workforce. Rise of professions.	Mechanisation. Adding machine. Electricity. Telegraph. Telephone. Typewriter. Copiers. Elevator. Air conditioning. Fluorescent light.
Corporate	**Part IV: 1945-2000**	**Process**. Globalisation. Scale. Bureaucracy. Command & Control. 'Corporate man'. Work landscapes. Birth of digital office.	Photocopier. Mainframe. System Furniture. Golf Ball Typewriter. PC. MS Office. Mobile phone. Internet. Email.
Digital	**Part V: 2000 onward**	**Knowledge**. Agile work. Data. Power.	Smart phone. The Cloud. Social Media. Augmented Reality. Automation. Artificial Intelligence. Robotics. Virtual Reality.

4 Banking business

The eighteenth century saw London's population rise from around 600,000 at the start to reach around one million by the close, making it by far the largest city in Europe. Much of this growth was absorbed in a westwards drift, away from the City of London, with the large-scale, intensive development of Westminster. The large estates – Berkeley, Burlington, Cavendish, Curzon, Harley, Marylebone, Portman and others – all witnessed enormous change as country estates were urbanised. By mid-century, the whole area between Oxford Street in the north and Piccadilly in the south, and between Park Lane in the west and Bond Street in the east had been built upon.

Some new districts were built for the growing aristocratic class, including Mayfair and St James', where the affluence of the new residents was reflected in the appearance of shopping and leisure. Bond Street was developed by Sir Thomas Bond: Old Bond Street to the south was built in the 1720s, and New Bond Street was developed to the north in the mid-1730s. Charles Asprey moved to 49 New Bond Street in the 1830s. Much of the Georgian architecture remains unscathed above and around today's glitzy shop displays. Indeed, the 'the West End' developed a character of its own – altogether more 'upmarket' than the City, and dominated by larger houses and shopping rather than by business.

In the later decades of the century the growing preoccupation with leisure can be seen in the opening of: the British Museum (1759); Hamleys toy shop (1760); the Orangery and Pagoda at the Royal Botanic Gardens (1761); Lord's Cricket Ground (1787); the rebuilt Theatre Royal Drury Lane (1794) and Hatchards bookshop on Piccadilly (1797). While department stores would not appear until the latter half of the nineteenth century, several large drapers set up shop earlier, including Dickins and Jones (1803); Swan & Edgar (1812); Shoolbred's (1820) and Marshall & Snellgrove (1837). The seventies and eighties saw these and other establishments expand into full department stores.[1]

Of course the emergent affluence was in sharp contrast to the level of despair that most Londoners endured daily; and the city was quite lawless. No one was safe in the streets of Georgian London: within the space of a few years in the 1770s and 1780s the Prince of Wales, the Prime Minister and the Lord Chancellor had all been robbed in broad daylight in the West End, while the Lord Mayor was held up at gun point on Turnham Green.[2] There was also major civil disorder.

In 1768, John Wilkes was imprisoned for writing an article for *The Northern Briton* that severely criticised King George III. His incarceration prompted a demonstration on St George's Fields in Southwark, in which troops shot dead seven people. Similarly, in September 1769, the Spitalfields Riots by silk weavers seeking pay rises resulted in troops firing and killing seven.

DOI: 10.1201/9781003582205-6

But the bloodiest uprising of the period occurred in June 1780, when London was rocked by the Gordon Riots. Led by Lord George Gordon, Protestants rose up against Roman Catholic emancipation, and particularly the Papists Act 1778, aimed at reducing discrimination against Catholics. A large and orderly protest rapidly degenerated into several days of widespread rioting and looting, including an attack on the Bank of England. The Government finally sent in the army and upwards of 300 people were killed; 450 arrested and, finally, around 25 executed.

Public safety and riots aside, the long-term differences between the mercantile City and the more refined, residential playground of the West End became more apparent. In the early nineteenth century, the City remained quite diverse in activities. The City retained shopping and leisure; for example, along Cornhill in 1847, on the southern side towards Gracechurch Street, could be found SW Silver & Co (shirt makers); Pretyman & Hobson (furnishing ironmongers); H&J Baldwin (woollen drapers and tailors); George Lowcock (watchmaker and cutler); and Maclaren & Son (boot makers).[3] However, it was also around this time that the drift to the west, the start of which was noted in the previous chapter, became a general trend, and the population of the City fell sharply. As the middle classes became more affluent, they found the city dirty and stuffy, and they began to look around for more pleasant and healthier dwellings. Some moved to Hatton Garden or Bloomsbury, others went as far into the country as Epsom and rode up to London every day.[4]

By 1829, omnibuses were running from Paddington to the City; and by 1837 their number had increased to over 200, serving various parts of the metropolis. Some 600 coaches daily conveyed more than 50,000 workers between the City and the outer residential districts of Camberwell, Chelsea, Clapham, Hackney, Hampstead, Islington, Kennington, Marylebone, Mile End, Paddington and Stepney. The fares were 6d (pennies) for short distances.[5]

Consequently, the City *felt* different. Kynaston cites Flora Tristan (French socialist feminist and grandmother of the well-known Parisian stockbroker Paul Gauguin), who wrote in her London Journal in 1840:

> *The inhabitants of this part of town [the City] are considered by those at the West End to be of pure John Bull stock. They are, for the most part, worthy merchants who are shrewd where their business interests are concerned and who care for nothing but those same interests. The shops, where many of them have made fortunes, are so dark, so cold, so damp that the West End aristocracy would disdain to have them as stables for their horses.*[6]

Of course, the 'shops' referred to in this quote are in fact what we would now call offices. At the time they remained largely domestic premises with a 'shop' or 'bank' on the ground floor. The quote underlines the primitive nature of many of the buildings occupied by the early banks; but also the functional separation of the City and West End that was to characterise London for another century-and-a-half.

We saw in Chapter 2 that the City began to lose its lustre as a place of residence in the early-eighteenth century, a process that gathered pace in the nineteenth century. David Morier Evans (1819–1874) was a financial journalist, one of the earliest of his kind, who published several books on matters pertaining to the City of London. In the extensively titled *The City; or the Physiology of London Business; with Sketches on 'Change and at the Coffee Houses* of 1865 (in which "Change" refers to the Royal Exchange), Evans stated that "*The City is not now much chosen for a residence. The old houses, in the best thoroughfares, are either let as offices, or given up to the occupation of housekeepers who have the charge of the premises*". He goes on to say that at the banks *the rule is, for the junior partner to reside on the premises; and a certain number of clerks also live in the house. Merchants and others, who formerly located in the suburbs have, in a great number of cases, chosen West End domiciles.*

It was the end of a venerable tradition, "*a dwelling in the City is a thing not now considered desirable – all move either towards the west, or emigrate to the suburbs, – the one for fashion, the other for economy and fresh air*".[7]

The City also fostered a business culture in which alcohol formed a central part for many. Bayne-Powell described the amount of alcoholic liquor drank at this time among businessmen: "*There were one-bottle men, who were considered very temperate, two-bottle men and three-bottle men, and there were others, like Lord Carlisle, who drank all the red wine within his reach. A few choice spirits, Sheridan, Lords Eldon and Stowell, and Professor Porson, could, it was said, drink six bottles in a sitting*". He was amazed that "*men could drink so deeply and yet be able to carry on their everyday business.*"[8]

Setting aside the occasional riot and the drinking culture, the City was becoming more and more concentrated on mercantile and financial activities; as was evidenced by the growing interest of the Government in the City to help it fund an ever expanding budget, not least in terms of being able to raise the vast sums of money that it required to fund over a century of war on the continent. Meanwhile, in the West End society was changing rapidly – with a far higher density of living, more affluence, more trade and rising social expectations – all of which demanded a more sophisticated business services underpinning.

Confident London and the first co-working space

Sitting at the heart of the heart of the financial system, alongside the Bank of England, is the Stock Exchange. We saw the emergence of the Stock Exchange in the eighteenth century in Chapter 2. When the brokers set up their own exchange in Sweeting's Alley in 1773, they changed the name from 'New Jonathan's' to the Stock Exchange. It was open to anyone so long as they paid the daily admission fee, covering expenses such as rent. In 1801, the Stock Exchange (Figure 4.1) became regulated for the first time; and in February 1812, the General Purpose Committee confirmed a set of recommendations, which later became the foundation of the first codified rule book of the Exchange.

Having outgrown Sweeting's Alley, a new building was constructed in 1801 in Capel Court. Opening in 1802, the new building was more spacious and designed for purpose. A gallery wrapped around the main hall, equipped with desks and bookcases, with clerks available to fetch and replace books. By the 1850s, and despite several incremental extensions in most directions, the Capel Court Exchange was becoming overcrowded and uncomfortable, losing much of its coffee house ambience. Consequently, the ruling council decided on a more radical solution and appointed Thomas Allason to design a new facility which opened in 1854.

By the mid-nineteenth century, London was a growing, confident, exuberant city; and this positivism was captured in a bold, albeit ultimately ill-fated, venture, which was a co-working prototype – the Hall of Commerce, designed as a prestige meeting place for the exchange of commercial news.

Entrepreneur and property speculator Mr Edward Moxhay (1787–1849) had previously been a cobbler who had by 1840 turned his talents to biscuit-making. As well as being an artisan, Moxhay was also an amateur architect. In this latter sphere, Moxhay thought big with his signature project.

Figure 4.1 The London Stock Exchange, 1847. © Illustrated London News Ltd/Mary Evans.

The Hall of Commerce was erected in Threadneedle Street between 1840 and 1843 (Figure 4.2). Not only did Moxhay design the edifice, but he and his son oversaw all aspects of the delivery. The exterior facades exuded simplicity and elegance, but it was the interior space that was quite remarkable. The main hall was a fine, overwhelming space 40m long, 13m wide and some 15m in height. Corinthian columns and pilasters, beautiful carvings, beautiful friezes and elegantly coved ceiling with large roof lights added up to a spacious and light interior space. The grand gesture cost Mr Moxhay the huge sum of £70,000.

The Hall of Commerce was designed as a mercantile club, with subscription members, and intended to concentrate under one roof multitude activities relating to trade and commerce. In addition to the large main hall, there was a generous reading room as well as office accommodation, which subscribers could book for transacting business of a confidential nature, or for the deposit of deeds. Drawers and cupboards were available to those unable to afford the rooms. The Hall hosted wool sales and auctioned railway shares. There was a grand bas-relief on the front by Mr Watson, a young sculptor of the time. The building was launched with some fanfare, including a lavish, inaugural banquet.

While the coffee houses had attracted merchants for information exchange and dealing, the Hall of Commerce was quite a distinct offering. Purpose designed, a selection of meeting rooms, sales and shows, store places for visiting members; a sense of experience: this was a genuine early version of a co-working or serviced office space – 150 years ahead of its time. And like many pioneering schemes, it was not a success.

The building was constructed while the Royal Exchange was being rebuilt and it is thought that Mr Moxhay had intended for the Hall of Commerce to compete for the custom of Lloyd's,

Figure 4.2 The Hall of Commerce, 1844.

the Baltic, Garraway's, the Jerusalem, and the North and South American Coffee Houses. But trading never really took off, and when the Royal Exchange reopened in 1844 it struggled to attract members from the coffee houses. Moxhay and his backers were forced to reduce the annual subscription of five guineas to £1 10s. 6d. But to no avail. From 1855, the building was occupied by a series of banks until its demolition in 1922. Moxhay died at 61 heavily in debt.

The Old Lady

When the Bank of England opened for business in the Hall of the Mercers' Company in Cheapside in 1694, it boasted a staff of seventeen clerks and two doorkeepers, and tables "*furnished with pens, inke, paper and sand*". But the accommodation was soon found to be inadequate, and before the year was out, the Bank moved to Grocers' Hall nearby. For some years it had been the wish of the Directors to house the Bank in its own building and, in 1724, they assembled an estate through the acquisition of neighbouring sites in Threadneedle Street. The site was then cleared to enable the Directors' ambitious solution. Designed by architect George Sampson (1718–1764), construction of the new building began in 1732 and, on 5 June 1734 business was transacted in the Bank's own premises for the first time.[9]

Thirty years later, the increasing work of the Bank made it necessary to acquire further adjoining properties to expand the property, between 1765 and 1770 under the guidance of Sir Robert Taylor (1714–1788).[10] Following Sir Robert's death, Sir John Soane (1753–1837) was appointed architect to the Bank, a position that was to keep him employed for 45 years. During that time the Bank's premises were effectively rebuilt: extended to the present boundaries of Threadneedle Street, Bartholomew Lane, Lothbury and Princes Street, and the area was increased from 0.4 to 1.4 hectares (1.0 to 3.5 acres). Except for minor alterations, Soane's work stood until a total rebuilding begun in 1925. A statue of him, by Sir William Reid Dick, stands in a niche in the Lothbury wall.[11]

As the City of London was assuming its European role, the Bank of England was the largest private employer of white-collar workers in Britain. In 1783, there were 321 permanent clerks working at the Bank, more than double the white-collar staff of the East India Company and ten times the numbers employed by the large insurance companies, the Royal Exchange and Sun Assurance companies. The Bank's workers were organized into two large departments: the Cashier's Office and the Accountant's Office and each of those departments was sub-divided into more than a dozen offices, some employing just a few men, while managing the Three Per Cent Consols Office made work for more than 50.[12]

The Bank's role and influence had grown during the Revolutionary and Napoleonic Wars which it was called upon to underwrite. Not only was the Bank the nexus of an increasingly complex and demanding domestic economy, it was also the lender of last resort and the manager of the national debt, which was coming to be accepted as a permanent and normal state of affairs. Indeed, while the public debt in 1750 amounted to £78,000,000, any "*serious attempts to pay it off had ceased. During the Seven Years War (1756–1763) 37 per cent of Government expenditure was raised by public borrowing, and at the end of the Napoleonic Wars (1815) the total national debt amounted to £670,000,000*".[13]

Figure 4.3 shows James Gillray's famous cartoon from which the Bank derived its nickname, the Old Lady of Threadneedle Street, from 1797. The cartoon shows Prime Minister William Pitt the Younger wooing an old lady (the Bank of England), while his real object is the get his hands on the Bank's gold reserves: the gold coins in her pocket and the money-chest on which she is seated, along with her dress made of £1 and £2 notes can all be seen. The image is set in the Bank's 'Rotunda', then the public area.

Figure 4.3 The Old Lady of Threadneedle Street, by James Gillray, 1797.

In managing debt on this scale, it is clear that capital was being raised from an increasingly engaged public, with creditors holding mainly Government, Bank, East India or South Sea Company stocks. Moreover, these investors "*were mostly drawn from the London mercantile bourgeoisie and petite bourgeoise – merchants, bankers, jobbers, brokers, lawyers, shopkeepers, craftsmen, and even domestic servants*".

Sheppard goes on to describe how the flotation of successive Government loans "*was heavily dependent on a small number of specialist London financial contractors who subscribed for large sums and quickly re-sold at a profit to 'permanent' fundholders*". By 1756, there were around 60,000 public creditors, nationally, with small investors (each with less than £500 each in the Bank of England and East India stock) forming around one-fifth of the total number of fundholders, holding less than one-tenth of the total stocks. "*An increasingly numerous new urban class of 'town' or 'city' gentlemen whose wealth and social position was not primarily based on ownership of land was beginning to emerge*".[14]

Figure 4.4 shows the Great Hall in the Bank just after the turn of the century and Figure 4.5 shows the exterior of the building in 1790, in relation to the Royal Exchange.

By the end of the eighteenth century the Bank of England was struggling with the rising costs of the Revolutionary and Napoleonic Wars (1793–1815). As well as providing escalating financial support to the estate, it was managing a growing workload.[15] For example, Clapham cites the 'vastly increased discounting business'. The average amount of bills of exchange

Figure 4.4 Bank of England, Great Hall, 1808. Augustus Pugin (1769–1832) & Thomas Rowlandson (1756–1827).

discounted annually had grown from a little over £2.5m in 1794 to over £13m by 1807.[16] As Sheppard noted

> *After the disruption of international commerce during the Napoleonic wars Amsterdam never recovered its former importance, while London's financial expertise had been sharpened by the great wartime flotations of Government debts and by the remittance of subsidies to Britain's continental allies.*[17]

Little wonder that the Bank's army of clerks grew to over 900 by 1813.[18] This undoubtedly made the Bank's staff the largest concentrated white-collar workforce in the world. Murphy compares the Bank's clerks to the civil service, in so far as they constituted a body of knowledge and experience that remained constant while directors and governors of the Bank came and went. Each operated, according to Murphy, in a specialised capacity in offices supervised by senior colleagues, and heavily coordinated with each other, thus making the organisation of work at the Bank akin to that in a large factory.

> *Individually, the jobs those clerks performed were mundane and repetitive but collectively the feat they achieved, that of managing the national debt and providing banking and*

Figure 4.5 The Bank of England and The Royal Exchange, c1790. William Marlow, 1740–1813. Credit: Bank of England.

> *discount facilities for London's business and financial community, was nothing short of extraordinary.*[19]

Here we have imagery of one the earliest corporate offices that is strikingly similar to imagery produced for the next century and more. Repetitive, mundane work within a complex organisational structure undertaken rather like a factory production line. A very similar picture was unfolding a stone's throw away in Leadenhall Street, where the East India Company established its own headquarters.

From its founding in 1694, through 1708 when it was granted a monopoly of joint stock banking in England and Wales, and through to the 1830s, the Bank of England dominated public finances through its symbiotic relationship with the state and through its vast capitalisation and volume of note issuing. Throughout this period, private banking evolved but remained small scale and, as a result, unstable in the face of economic crisis. Both London and provincial banks were restricted to private partnerships, with a maximum of six partners. Little wonder that most banks remained small in scale and narrow in focus. By the 1830s, the financial system was under strain in the context of a growing economy and increasingly affluent population. The financial crisis of 1825–1826, resulting in the failure of over 80 banks, ultimately led to the emergence of joint stock banks, which through their wider capital base and more aggressive approach to banking transformed the culture of banking and business fundamentally.

The Bankers' Clearing House. The origins of the bank clearing system were described in a delightful book by a former chief Inspector of the London Bankers' Clearing House, Philip

Matthews.[20] Until the 1770s, bank clerks from each bank visited all the other City banks to exchange with one another the cheques drawn upon and bills payable at their respective houses during the day.

> *... the clerk from Bank A met the clerk from Bank B at a given place and the clerk from Bank A informed the clerk from Bank B that he had drafts representing so much to present for payment at Bank B, and the clerk from Bank B informed the clerk from Bank A that he held drafts representing so much to present for payment at Bank A they then struck a balance between the two and agreed to meet for settlement later in the day.*[21]

Matthews states that this arrangement reduced the amount of currency required for the settlement as, if the presentations had taken place at the offices of the Banks A and B, the full amount of each charge must have been paid in cash. The system of clerks visiting every other banker in London was replaced with a central facility, or clearing house, and single interbank transfers. By 1921, Matthews was able to describe the Bankers' Clearing House as the hub of city finance, without which the huge daily financial transactions would be impossible. The knowledge that a "*final settlement and payment ... will be made at a given hour, of cheques issued on clearing bankers and paid in to other clearing bankers to meet the requirements of the business in hand is invaluable*".[22]

The first venue for the Clearing House was a private, rented room in a tavern! The earliest records, of 1773, refer to the Five Bells in Lombard Street. Soon this was found to be too small, and a larger room was taken at Mrs. Irving's, a private house next door to the Five Bells. A dedicated facility was built in 1833 by Sir John Key. The site was small, and

> *precluded the possibility of a pretentious or commodious building. There was no desire to make the building attractive from the exterior; the bankers were anxious that their clerks, carrying as they did securities of great value, frequently Bank of England notes for large amounts, should not be observed by the public.*[23]

Thus the clerks could travel to and fro for the most part unobserved.

The 39 member bankers acquired the freehold of the site in 1834. Several of these disappeared over time and the remainder were joined by joint stock banks after 1853. The balance of leavers and joiners numbered 31 banks in 1854. In the same year, Matthews indicates a further round of structural alterations to the building, necessitating the decamping of the Clearing House to the Hall of Commerce. Further purchases of neighbouring properties followed, including the adjoining premises, 3 Abchurch Lane, the basement of which was let to a wine merchant whose cellars extended underneath the Clearing House to Post Office Court (the upper floors were let out as offices). "*These premises were a valuable acquisition and enabled the authorities to take in from time to time such space as their growing business required*". In 1892, extensive alterations were taken in hand, and the whole of the upper floors of 3 Abchurch Lane were taken into use. The wine merchant, however, retained his cellars until 1915.[24]

In 1902, the adjoining premises, Nos. 84 and 85 King William Street, became vacant and a lease was secured by the Clearing House Committee. The alterations which then took place rendered possible the introduction of the Burroughs adding machine into the Clearing House and the working of the Clearing System underwent considerable alteration.[25]

Private bankers and merchant bankers

The distinctive roles of the private and merchant banks and bankers, their social position and their relations with customers are fundamental to the story being told here. From the very beginning, banking reflected two contrasting cultures. First, and as we saw in Chapter 3, many of the early bankers grew out of goldsmithing and these bankers typically built a clientele among the aristocracy. Secondly there were merchants who diversified their businesses in order to finance their core activities. The latter part of the eighteenth century and the start of the nineteenth saw private bankers rise and expand, before they started to face growing competition in the 1830s from joint stock banks that were highly capitalised and more 'commercial'.

The earliest of the City's private bankers included Sir Thomas Gresham, also founder of the Royal Exchange, who had acted on behalf of King Edward VI, and by the 1770s, there were several firms, including Barnett, Hoare & Co and Willis, Percival & Co, both of Lombard Street, that had been bankers in the City since the reign of Charles II.[26]

The latter half of the eighteenth century saw both an explosive growth in banking activity and restructuring of the banking system. Before 1750, only about 12 banking shops existed outside of London while by 1795 they were in almost every market town. In fact they grew from 12 in 1750 to 100 in 1775 and to 370 in 1800.[27] Over the same period the number of banks in London grew from 20 to 70.[28]

By the end of the eighteenth century there were around 70 banks in London, and these broke down into two distinct markets, differentiated both by their location and by the nature of their business: City banks and West End banks.

The City banks, or merchant banks, initially centred around Lombard Street in the City, led the drive towards international finance, as London developed its pre-eminent position as a world centre for the finance of international trade and the export of long-term capital overseas. Currie's, Gosling's and Martin's Bank were just three of the more well-known banks.

City banks were defined as a distinct business grouping by a number of common activities, which encompassed trading goods on their own account or a joint account, or obtaining consignments of products for themselves, their agents, or friends; trading in securities (on own account or on commission), including floating them in the European markets; and accepting (granting of advances) or financing of international trade and even financing the trade of securities. Other activities included insurance (as underwriters or brokers); exchange (negotiating foreign exchange rates or getting acceptance of bills of exchange on behalf of other merchants resident in other markets); shipping (owning of vessels or shipping brokerage); and issuing of public debt.[29]

The practice of banking was still carried on alongside other trading ventures, and too many of its practitioners paid the penalty for rash speculation in commodities, like Sir George Colebroke's unsuccessful attempt to corner alum supplies, or else lacked the expertise necessary to distinguish between various types of assets, or to keep the requisite reserves of cash. This combination of other interests and inexperience probably accounts for the high mortality among firms, which came into existence from I769 to I773.[30]

Black identifies the Smithfield bankers, specialising in financing the droving trade; the Southwark bankers on the City's fringe, specialising in financing the trade in hops, and a further cluster on the western edge of the City along Fleet Street and around Temple Bar.

West End bankers moved away from mercantile dealings, and chose instead to develop banking services with landowners and other wealthy or aristocratic private clients. These private banks of the late-eighteenth century provided short-term loans and mortgages, transfers of cash between country and town, and the management of investments in Government stocks. They were also involved in deposit taking, bill discounting and investing in stock on behalf of their clients. In 1865, Evans noted that

> *Private bankers of London number sixty firms, with two hundred and seventeen partners. The leading houses carrying on business in the midst of the City, amount to thirty-two; the*

> *remainder prosecute their operations from Cheapside, to the extremity of Charing Cross and Old Bond Street, westward.*[31]

We met Asgill's and Drummond's earlier; but Child's Bank, Coutts' and Hoare's Bank were all very successful, and by 1785 there were around 15 banks in this area.[32] These banking families "*became socially indistinguishable from their customers*"[33] as evidenced by the following quote.

> *The West End private bankers enjoy very stable reputations. Principally they are banks of deposit, whose customers do not overdraw their accounts or require advances. Coutts & Co., Child's & Co., and Cockburn & Co., with others, are all know to possess very large balances on current accounts; and thus they are relieved in a great measure from the responsibility attaching to business which requires either bill or loan accommodation.*[34]

Typically the West End banks focused on servicing the needs of wealthy individuals who resided in London on a permanent or temporary basis.[35] Many of the latter were drawn to London for their roles in parliament and in the courts, to sort financial and legal affairs business reasons and to be a part of 'the season' of balls, dinner parties and cultural and leisure activities. They also came to make arrangements for the marriages of their sons and daughters. It was a small step to entrust the London private bankers with the security of their income and wealth.

Their role at the fulcrum of London's financial community gave private bankers a reputation that was reflected in their social and economic status. Whether City or West End bankers, they affected an air of solidity, trust and respectability; they exuded social gravitas and were extremely well-connected. The private banker was perfectly encapsulated.

Daniel Hardcastle, writing in *Banks and Bankers* in 1843, described the London private banker thus

> *... a man of serious manners, plain apparel, the steadiest conduct, and a rigid observer of formalities. As you looked in his face you could read ... that the ruling maxim of his life ... was, that he who would be trusted with the money of other men, should look as if he deserved the trust, and be an ostensible pattern to society of probity, exactness, frugality, and decorum. He lived – if not the whole of the year, at least the greater part of the year – at his banking house; was punctual to the hours of business, and always to be found at his desk.*[36]

No less an authority than Bagehot linked the success of the London private banks closely to the personal character of their bankers:

> *a man of known wealth, known integrity, and known ability is largely entrusted with the money of his neighbours. The confidence is strictly personal. His neighbours know him, and trust him because they know him. They see daily his manner of life, and judge from it that their confidence is deserved.*
>
> *the name "London Banker" had especially a charmed value. He was supposed to represent, and often did represent,' a certain union of pecuniary sagacity and educated refinement which was scarcely to be found in any other part of society.*[37]

This 'social' or 'cultural' backdrop to the early development of banking, played an enormously important role in how they grew as businesses, particularly here concerning social relations in the office.

C. Hoare & Co. The story of Hoare's is emblematic. C. Hoare & Co is the UK's oldest independent bank, founded by Richard Hoare (1648–1719) in 1672. In the seventeenth century there were no street numbers, and businesses were identified by signs. Hoare traded under the 'Sign of the Golden Bottle'. Richard Hoare completed his apprenticeship as a goldsmith and, as goldsmiths often did in those days, set up a business safeguarding gold and valuables, in Cheapside. In 1690, the business moved to its long-term location in Fleet Street, where it continued to prosper. Richard was knighted by Queen Anne in 1702, and appointed Lord Mayor of London 10 years later.

The business gradually moved away from goldsmithing, pawnbroking, dealing in diamonds and speculation in ships, towards a specialisation in providing banking for the aristocracy and gentry. By 1720, the goldsmith-banking firm had shed its active goldsmith's business and "*rapidly became one of the great banks for the aristocracy and gentry*".[38]

In this more specialist role, Hoare regularly received rents from estates all over the country, with large balances held in London and drawn on by note or by bill when required; while also handling investments in stocks and collecting dividends for their customers. With its clientele among the richest group in the kingdom, the business grew to a considerable size.

> In 1725 the liabilities, which were mainly deposits, totalled £209,074; on the asset side their loans and investments (mainly loans) stood at £90,868 while the liabilities were backed by a very healthy reserve of cash, £89,474. By 1785, after more than half a century of growth interrupted only by the wars, deposits and other liabilities stood at £821,002; total loans, advances and investments were £599,220; overdrawn accounts came to £47,116; with again a substantial though lower cash reserve of £156,363.[39]

In 1829, the Fleet Street premises were rebuilt (at what is now 37 Fleet Street; see Figure 4.6) as one of the earliest purpose-built bank buildings, thereby breaking with the town house tradition. Designed by Charles Parker, the building is elegant and well-balanced, shunning any thought of ostentation. The building is dressed in Bath Stone, behind which a basement and three storeys are massed. Three rows of five windows are bookended by projecting balconies and windows topped by pediments; and the ground floor windows are all arched. Otherwise simple lines and modest relief give the building a rather austere look – perfectly in tune with the discretion and probity of the occupants!

The interior retains many Grecian Revival features dating from the construction of the building.

> *The Banking Hall still has its original, but altered, oak counter with low bronze screen, and a columnar bronze stove surmounted by a lamp of the same date. Each of the balusters of the fine staircase, leading from the ground floor to the Partners' rooms on the floor above, takes the form of a bronze Doric column. The pedestal of the lampstand on the landing is decorated with scagliola, its drum girdled with a bronze frieze and surmounted by a bronze candelabrum. The suite of rooms on the first floor retain all their early-C 19 cornices, fireplaces, doorcases and doors.*[40]

The early-nineteenth century also saw a major European influence on London banking. The turmoil of the Napoleonic and Revolutionary wars in mainland Europe during the latter part of the eighteenth century led many merchants to flee their home cities for a safe haven in Britain and London, where they could trade freely, often diversifying into banking. While there was a

Figure 4.6 Hoare & Co, 37 Fleet Street.

'push' in the form of continental Europe becoming an untenable base for business, there was also a 'pull' from London in the form of a critical mass of support facilities existing in London to maximise the trading or trade-financing potential of distinct areas of local knowledge.[41]

Llorca-Jana refers to the exceptional degree of mixing of merchants of different nationalities in London in the early years of the century. Among the foreign merchants moving to London during this time, he notes Frederick Huth, the founder of Huth & Co., who arrived in London in 1809 via Spain (although originally German), and John Henry Schroder, the founder of J. H. Schroder & Co. of London, who arrived in England from Hamburg in 1802, in the midst of the Napoleonic Wars. "*The cosmopolitan atmosphere, with so many European merchants residing in London, was a new phenomenon at the beginning of the nineteenth century, and it certainly shaped merchant bankers' activities*".[42] Other merchant banks arriving around this time – many of whose names echoed through the city for the following two centuries – included Baring, Hambro, Kleinwort, Lazard, Rothschild and Seligman from Europe, while others came from the USA. One such was George Peabody who emigrated from Baltimore in 1831 to found the firm that was to grow and split into JP Morgan and Morgan Grenfell. Many diversified into banking and loaned money overseas, providing capital for the expansion of the British Empire. As their descendants assimilated into British society, their banks became decidedly British institutions.

Taking Schroders as an example, the brothers Johann Friedrich Schröder (1780–1852) and Johann Heinrich Schröder (1784–1883) were the sons of a successful Hamburg merchant, Christian Matthias Schröder (1742–1821). Johann Friedrich set up a London-based merchant firm in 1800, J. F. Schröder & Co. In 1804 Johann Heinrich became a partner in his brother's firm. The company prospered, focusing on financing trade between America and Europe, particularly in the tobacco, cotton and sugar markets. The firm also acted as an acceptor of bills of exchange.

After his elder brother's retirement in 1817, Johann Heinrich set up his own firm, J. Henry Schröder & Co. in 1818, and maintained close ties with Germany through his Hamburg firm of J. H. Schröder & Co., set up in 1819. The new company continued the family's traditional trading activities while gradually diversifying into merchant banking. The rise of international trade during the nineteenth century brought a need for more secure payment method, and banks such as Schroders capitalised on the opportunity to provide trading houses with guaranteed payments. The appearance of merchant banks was to play a central role in encouraging the growth of international trade. In 1853, the firm began issuing bonds, at first for overseas railway construction (starting with the Matanzas and Sabanilla Railway in Cuba) and from 1863 for sovereign states.

Bonds were a critical development in the evolving financial services market. It is worth summarising the difference between bonds and loans, which boils down to two key features. The first key difference is that the cost of loans fluctuates by being pegged to a central bank interest or some other financial benchmark; while bonds normally pay an unchanging rate. This feature gives bonds their other name: fixed income. The second key difference is that bonds are designed to be traded, loans are not. Trading allows creditors to move risk around. The scale of the market can be appreciated by highlighting that the global bond market was worth $141tn in 2022 (by comparison, the global equities market were valued at around $42tn in 2022).

Baring Brothers. There are few businesses whose physical presence in the City (or anywhere else) can be tracked over centuries, but this is the case with Baring Brothers bank. Co-founder Francis Baring (1740–1810) lived almost across the entire span of this chapter's focus and his bank's office history can be traced through to 2012.

In the early-eighteenth century, Exeter was a thriving town of 12,000 people and a key port in the West Country. The region's wool trade sent woollen goods to Holland, Flanders, Germany, Spain and Portugal. The market attracted merchants from wide and far, and one of the earliest to arrive was John Baring from Bremen, Germany, in 1717. At the tender age of 20, John entered an apprenticeship to learn the Exeter merchanting trade, with a view to acting as an agent for the family business in Germany. John married Elizabeth (daughter of another successful merchant) and they had had three sons. John was very successful, becoming an eminent merchant and landowner.[43]

John senior was followed into the business by his sons John, Charles and Francis, and in 1762 the family reorganised the business, with a view to establishing a London house (as it was referred to). On New Year's Day, 1763, John & Francis Baring & Co opened for business on Queen Street, Cheapside in the City of London, with Francis running the house. Francis built a strong business, gradually diversifying from wool into many other commodities, providing financial services for the rapid growth of international trade, and moving into accepting bills of exchange very successfully.

Towards the end of the century, a few of the leading London merchants, above all Barings, were taking on a 'finance' function and becoming what would eventually be termed merchant banks – or, more narrowly, accepting houses – to service the international trading community. Francis Baring's acceptance business commenced in 1768. It was a profitable business, done on a commission basis; but since it involved guaranteeing bills of exchange that would eventually be sold in the London bill market centred on the Royal Exchange, it was one that demanded the nicest possible judgement of clients, of trades, and of countries.[44]

In 1769, the business moved to a new office at 6 Mincing Lane. Francis was becoming a senior city figure: in 1771, he was appointed a director of the Royal Exchange Assurance, and in 1779 a director of the East India Company (later to become Chair). In 1784, he became a Member of Parliament, and was created a baronet in 1793. In 1767, Francis married Harriet Herring, with whom he had twelve children – 11 of which were born in the family's quarters above the Mincing Lane office.

European war and strife afflicted the continent's trading patterns badly in the closing years of the eighteenth and early years of the nineteenth centuries. British merchants benefited, and London toppled Amsterdam as the leading financial centre, directly benefiting key London firms, including Baring. Francis also began to trade successfully with North America. The house of Baring also found new trade in financing the needs of European governments.

Over the last quarter of the century, Barings flourished, becoming a dominant force in the City. In 1790, total assets were about £385,000, by 1795 £765,000 and in 1797 £1.66 million. In terms of total asset size, only the largest banks, such as the Smith, Payne & Smith or Hoare's, reached a similar scale.

All this success was reflected in a move from the cramped Mincing Land premises to a more palatial setting at 11 Devonshire Square. The new premises, however, also served as a residence, before Francis moved to Hill Street in the West End in 1802. The business moved again in 1805 to "*a commodious house fronted by an open courtyard with street access through a carriage arch*".[45] In 1806, the firm moved to 8 Bishopsgate, where it remained – with multiple modifications – for nearly two centuries. Sir Francis died in 1810, aged 70.

In 1853, the well-known building firm, William Cubitt, with architect Lewis Cubitt, partially filled in the courtyard to create a banking hall and partner's room. Then, in 1885, more modifications took place under the eye of Richard Norman Shaw, including a new building in the Bishopsgate frontage. Following further growth, the bank bought neighbouring plots to facilitate in situ growth. In 1887, 13 Bishopsgate was bought; and in 1906 further acquisitions were made of 12, 14 and 16 Bishopsgate. Soon afterwards Nos. 10 and 11 Bishopsgate were also added. The composite plot was extended to the rear in 1945 through the acquisition of 145 Leadenhall Street.

The later-nineteenth and twentieth century history of Baring Brothers' offices is summarised by Harrow[46] who notes the arrival of modern amenities, including a water closet and gas lighting in 1853; electricity in the 1880s; the first telephone in 1904; and in 1913 the first lift. Harrow cites a journalist writing for *The Daily News* in 1886:

> *For myself, I have no objection to its high narrow windows and stony front for that means the safety of valuable property; nor can you cavil at the low, arched doorway, with its two swinging doors, through whose glass panes you look up the long passage into the serene and mystic interior, where gentlemanly clerks deal with heaps of gold as children deal with a handful of hazelnuts.*

Harrow also cites the first-hand account of a Baring clerk, J Walter Wood:

> *The silence was impressive, and the great banking room, with its mellowed features seen through the soft light was very dignified. No-one seemed in a hurry. Business did not really begin 'til about 11 o'clock and was usually over by 4 o'clock. For those clerks who remained at work until 5 o'clock, tea with thin slices of buttered bread was served by the liveried 'walkers' [bank messengers].*

There was then the partners' room that adjoined the general office and retained many domestic features and comforts, such that it was

> *more like a gentleman's library, with a cheerful open fire in a marble mantelpiece at the end of the room, and a soft red and blue rug. The mahogany desks were massive, and the chairs deeply upholstered in dark leather. In the centre of the room was a large table with a limited number of necessary reference books, and from the walls deceased partners of the House looked down from mellowed gilt frames.*

Harrow reports that, by 1973 the building had become a rabbit warren of small offices and that plans to widen Bishopsgate meant that the site needed to be vacated and the bank therefore relocated to 88 Leadenhall Street seizing the opportunity to erect a new 20-storey building that was occupied from October 1981. The business moved once again in 2016 to ING's new premises at Ten Moorgate.

The story of Barings' bank and its buildings illustrates in microcosm the story of banks and their buildings generally over two centuries and more. Starting as merchants and based in a mixed use building comprising shop and domestic accommodation; the bank then grew in stature and size, necessitating a series of moves to larger, more complex buildings, before adapting to global trading in an open plan office with a large trading floor. In a single company – one that unusually resisted the widespread practice of merger and acquisition – the whole history of the modern office can be traced.

Towards an organised industry. The banking industry was growing quickly and as it did so interbank payments increased very rapidly. As we saw earlier, this growing interbank activity led to the creation of the Bankers' Clearing House, which was used by City banks, who needed to formalise the previously haphazard private exchanges by clerks. The West End bankers chose to develop their own method of exchanging bills and notes via direct exchanges by peripatetic clerks.[47]

The creation of the Clearing House was a sign of things to come. London was growing and, with it, the banking industry was growing and diversifying. The private bankers were generally successful but, with the benefit of hindsight, their limitations were becoming apparent. Private bank clients were, by and large, financially sound, making only limited demands on their banks, and the bankers themselves worked within themselves in terms of their skills and services. However, later years

> of the 18th and the first quarter of the 19th century were traumatic ones for British banking. The industrial expansion of the country was punctuated by a number of commercial crises which produced business failures and runs on the banks. In 1825 the crisis was particularly acute.[48]

The turbulent times exposed weaknesses of the banking system and a widespread

> *outcry against the dangers of and disorder in private banking gradually solidified into legislation designed to bring a little more stability into it. Most important was the Act of 1826 by which the Bank of England's monopoly of joint stock banking in the country was terminated.*[49]

Private banks were always precarious because their very size limited their ability to fend off and see through difficult periods. Even a rumour of some external crisis could cause customers to withdraw their cash in haste in order to 'keep it safe' themselves. There was one such crisis in 1825–1826 when economic turbulence precipitated a run on the banks. The impact was unprecedented, leading to the failure of 93 banks in England and Wales during the year.[50]

Small banks were fine for a small London. But a large London, with rapidly growing numbers and types of customers requiring a burgeoning range of financial services, not to mention the international trade dimension, required something altogether different. Bagehot summarised the situation well.

> *It is painful to have to doubt of the continuance of such a class, and yet, I fear, we must doubt of it. The evidence of figures is against it. In 1810 there were 40 private banks in Lombard Street admitted to the clearing house: now there are only three.*[51]
>
> *The truth is that the circumstance which originally favoured the establishment of private banks have now almost passed away. The world has become so large and complicated that it is not easy to ascertain who is rich and who is poor.*[52]

Joint stock banks

As was shown in Chapter 2, joint stock companies began to spread during the seventeenth century, particularly in response to the increasing costs and risks attached to long-distance trading. Exporting and importing goods to and from far-distant lands began to flourish, but safe transit was far from guaranteed. In allowing individuals to purchase in interest or a 'share' in companies, making them joint owners with the other shareholders, or investors, joint stock companies spread costs and risks (and of course, reward) among a large number of individuals. Shareholders were paid in divisions, or dividends, which distributed the profits of a voyage in proportion to the number of shares held.

Despite the growth of JSCs, the banking sector was unaffected as long as the Bank of England held its monopoly position over joint stock banking between, and the private banks were legally limited to private partnership status. Black observes that banking in the 1820s remained a closed and private world, one which was

> *jealous of its privileges and exclusivity. Such privileges, enjoyed for over a century, had produced a private landscape of capital reinforced by the micro-geographies of the private banking houses and the exclusionary effects of Soane's windowless walls at the Bank of England. It was this landscape and the cultural values it sustained, legitimated and reproduced that was to face a sustained challenge following the joint-stock legislation of 1826 and 1833.*[53]

In May 1826, the Banking Co-partnership Act authorised the establishment of joint stock banks, in England and Wales (outside London), along the model earlier adopted in Scotland. It was not until 1833 that they were permitted in London, and even then for deposit only, not note issue, because the Bank of England retained its monopoly on note issuing within a 65 mile radius of Charing Cross. Banks could operate within this area on condition that they relinquished their note-issuing.

Despite this constraint, joint stock banks – even without limited liability – were able to conduct business on a much deeper capital base, managed rather than owned, by a Board of Directors. And they brought with them a new banking culture: they were more aggressive and more growth orientated; they acted on behalf of shareholders, with transparent annual reports and accounts.

The private bankers – who built their businesses based on both their personal wealth and their intimacy with the aristocracy and gentry – saw the new comers as a severe threat; they were regarded with a mixture of suspicion and contempt. This expressed itself in a number of ways, perhaps most notably in their collective refusal to grant joint stock banks permission to join the London Clearing House; a position they maintained until 1854. Bagehot noted a widespread fear that the emergent banks "*would fast ruin themselves, and then cause a collapse and panic in the country. The whole of English commercial literature between 1830 and 1840 is filled with that idea*".[54]

Indeed, between the 1826 and 1833 Acts relations between the established banks and the upstart banks were fractious as competition for clients intensified. Booker describes banking during this time as "*immature, petty and acrimonious*" as private bankers in London sided with the Bank of England in opposing the extension of joint stock banking into the metropolitan area.

> *With great suddenness, English banking, so long the scene of complacent privilege, was electric with a competition which could not fail to find expression in the choice of places of business.*[55]

The first joint stock bank to open in the capital following the legislation of 1833 was the London and Westminster Bank, which opened offices in both the City (38 Throgmorton Street) and West End (9 Waterloo Place), on 10 March 1834. The new bank was viewed with scepticism by both the Bank of England and the private banking community; it would not be until 1854 that London and Westminster was admitted to the Clearing House. Despite the opposition, it was not long before the bank had outgrown its City office and had

> *assembled a substantial freehold plot in the heart of the City opposite the northeast corner of the Bank of England. The symbolism of the location, immediately across the road from its great rival, could hardly be missed.*[56]

The new plot was on neighbouring Lothbury, and both premises eventually consolidated into a single building, known as 41 Lothbury. Black notes that the bank's new headquarters opened on 26 December 1838, at a cost of £37,000, including freehold, building, materials and architects' fees!

By 1835, a further 53 banks were registered, and in an unprecedented upsurge in demand in 1836, no less than 45 were registered.[57] The Surrey, Kent & Sussex Banking Company (later renamed the London and Country Bank) and the London Joint Stock Bank (later Midland Bank) both incorporated in 1836. The Union of London Bank and Smiths Bank opened in 1839, while the Commercial Bank of London opened in 1840.

The Union Bank of London was established at 8 Moorgate Street and developed a network of branch offices across the capital. The head office moved to Princes Street in 1844 (Figure 4.7), where, in 1887, a new head office was built. The detailing, window design and columns all give the bank a look of financial and corporate strength and demonstrate the additional funds that they allocated to creating impressive premises. At the same time, the building retained a domestic quality.

It was not until the Joint Stock Companies Act of 1844 that the 65 mile rule was abolished and when incorporation without specific legislation was allowed; when any company could, through a simple registration procedure, incorporate. Further 1854 legislation permitted joint stock banks to join the London Clearing House. The Bank of London and City Bank both opened in 1855; while the 1860s saw a further seven London openings: Alliance Bank, East London Bank, Imperial Bank, London and Middlesex Bank, London and Provincial Banking Company, Metropolitan and Provincial, North London and South London Bank.[58]

At this stage, incorporated companies did not have limited liability, but it became common for companies to include a limited liability clause in their internal rules. A critical further step was the Limited Liability Act 1855, which allowed investors to limit their liability in the event of business failure to the amount they invested in the company. These two features – a simple registration procedure and limited liability – were subsequently codified in the first modern company law Act, the Joint Stock Companies Act 1856. The Vice President of the Board of Trade at the time, and Liberal Member of Parliament for Kidderminster, Robert Lowe, is often referred to as the 'father of modern company law' for his role in getting the 1856 Act through Parliament.

By the late-1850s, joint stock banks were common in London, but still faced prejudice from the financial elite. For example, the London and Westminster Bank was refused an account at the Bank of England, along with all the endorsement that holding such an account could bring. But there was something inevitable about the progress of joint stock banks and, by 1867 they outnumbered private banks in London. Bagehot himself remarked how they had been "*a most remarkable success*". He went on to argue that the private banker of past times lent to private individuals, and he could form a "*judgement of the discretion, the sense, and the solvency of those to whom he lent*". But this was a time when London was a small city where "*everyone stuck to his proper business*", but "*now that London is enormous and that no one can watch anyone, such a trade would be disastrous*".[59]

Financial journalist David Evans observed that the joint stock bank system had made good progress over the previous 15 to 20 years, producing "*five establishments of great pretentions in the metropolis*". The chief of these was the London and Westminster Bank, followed by the London Joint Stock Bank, the Union Bank of London, the Commercial Bank of London and the London and County Banking Company.

> *All these banks pay a dividend varying from five to six per cent, and report good reserve funds, judicious management having enabled the directors (at least so they say at every*

Figure 4.7 Union Bank of London, Princes Street, 1857. © Illustrated London News Ltd/Mary Evans.

> *yearly or half-yearly meeting) to keep their business within the limit of a small amount of bad debt, while profits are well supported.*[60]

Evans highlighted some of the key differences between the emergent joint stock banks and their more conservative private counterparts. In particular, he discussed scale and culture. In terms of scale, Evans pointed out that the new banks "*have a large staff of officers in the character of managers, sub-managers, and secretaries, beside their cashiers and clerks as in ordinary banking houses*". He then suggested that such employment is expensive "*because it is necessary the several gentlemen holding them should be well paid to insure honesty and strict integrity of conduct*". Chief managers' salaries ranged from £1,000 to £2,000 a year, and the secretaries rarely received less than from £500 to £800 a year.[61]

In terms of cultural differences, Evans pointed to the great difference in class between people employed in joint stock banks and those employed in private banks. Whereas in private banks customers met "*cashiers and clerks peering through spectacles with a steady and staid appearance, whose only inquiries are respecting the weather and the prospects of business*", in the new joint stock banks customers met,

> *sprightly young gentlemen, who talk about new operas and all the other amusements of the town with all the ease of connoisseurs in high life; and whose chief study is to give effect to chequered neckerchiefs, showy chains, and mogul pins.*

Evans then argued that while the change in culture might reflect progress,

> *to the quiet man of business, the times, in this respect, are scarcely so acceptable as the old days of white ties, venerable faces, and tranquil attention to the wants of customers. The modern improvements do not facilitate the counting or weighing of sovereigns, crossing cheques, or balancing ledgers.*[62]

For others, joint stock banks brought with them what we would today refer to as professionalism, and an approach which improved on the family bankers, because they had

> *at the head of the executive a general manager who was tried in the detail of banking, who is devoted to it, and who is content to live almost wholly in it. He thinks of little else, and ought to think of little else. One of his first duties is to form a hierarchy of inferior officers, whose respective duties are defined, and to see that they can perform and do perform those duties.*[63]

The emergence and expansion of the joint stock banks is a critical step on our history here. It represented the formal separation of ownership and management; the 1826 Act directed that the new banks were to be led by boards of directors, who appointed professional managers. The agency of the manager, operating between owners and workers had been established. And yet most businesses were small by modern standards. In 1815, about 8,500 firms were operating in the City of London (alongside a residential population of some 122,000). The overwhelming majority of these firms were small, specialised and family owned. Glyns, for example, was one of the larger banks, but still only had 36 staff in 1815; while in the same year Antony Gibbs employed only five men in its London office in addition to three partners.[64]

The picture was very similar in the United States where merchants ran their small companies as family businesses, normally with just a handful of clerks. They served local markets, importing, exporting, storing, buying, selling and financing. The owner-merchant managed every aspect, with the clerks keeping the accounts in order and writing correspondence. The early businessman was a diversified economic type: merchant, moneylender, speculator, shipper, cottage manufacturer.

Summary. By the end of the eighteenth century, two of the largest white-collar organisations in London were the Bank of England and the East India Company. Both exhibited complex corporate structures; both occupied impressive office buildings and both employed dense (for the time) concentrations of clerks. Commercial banks and insurance companies, in particular, were following a similar path of development.

London's role as the European focus for trade grew as continental conflict severely curtailed growth in previously important cities such as Amsterdam, Brussels, Florence, Paris and Venice. Conflict on the continent was being funded by a ballooning government debt in London. At the same time, JSCs were increasingly popular for investment among the bourgeoisie, including merchants, bankers, jobbers, brokers, lawyers, shopkeepers and craftsmen.

As organisations and concentrations of clerks grew, so their working conditions overall began to deteriorate. Division of labour led to repetitive and monotonous work, working days lengthened and hierarchies began to spread. Buildings were not designed to cope with the high densities, lacking appropriate sanitation and heating and cooling.

The distinctive roles of the private and merchant banks, their social position and their relations with customers are fundamental to the story being told here. From the very beginning, banking reflected two contrasting cultures. First, and as we saw in Chapter 3, many of the early bankers grew out of goldsmithing and these bankers typically built a clientele among the aristocracy. Secondly there were merchants who diversified their businesses in order to finance their core activities.

The social differences between private and merchant bankers were compounded by the arrival of joint stock banks, the first of which opened in 1833. The key point was that joint stock banks could were able to conduct business on a much deeper capital base, and they were managed rather than owned, by a Board of Directors. They introduced a different banking culture, what we might call today, more commercial compared to the personal service offered by private bankers. Despite resistance from the established banking community, by the late-1850s, joint stock banks were common.

Notes

1 Olsen DJ (1976) *The Growth of Victorian London* Penguin Books, London p124.
2 Gray R (1978) *A History of London* Hutchison, London p227.
3 Kynaston D (1994) *The City of London Volume One: A World of its Own 1815–1890* Chatto & Windus, London p141.
4 Bayne-Powell R (1937) *Eighteenth Century London Life* John Murray, London p67.
5 Clunn HP (1935) *The Face of London: The Record of a Century's Changes and Development* 6th Edition Simkin Marshall, London p20.
6 Kynaston D (1994) *The City of London Volume One: A World of its Own 1815–1890* Chatto & Windus, London pp141–142.
7 Evans DM (1865) *The City, or The Physiology of London Business with Sketches on 'Change, and at the Coffee Houses* Baily Brothers, London pp189–190.
8 Bayne-Powell R (1937) *Eighteenth Century London Life* John Murray, London p50.
9 Bank of England (1970) *History and Functions Bank of England Archive* (G15/634) p7.
10 Bank of England (1970) *History and Functions Bank of England Archive* (G15/634) p7.
11 Bank of England (1970) *History and Functions Bank of England Archive* (G15/634) p8.
12 Murphy AL (2017) Clock-watching: Work and Working Time at the Late Eighteenth Century Bank of England *Past & Present* Vol 236 Issue 1 pp99–132 (August 2017).
13 Sheppard F (1999) *London: A History* BCA Books, London p226.
14 Sheppard F (1999) *London: A History* BCA Books, London p227.
15 Murphy AL (2015) Writes a Fair Hand and Appears to be Well Qualified: the Recruitment of Bank of England Clerks *Financial History Review* Vol 22 No 1 2015 pp19–44.
16 Clapham JH (1945) The Bank of England: A History 2 Vols Cambridge University Press Cited in Murphy, Writes a Fair Hand and Appears to be Well Qualified.

17 Sheppard F (1999) *London: A History* BCA Books, London p235.
18 Kynaston D (1994) *The City of London Volume One: A World of its Own 1815–1890* Chatto & Windus, London p30.
19 Murphy, Writes a Fair Hand and Appears to be Well Qualified.
20 Matthews PW (1921) *The Bankers' Clearing House: What it is and what it Does* London, Sir Isaac Pitman & Sons LTD.
21 Matthews PW (1921) *The Bankers' Clearing House: What it is and what it Does* London, Sir Isaac Pitman & Sons LTD p2.
22 Matthews PW (1921) *The Bankers' Clearing House: What it is and what it Does* London, Sir Isaac Pitman & Sons LTD p3.
23 Matthews PW (1921) *The Bankers' Clearing House: What it is and what it Does* London, Sir Isaac Pitman & Sons LTD p12.
24 Matthews PW (1921) *The Bankers' Clearing House: What it is and what it Does* London, Sir Isaac Pitman & Sons LTD pp13–14.
25 Matthews PW (1921) *The Bankers' Clearing House: What it is and what it Does* London, Sir Isaac Pitman & Sons LTD p14.
26 Kynaston D (1994) *The City of London Volume One: A World of its Own 1815–1890* Chatto & Windus, London p14.
27 Turner GD (2015) *English Banking in the Eighteenth Century: Bankers, Merchants and the Creation of the English Financial System* Durham theses, Durham University Available at Durham E-Theses Online: http://etheses.dur.ac.uk/11297/ p 171.
28 Cameron R (1967) *Banking in the Early Stages of Industrialization, A Comparative Study* In Turner, *English Banking in the Eighteenth Century* p171.
29 Llorca-Jana M (2014) Shaping Globalization: London's Merchant Bankers in the Early Nineteenth Century *Business History Review* Vol 88 (Autumn 2014) p471–472 doi:10.1017/S0007680514000403.
30 Joslin DM (1954) London Private Bankers, 1720–1785 *The Economic History Review* New Series Vol 7 No 2 pp167–186.
31 Evans DM (1865) *The City, or The Physiology of London Business with Sketches on 'Change, and at the Coffee Houses* Baily Brothers, London p8.
32 Joslin DM (1954) London Private Bankers, 1720–1785 *The Economic History Review* New Series Vol 7 No 2 pp167–186.
33 Sheppard, *London: A History* p228.
34 Evans DM (1865) *The City, or The Physiology of London Business with Sketches on 'Change, and at the Coffee Houses* Baily Brothers, London p12.
35 Black IS (2003) Private Banking in London's West End, 1750–1830 *The London Journal* Vol 28 No 1 pp29–59.
36 Hardcastle D (1843) *Banks and Bankers* Whittaker & Co, London pp22–23.
37 Bagehot W (2009) *Lombard Street: A Description of the Money Market* Seven Treasures Publications (Original work published 1873) p118.
38 Joslin DM (1954) London Private Bankers, 1720–1785 *The Economic History Review* New Series Vol 7 No 2 pp167–186.
39 Joslin DM (1954) London Private Bankers, 1720–1785 *The Economic History Review* New Series Vol 7 No 2 pp167–186.
40 Hoare's Bank, A Grade II* Listed building in the City of London https://britishlistedbuildings.co.uk/101192633-hoares-bank-farringdon-without-ward
41 Kynaston D (1994) *The City of London Volume One: A World of its Own 1815–1890* Chatto & Windus, London p26.
42 Llorca-Jana M (2014) Shaping Globalization: London's Merchant Bankers in the Early Nineteenth Century *Business History Review* Vol 88 (Autumn 2014) p471–472 doi:10.1017/S0007680514000403.
43 This brief history of Baring Brothers is summarised from Orbell J (1985) *Baring Brothers & Co, Limited: A History to 1939* Baring Brothers & Co, London.

44 Kynaston D (1994) *The City of London Volume One: A World of its Own 1815–1890* Chatto & Windus, London p12.

45 Orbell J (1985) *Baring Brothers & Co, Limited: A History to 1939* Baring Brothers & Co, London p20.

46 Harrow C (2016) Barings and its Buildings *Architecture & Finance Bulletin* pp169–170 European Association for Banking and Financial History Available at: https://bankinghistory.org/wp-content/uploads/Barings.pdf

47 Joslin DM (1954) London Private Bankers, 1720–1785 *The Economic History Review* New Series Vol 7 No 2 pp167–186.

48 Clarke P (1973) *The First House in the City: An Excursion into the History of Child & Co to Mark its 300th Year of Banking at the Same Address* Perivan/Williams Lea Group, Southend-on-Sea p31.

49 Clarke P (1973) *The First House in the City: An Excursion into the History of Child & Co to Mark its 300th Year of Banking at the Same Address* Perivan/Williams Lea Group, Southend-on-Sea p32.

50 Collins M (1988) *Money and Banking in the UK: a History* Croom Helm Ltd, London & New York.

51 Bagehot W (2009) *Lombard Street: A Description of the Money Market* Seven Treasures Publications (Original work published 1873) p118.

52 Bagehot W (2009) *Lombard Street: A Description of the Money Market* Seven Treasures Publications (Original work published 1873) p119.

53 Black IS (2000) Spaces of Capital: Bank Office Building in the City of London, 1830–1870 *Journal of Historical Geography* Vol 26 No 3 p357.

54 Bagehot W (2009) *Lombard Street: A Description of the Money Market* Seven Treasures Publications (Original work published 1873) p111.

55 Booker J (1990) *Temples of Mammon: the Architecture of Banking* Edinburgh University Press p44.

56 Black, Spaces of Capital p359.

57 Klingender FD (1935) *The Condition of Clerical Labour in Britain* Martin Lawrence, London p1.

58 Orbell J; Turton A & Pressnell LS (2001) British Banking: A Guide to Historical Records, Studies in British Business Archives Ashgate, Aldershot Cited in Barnes V & Newton L (2019) Symbolism in Bank Marketing and Architecture: the Headquarters of National Provincial Bank of England *Management & Organizational History* Vol 14 No 3 pp213–244.

59 Bagehot, *Lombard* p111.

60 Evans, *The City, or The Physiology of London Business* pp13–14.

61 Evans, *The City, or The Physiology of London Business* p14.

62 Evans, *The City, or The Physiology of London Business* p15.

63 Bagehot, *Lombard Street* p122.

64 Kynaston, *City of London Volume One*p30.

5 Clerking

Laborious and monotonous

Adam Smith speculated on the division of labour within industrialism and its role in driving productivity. The specialisation and concentration of workers in their sub-tasks, he argued, led to greater skills and productivity than would be achieved with the same number of workers each undertaking the original broad task. He famously cited pin-making to make his point:

> *A workman not educated in this business ... nor acquainted with the use of the machinery employed in it ... could scarce, perhaps, with his utmost industry, make one pin in a day, and certainly could not make twenty.*[1]

To achieve an output of 20, Smith observed that the pin-making process is divided into 18 distinct stages. He also recognised the role of machine capital in the manufacturing process. In short, he was recognising the blending of human skills and technology, in the context of a wider organisation (the firm) to enhance productivity. This becomes highly relevant when we consider the 'invisible glue' that today holds knowledge economy firms together in tight business ecosystems, or business clusters, in our cities.

This division, or specialisation of labour is seen in the early years of the office; the degree of specialisation that separated people and sub-tasks in Smith's 'production' process were visible in offices as soon as they began to scale-up from small, owner-managed businesses and moved towards corporate structures. In the former, clerks are often portrayed as 'masters of the whole business', handling all aspects of administration. But this mastery broke down when a Smith-like production process, albeit paper production process, got underway. The early offices were the original 'paper factories'.

It is perhaps difficult to appreciate from the perspective of the twenty-first century that the emergence and spread of clerks represented a huge shift in the structure of the workforce; they not only formed a new economic function, but also eventually the core of a new social class. In the pre-Industrial Revolution eighteenth century, there was a gulf between the agricultural and labouring workforce and the gentry and aristocracy. The merchant class was growing in the larger cities, and particularly London, but their incomes and lifestyles – many were landowners – allied them more closely with the gentry than with labourers.

The gulf between gentry and labourer carried deep cultural and socio-economic meaning. The emergence of people whose role sat somewhere between these two worlds (i.e. managers) was in many senses revolutionary. This is important because it shaped so much of what was to follow in terms of the rise of the middle class, the professional class and the office economy.

At its most basic, certainly in the early years, the role of the clerk was to undertake manual work, but in an indoor environment where management of paperwork was the main focus,

DOI: 10.1201/9781003582205-7

requiring literary and numeracy skills not common in large sections of society. Clerks were, as we see elsewhere here, manual workers undertaking mundane and repetitive tasks, but they were doing so in a context from which they could progress economically and socially.

Four pent walls, without relief

The East India Company was chartered as a joint stock company in 1600, and granted exclusive rights to all trade with Asia. The company occupied a suite of rooms in the mansion of its governor, Sir Thomas Smythe, in Fenchurch Street but having outgrown this space, in 1621 it moved to Crosby Hall on Bishopsgate. For reasons that are not clear, it then made a further move, into the house of its current governor, Sir Christopher Clitherow, in Leadenhall Street in 1638. Once again it outgrew its accommodation and, in 1648, took a leasehold on a headquarters at Craven House in Leadenhall Street, on the site of what is now the Lloyd's of London building in Leadenhall Street.

In 1710, East India Company purchased Craven House. The business continued to adapt and expand the property, but the building was, overall, in a very poor condition. Eventually, the business took temporary accommodation in Fenchurch Street in 1726 to allow a complete rebuild using the site of Craven House and neighbouring properties. East India House was completed in 1729 to a design by Theodore Jacobsen. This was the first purpose-built corporate headquarters.

Figure 5.1 East India House, c1800.

In 1796, the company purchased an additional plot of land in order to extend its premises in a project started by Richard Jupp and completed by Henry Holland in 1799 (Figure 5.1). Like many modern global corporations, the East India Company created a headquarters to reflect its dominant trading position, housing people and work settings for generating, processing, receiving, despatching and filing vast quantities of paper records. As with the Bank of England, the large numbers of workers created a large and complex bureaucracy.

Externally the neo-classical architecture was extravagant, finished off with a dominating six-column portico. The large tympanum that sat atop depicted King George III defending the commerce of the East, with a statue of Britannia crowning the top, and statues of Asia and Europe to the east and west, respectively. Internally, the building was extraordinary.

> *Reflecting the great wealth of the East India Company, no expense was spared on the interior decorations. Palladian architectural features, including Corinthian pilasters and heavy moulding, continued throughout the interior. Lavish Georgian furnishing including ornately carved boardroom tables and velvet upholstered chairs for the Chairman and Vice-President, were commissioned.*[2]

Figure 5.2 shows the sale room in East India House in 1809, emphasising the sense that it operated rather like some of the merchant exchanges.

Figure 5.2 East India House, sale room, 1809.

Charles Lamb, essayist and clerk. A curious number of literary folk passed through the front doors of East India House to supplement their meagre earnings from writing. One of the longer-serving, and perhaps the most prolific, commentators on working conditions was essayist Charles Lamb (1775–1834).

Lamb's most famous works include *Essays of Elia* and *Tales from Shakespeare*, the latter a children's book co-authored with his sister, Mary. He enjoyed a sophisticated social life, rubbed shoulders with the likes of Coleridge, De Quincey, Hazlitt and Wordsworth and was a mainstay of the literary scene in London. However, writing and literature did not pay the bills, and so Lamb was also a wage-earner. While still not 15 years of age, Lamb left school to become a clerk for London merchant Joseph Paice. A year later, he held a brief 6-month tenure at South Sea House as a junior clerk in the Examiner's Office, before moving to become a clerk in the Accountant's Department of the East India Company. Here he would remain until he retired 33 years later.

As was customary at the time, Lamb's employment was based on a £500 surety and two sponsors – a measure designed to serve as a surety for good and honourable behaviour. Lamb also had to serve a 2-year probationary period for which he was paid a gratuity of £30 per year. Only in his third year of service did he receive a regular salary, starting at £40.

The superficial grandeur of East India House disguised what, by most accounts, were grim working conditions – both in terms of the physical environment and the pattern of work. The atmosphere in the building was captured well by De Quincey while paying a visit to Lamb in the winter of 1804–1805; De Quincey himself was just 19 years old. Years later he gave an account of this first meeting in his London Reminiscences. He describes how he was

> *shown into a small room, or else a small section of a large one in which was a very lofty writing-desk, separated by a still higher ceiling from that part of the floor on which the profane – the laity, like myself – were allowed to approach the clerus, or clerkly rulers of the room.*[3]

He then wonders what might have been the purpose of the lofty furniture, which was so absurdly high "*that I can imagine no possible use or sense in such an altitude, unless it were to restrain the occupant from playing truant at the fire by opposing Alpine difficulties to his descent*".[4] De Quincey approached one of the clerks

> *I walked, therefore, into one of the two open doorways of the railing, and stood closely by the high stool of him who occupied the first place within the little aisle. I touched his arm, by way of recalling him from his lofty Leadenhall speculations to this sublunary world; and, presenting my letter, asked if that gentleman (pointing to the address) were really a citizen of the present room; for I had been repeatedly misled, by the directions given me, into wrong rooms. The gentleman smiled; it was a smile not to be forgotten. This was Lamb.*[5]

He then described how Lamb descended from his writing desk as an

> *act of descending from his throne, a very elaborate process, with steps and stages analogous to those on horseback … to all intents and purposes, the same thing as dismounting from a great elephant of a horse.*[6]

It must have been quite a sight. And add to this the probable physical environment: De Quincey hints at the labyrinthine room layout, each of which would have been very quiet; with little or no air movement; dusty and musty and poorly lit. All in sharp contrast to the superficial grandeur of the external architecture and the stately porticos; imposing staircases; grand court and committee rooms, and offices of the Directors.

Lamb was damning of the working conditions at East India House. During his early years in the Accountants' Office, he often worked hard and late, to the point that he arrived home every night "*o'er wearied and quite faint*". He found the "*strain unbearable at times and he suffered from what today would be regarded as work-related stress*". In 1815, he wrote to William Wordsworth

> *On Friday I was at office from 10 in the morning (two hours dinner except) to 11 at night – last night till 9; my business, and office business in general, has increased so.*[7]

Work conditions in offices of this period were often cramped, poorly lit and poorly ventilated; and the work was often repetitive and physically demanding. On 20 March 1822, Lamb was morose in his latest letter to his dear friend Wordsworth:

> *I grow ominously tired of official confinement. Thirty years I have served the Philistines, and my neck is not subdued to the yoke. You don't know how wearisome it is to breathe the air of four pent walls, without relief, day after day, all the golden hours of the day between ten and four, without ease or interposition. Taedet me harum quotidianarum formarum, these pestilential clerk-faces always in one's dish. Oh for a few years between the grave and the desk! They are the same, save that at the latter you are outside the machine.*[8]

The Latin element of this quote translates as *I am tired of these everyday forms*. It is also clear that as well as the long hours, management–worker relations were frayed. For example, Lamb fumed at a top-down initiative to cut costs, including retraction of the annual £10 holiday allowance; turning Saturday from a half-day to full working day; taking away holidays on Saints' Days; taking away the perk of sending private letters postage free from India House and, perhaps most resented of all, the introduction of a clocking in/out system.

Noting that 36 years had elapsed since becoming a clerk, since leaving behind the freedoms of a 14 year old with abundant playtime and holidays, for the 8, 9, and sometimes 10 hour days of work at the counting house, Lamb is melancholic.

> *It is true I had my Sundays to myself.... But besides Sundays I had a day at Easter, and a day at Christmas, with a full week in the summer .... Before I had a taste of it, it was vanished. I was at the desk again, counting upon the fifty-one tedious weeks that must intervene before such another snatch would come ... . I was fifty years of age, and no prospect of emancipation presented itself. I had grown to my desk, as it were; and the wood had entered into my soul.*[9]

The pain expressed by '*the wood had entered my soul*' is almost tangible. Lamb also describes how his work followed him home and weighed on his mind into the small hours.

> *Besides my daylight servitude, I served over again all night in my sleep, and would awake with terrors of imaginary false entries, errors in my accounts, and the like.*[10]

After 33 years of what he variously referred to as servitude, slavery and captivity, Lamb reached the point where he could bear no more: he had had enough; he was spent. So much so that he decided to quit. In February 1825, Lamb wrote to the Board of Directors requesting that he be permitted to retire, citing his ill health as the main reason. For reasons that are not entirely clear, it took the Board over 7 weeks to respond. On 29 March the Directors resolved that

> *the resignation of Mr. Charles Lamb of the Accountant General's Office, on account of certified ill health, be accepted, and it appearing that he has served the Company faithfully for 33 years, and is now in the receipt of an income of £730 per annum, he be allowed a pension of £450 ... per annum ... to commence from this date.*[11]

The contrast between Lamb's two pre-occupations – literature and ledgers – could not have been starker. Little wonder he was so demoralised by the latter. Within days of his retirement Lamb was writing to Wordsworth once again: "*Here I am then after 33 years of slavery, sitting in my own room at 11 o'clock the finest of all April mornings, a freed man*".[12] A little later in that same letter he says: "*I came home for ever on Tuesday in last week. The incomprehensibleness of my condition overwhelm'd me. It was like passing from life into Eternity*".[13]

Lamb likened his retirement in 1825 to "*emancipation from slavery*". In *Essays of Elia*, 'The Superannuated Man' Lamb described his experience with pathos.

> *If peradventure, Reader, it has been thy lot to waste the golden years of thy life thy shining youth – in the irksome confinement of an office; to have thy prison days prolonged through middle age down to decrepitude and silver hairs, without hope of release or respite; to have lived to forget that there are such things as holydays, or to remember them but as the prerogatives of childhood; then, and then only will you be able to appreciate my deliverance.*

Figure 5.3 shows clerks at work – one showing clear signs of fatigue! – in an illustration from the 'Superannuated Man', The Last Essays of Elia in the *Works of Charles Lamb* Volume II, edited by William MacDonald (1903), with illustrations by CE Brock.

As the eighteenth century moved through its final decades, large offices like East India House remained a rarity; small owner-managed businesses continued to dominate the economic landscape. And it was in these businesses, where the clerk multitasked across the business, earning a high level of trust from the owner, that the more typical early office culture was evident. Clerking was at this stage an aspirant role; one that promised economic and social progression for incumbents, even the prospect of ownership one day. This was before business more generally began to replicate the structure of the East India Company, turning into 'clerical factories' with specialism, hierarchy and repetitive and mundane work.

Worsening working conditions. It has already been stated that clerks and clerking were enormously varied; and it is almost as problematic to describe a typical eighteenth or nineteenth century clerk as it is to describe a typical office worker today. Having said that, there was a distinct deterioration in working conditions towards the end of the eighteenth and into the nineteenth centuries – a time when businesses typically grew in size, with more bureaucracy and specialisation of work roles.

A particularly stark description is provided by journalist and politician William Cobbett (1763–1835), who worked as a copying clerk in a lawyer's office in 1783, describing it as the least joyous part of his entire life.

> *No part of my life has been totally unattended with pleasure, except the eight or nine months I passed in Gray's Inn. The office (for so the dungeon, where I wrote, was called) was so dark, that on cloudy days, we were obliged to burn candles. I worked like a galley-slave from five in the morning till eight or nine at night, and sometimes all night long.*[14]

Figure 5.3 Eighteenth-century clerking, by CE Brock.

Source: 'Superannuated Man' by CE Brock from Charles Lamb, *The Last Essays of Elia* ed. William Macdonald (1907).

He went on to describe vividly how wretched the experience made him feel.

> *When I think of the saids and soforths, and the counts of tautology that I scribbled over; when I think of those sheets of seventy-two words, and those lines two inches apart, my brain turns. Gracious Heaven! if I am doomed to be wretched, bury me beneath Iceland snows, and let me feed on blubber; stretch me under the burning line and deny me thy propitious dews; nay, if it be thy will, suffocate me with the infected and pestilential air of a democratic club-room; but save me, O save me from the desk of a pettifogging attorney!*[15]

For many, including those working at the Bank of England and the East India Company in the early years, office work was repetitious, mundane and unhealthy. Unfortunately, it was a work experience that was becoming established across emerging office organisations. Charles Dickens (1812–1870) was an astute observer of the details of office life at this time. Figure 5.4

Mr Stryver at Tellson's Bank

Figure 5.4 Mr Stryver at Tellson's Bank, from Dickens' *A Tale of Two Cities*.

shows the detail of a clerk at Tellson's Bank in Tales of Two Cities, while he described his own downtrodden experience in 1836–1837 in *Sketches by Boz*:

> *The dingy little back office into which he walks every morning, hanging his hat on the same peg, and placing his legs beneath the same desk: first, taking off that black coat which lasts the year through, and putting on which did duty last year, and which he keeps in his desk to save the other. There he sits till five o'clock, working on, all day, as regularly as the dial over the mantel-piece, whose loud ticking is as monotonous as his whole existence.*

As monotonous as his whole existence. Quite a simile. Indeed the relationship between work and the clock became deeply embedded around this time. Both in factories and in offices, the clock dictated working patterns, breaks, the start and end of the day, the amount of work to be completed and so on.

As can be gleaned from the foregoing quotes, poor working conditions were perhaps most often associated with the nature of the work itself. Mundane, repetitive tasks undertaken in poor environments were common. Attewell underlined the routine and the repetitive nature of much work:

> *The bulk of traditional office-work was repetitious and deadening because so much of it consisted of writing, copying, and crosschecking. Not only was everything done by hand, but multiple copies were usually necessary. Letters often went through several drafts, and it was standard practice to keep a record of the final product, so outgoing letters were hand-copied into a correspondence book. This generated such a volume of work that even merchant-owners weren't exempt from copying.*[16]

Even many of the higher-level tasks, such as accounting and correspondence, were full of routine and monotony. Attewell describes how

> *bookkeeping involved multiple ledgers, each of which contained column upon column of figures. Any purchase or sale had to be written up in a daybook or scratchbook, then copied into a journal, and then into the general ledger, then again under the specific person or firm's account, sometimes also in the cashbook, and so on.*[17]

While routine and monotonous, employers called for accuracy and precision in everything: handwriting, writing style, spelling and arithmetic were all critical skills. Multiple transactions of the same kind, or multiple letters containing repetitious phrases all had to be written and copied with complete precision.

Indeed, quality of penmanship was of overriding concern to employers. Good handwriting was viewed as an essential tool for a successful banking career, just as poor handwriting could even lead to dismissal. Such careful penmanship not only improved legibility but it also avoided the correction of errors, which could be used to hide fraud.

The situation in American offices of the time was little different to those in Europe. Zakim observed that cashiers, bookkeepers, penmen and sales agents worked in close or adjacent *quarters*,

> *increasingly in "offices" physically designed for the administration of business, documenting, arranging, and moving merchandise and information through a matrix of journals, ledgers, cash books, invoices, commission accounts, receipts, inventories, expense records, accounts current, copy books, and memoranda.*[18]

Zakim reinforces the point about the monotony of copying: lawyers typically required four copies of every real estate transaction they brokered (one each for the seller, the buyer, the public records office, and for their own files) while businesses usually made three copies of all their correspondence in order to post two of the same letter in case one was lost in the mails. And that such record keeping required workers "*to bend over a desk and scratch from 'morn till dewy eve' without intermission from day to day*".[19]

A thick and fetid atmosphere

Nineteenth century offices were, generally, uncomfortable and unhealthy places. The fact that the incidence of phthisis among clerks was particularly acute tells the story. Anderson observed that

> *... clerks were susceptible to that great scourge of Victorian society, phthisis. It was a disease which obsessed the Victorians and Edwardians, to the extent that there were several Royal Commissions on its causes and a programme of sanatoria was initiated to combat it.*[20]

Phthisis, otherwise known as consumption, was of course symptomatic of the broader social conditions of the period (sanitation, diet and housing conditions), but its prevalence among clerks was notable. Early American offices were no less unhealthy. Zakim refers to the somatic expressions of ill-health in the clerk's "*pale and pimply face untouched by the sun, in his pasty complexion and gouty inertia*", and in the characteristic "*curving forward of the chest of thousands of young men engaged from morning till night in the occupation of writing down figures, sitting in uncomfortable positions, breathing an unwholesome atmosphere, unrelieved by relaxation*". Such inclination toward disease, according to Zakim, was symptomatic of the pathological nature of the clerk's employment in general, of the ill effects produced by habits of business.[21]

Despite these vivid descriptions of poor working conditions, they were not universal, and seem to be more typical of situations where comparatively large organisations were growing quickly without proper planning of the workplace. Many clerks experienced work in much smaller organisations and in these situations, working conditions were often more bearable, as shown by the following quote.

> *Some stroll into their cosy, well-ventilated offices at ten, then leave them at five, after being out 'for dinner' an hour and a half; others, seated in their places at nine, toil diligently for ten or even eleven hours, with scarcely any cessation, in a thick and fetid atmosphere which is certainly and rapidly poisonous.*[22]

But even if conditions were better for some than others, the overall experience was evolving from a focus on small owner-managed businesses where clerks were held in some esteem by their employers – when clerking remained an aspirational occupation for gentlemen – to a situation where employment in larger firms with more job specialisation meant more widespread relentless, mundane and repetitive work. Added to poor working conditions – uncomfortable furniture, lighting, ventilation, over-crowding – and a general decline in the attractiveness of clerking can be detected as the second half of the century got underway. A similar trend can be detected in pay, as more routine tasks brought with them a wider spectrum of pay levels.

Lockwood observes that, in the usual course, a bank clerk at 30 earns about 200 pounds a year, working hours being from 9.30 to 4 o'clock. "*These are excellent average salaries for superior clerks; they are made still more attractive by the social status which accompanies them and by the ease with which anyone leaving a bank secures employment elsewhere*".[23] Similarly, the philosopher and political economist JS Mill (1806–1883) was kept in the office only from ten till four, and he wrote much of his great works, *System of Logic* and *Principles of Political Economy*, in office hours. He explained in his autobiography that he found office duties

> *an actual rest from the other mental occupations which I have carried on simultaneously with them. They were sufficiently intellectual not to be a distasteful drudgery, without being such as to cause any strain upon the mental powers of a person used to abstract thought, or to the labour of careful literary composition.*[24]

It should also be noted, for balance, that working conditions in offices were almost universally better than in many manual and agricultural roles. The clerk worked fewer hours for his income than the wage-earner and enjoyed paid holidays much earlier than the mass of manual workers.[25] JS Mill was clear in contrasting the differing fortunes of clerk and labourer.

> *A clerk from whom nothing more is required than the mechanical labour of copying, gains more than an equivalent for his mere exertion if he receives the wages of a bricklayer's labourer. His work is not one tenth-part as hard, it is quite easy to learn, and his position is less precarious, a clerk's place being generally a place for life.*[26]

There have been precious few systematic studies of clerical workers as a class – a surprising fact given their important role. One of the earliest was Benjamin Orchard's wonderfully titled *The Clerks of Liverpool: Being Ten Chapters on their Numbers, Distribution, Salaries, Grievances, Marriages, etc.* of 1871.[27] Writing at the time rather than as history, Orchard takes a more robust view of the working conditions of clerks. He argues that there was no 'standard' for employment conditions and that the fortunes of clerks "*are as various as are the natures of those who hire them*", and that there are "*extreme cases of benevolence as well as of hardship, though the latter are what come most into notice*". He goes on to suggest that evidence

> *does not corroborate the angry complaints which are often heard; and perhaps if those clerks who write to newspapers and otherwise persistently grumble were cross – examined, it would be found that a very large proportion of these ... complaints are really not founded on reason, and proceed from a very few persons.*[28]

He then describes many complaints as egregiously absurd; and that on the whole

> *in this portion of the labour market, as in many others, stupidity and intelligence, frail morality and trustworthiness, industry and slothfulness, fickleness and steadiness, experience and inexperience, find that the great eternal laws of justice which pervade the universe are not less effective when acting upon them.*[29]

Lockwood and Orchard both offer some balance to the overall picture of poor working conditions, but the weight of evidence suggests otherwise. We have at least two sources Thackrah and Parsons – that one might expect to offer an objective, or accurate picture; and neither of these would support either Lockwood or Orchard.

One of the first attempts to describe the medical implications of clerks' working conditions was that by Charles Turner Thackrah (1795–1833), a surgeon and a pioneer in the field of occupational health and a more humane workplace: he believed that employment of children in any labour was wrong. Ironically, he died very young, at 38, from tuberculosis. His principal work, published in 1832, was *The Effects of Arts, Trades, and Professions, and of Civic States and Habits of Living, on Health and Longevity*.[30]

He described how clerks, bookkeepers and accountants "*suffer from confined atmosphere, a fixed position, and often also from long days*" and often their "*muscles are distressed by maintenance of one posture; and they complain frequently of pains in the sides of the chest*". Furthermore, in clerks and bookkeepers the digestive organs suffer the most:

> *In clerks and bookkeepers, the digestive organs suffer most; a fact apparent even from the countenance of the tongue. The circulation is imperfect; the head becomes affected; and though urgent disease is not generally produced, yet a continuance of the employment in its full extent, never fails to impair the constitution, and render the individual sickly for life.*[31]

Thackrah also identified the legal profession for specific attention. He suggested that most branches of professional law practice were sedentary and that solicitors and other clerks "*are kept, from morning to night, in a bad position, with the limbs fixed, and the trunk bent forward*".[32]

In 1876, Charles Edward Parsons wrote *Clerks; Their Position and Advancement*. The book was further sub-titled '*Addressed to Parents, Employers and Employed*'. Parsons argued that amidst wider economic progress and growth in wages, there was "*one large section of the community whose claim to proportionate participation has been utterly ignored*". His aim in publishing the book was to set out a framework for redressing the clerks' situation and to "*prove that, by the adoption of the system about to be submitted, employers and employed may meet each other on a fair and clearly-defined basis, and to show how each may be considerably benefited by a mutual understanding*".

In preparing his case for improvement, Parsons set out his perception of the clerks' lot in graphic detail. He begins with a description

> *Underpaid, with a great deal of very laborious and monotonous work, in addition to heavy responsibilities and anxieties as his position gradually "improves," his constitution frequently impaired (often ruined) by sedentary duties and confinement in ill-ventilated back offices, without either time or means to enjoy the recreation necessary to everyone so employed, he grows up an unhealthy, dissatisfied man – oppressed by a consciousness that he would have done better had he but had opportunities – his nervous susceptibilities highly wrought upon by the worries of business, the fear of losing his precarious situation, and, lastly, the anxieties of a home-life often destitute, as he increases in years, of the comforts which render middle-age enjoyable to other classes.*[33]

Parsons highlights both the impact of physical privations on physiological wellbeing and also the psychological impact, highlighting worries and anxieties, as well as 'being oppressed by a consciousness'. These were radical thoughts for 1876. But he then went further to speculate on the mental state of clerks in their dying days.

> *He commences life as an ill-paid clerk, his ambitions are never encouraged, all the hopes of his manhood are thwarted, his employer can get numberless others to replace him at even less than he earns; and so he continues at the only calling he is capable of following, and with nearly every feeling soured by adversity – he dies: a hard-worked, ill-paid clerk from beginning to end.*[34]

Working conditions, like many aspects of clerking life, were extremely variable. Buildings were simply not designed for purpose and often domestic premises were being pressed into service for activities for which the buildings were inappropriate. Even in more designed workplaces, such as the Bank of England and East India Company, more attention was given to the image of the exterior and public spaces than was given to the welfare of the scribes who laboured there.

It was not as if the dangers associated with coal and gas pollution were not understood. In his 1865 *Practical Treatise on Banking*, Gilbart drew attention to the importance of ventilation. Drawing a comparison with the impure air of towns and cities, he asked plaintively

> *what must be the state of those offices or rooms where twenty or thirty persons are breathing close together during the whole of the day, and gas lights are burning during the evening. In such cases we are told that a person afflicted with consumption of the lungs may*

> *communicate the complaint to others, as they must inhale a portion of the atmosphere which he has breathed out.*

He went on to add that not only the proximity of people and burning of gas renders air impure in offices, but that it

> *also contains very frequently particles of dust arising from the floor, through the number of people constantly walking in and out. It is almost impossible for persons so circumstanced to enjoy for a length of time even moderate health. A portion of this evil may be mitigated by a good system of ventilation. To obtain this should be regarded as an object of the first importance.*[35]

Locker describes some of the practical difficulties of heating and ventilating Victorian offices. The number of clerks in a bank or insurance office could easily number 25 to 50, "*sufficient to produce considerable amounts of 'vitiated' air*". Ceilings were generally high; spaces were generally too large to be effectively served by one or even several fireplaces. "*Large offices, and banks in particular, had an additional problem. The large glass areas, often in the roof, produced cold radiation and convection currents referred to as the banker's draught*".[36] He pointed out that systems developed for dealing with such issues in public buildings "*required a considerable amount of space for ducts, flues, and equipment*" and that the "*cost and spatial inefficiencies of such systems were clearly in excess of what even the banks and insurance companies were able or willing to afford*".[37]

Locker then describes what was possibly the "*most extensive, and certainly the best documented heating system*" installed by John Gibson in National Provincial Bank's Bishopsgate building in 1863. The building was a large one, housing in excess of 200 clerks, with extensive basement vaults spreading under most of the clerical office.

> *Gibson placed a closed stove and a 'ventilating machine' at one end of the vaults. Hot air was blown through spaces built between the vaults and the banking hall above. The ground floor was supplied by open grates, but the basement was supplied by an ingenious arrangement of hollow columns with openings at the floor and the vault line. Gibson maintained the traditional cheerfulness of the open fireplace by placing one in a prominent position, visible from the entry to his grand office.*[38]

As the Victorian era ended, so too did this universal reliance on coal fires for heating and ventilation, together with the piecemeal adoption of central warming systems; yielding to the "*unequivocal integration of modern heating and ventilating systems in offices*".[39]

Employer and clerk: a case of mutual dependency

In the later-eighteenth century, there remained a sharp division between owner and employee; there was at this stage no concept of a 'manager' class – just clerks to provide the owner with the means to execute his business. In return, the clerk enjoyed live-in accommodation, a secure job and income and a career that offered the potential for social and economic advancement. Owner and clerk were, in many ways mutually dependent.

The clerk got to know everything about a business and was often in a position of trust – having access to the firm's money and assets. Owner and clerk were often working side-by-side, all day and most days; their relationship was often a very close one. There was a mutual interest in nurturing a cordial relationship. In such circumstances, interpersonal relationships assumed major importance. As Lockwood observed the value of a clerk's services depended

> *closely and somewhat curiously on relations with the employer, that is to say, upon possibilities of combination in action between men who have learnt to know each other's ways and who suit each other. Such relations are usually formed gradually and are the essence of all high value in clerk's work.*[40]

Marxist and historian Francis Klingender (see Chapter 8) noted that in the early nineteenth century "*the requirements of banking, commerce, or industry did not exceed the resources of family concerns or small partnerships*", which limited "*the extensive development of clerical labour*". He suggested that within this cottage industry style economy

> *there was an almost feudal relationship between the small number of clerks to be found in such offices and their employers. The clerk was more a family servant than a wage labourer.*

There is an element of truth in this view because clerks were seen to some degree as an extension of other domestic employees, such as butlers, cooks and chamber maids. The close relationship between owner and clerk was frequently reinforced by physical proximity: clerks often lived 'over the shop', normally on the first floor, and with their employers and their families on the upper floors. While the practice lasted, the sharing of space by employer and employee was a convenient arrangement, but one which must have also been a cause of tension – for both parties. In some ways it replicated the arrangement of wealthy owner and live-in staff, for whom there was often a master–servant relationship.

However, the experience of clerks changed rapidly over time and was, at any single point, extremely varied. As the nineteenth century progressed and as businesses grew in size and complexity, feudal relationships became less common. Indeed, both clerk and employer were replaced in the later-nineteenth century by managers, particularly in the expanding joint stock banks.

Whatever the class of the clerk, the relationship with the owner was normally a paternalistic one. Many clerks, for example, were reliant upon receiving gifts, Christmas boxes and gratuities as a physical sign of their employers' paternalism.[41] But they were mutually dependent upon one another: the owner was heavily invested in the honesty and ability of the clerk.

In the early days, social compatibility was made easier by the fact that many clerks were selected based on a knowledge and acceptance of their background. JS Mill made the point that until universal education had been made available, any jobs requiring even basic reading and writing skills "*could be recruited only from a select class, the majority having no opportunity of acquiring those attainments*". In these circumstances, claimed Mill, most such jobs were overpaid compared to the wider labour market; but the owner was happy with this relationship because it guaranteed a certain level of commitment and behaviour.

Most clerks found their positions, in these early days, through personal recommendation, often through wider family or other social connections. For larger employers, they simply sought a 'particular kind' of clerk; for example, hiring from among the middle classes was a prime priority for the Bank of Scotland.[42] As a result of these social influences, to a certain degree, like attracted like – people tended to work for people who were similar to themselves in a social context. This made the forging of close working relationships easier as employer and clerk shared social context.

Even with personal recommendations and hiring within social networks, employers often required a surety as a part of recruitment. While such a guarantee offered the employer some financial protection from staff embezzlement, it also acted as a further indicator of the personal reputation of the candidate, since financial stability could be equated with good character. But what did this mean?

Galassi and Newton provided some insights into what was meant by good character or reputation. They found that factors such as social standing, religion and kinship were more important in the nineteenth century banker's measurement of reputation than more observable and tangible economic information on the individual. Certainly, activities such as drunkenness and indebtedness were particularly feared within the banking community. For example, the Glamorganshire Banking Company dismissed one of its clerks who had been fined for drunk and disorderly behaviour.[43]

While the quality of school education available at the end of the eighteenth and start of the nineteenth century remained inadequate for providing the writing and numeracy skills necessary for clerical work, most young people were excluded from considering such a career. As a rule, clerks were comparatively well-paid, adding to the allure of the vocation, by reinforcing its aspirational qualities. However, once universal education was available, the general level of writing and numeracy skills was lifted, opening the way for many more people to pursue a clerking career, ultimately resulting in great competition for positions and downward pressure on salaries.

For those fortunate to find a clerking tenure, the early stages often involved an apprenticeship. Anderson described the apprentice as one whose duties included outdoor work such as the distribution of circulars, the carrying of notes and telegrams and the collection of receipts and invoices.[44] Indoor work consisted largely of menial and repetitive tasks, including the routine of copying letters, addressing envelopes and checking invoices, all under the gaze of a disciplinarian boss. Indeed, at the start of their careers most clerks found their position to be similar to the school that they had recently left behind; only they no longer had the short hours and long holidays. Charles Lamb was one such young clerk; he found the transition "*at fourteen from the abundant playtime, and the frequently-intervening vacations of school days, to the eight, nine, and sometimes ten hours' a-day attendance at a counting-house*" very difficult to accept.[45]

Apprenticeship was seen as the result of business prudence, based on the supposition that to learn anything thoroughly one must be kept at it for a sufficient period. It was also thought that the apprenticeship helped to lay the basis of mutual trust and affection between clerks and employers.[46]

But even with this adjustment in salaries, Mill argued that there remained a "*much greater disparity than can be accounted for on the principle of competition*".

> *A clerk from whom nothing more is required than the mechanical labour of copying, gains more than an equivalent for his mere exertion if he receives the wages of a bricklayer's labourer. His work is not one tenth-part as hard, it is quite easy to learn, and his position is less precarious, a clerk's place being generally a place for life.*[47]

Mill speculated as to why the disparity between clerk incomes and other jobs continued. As part of the solution, he suggested "*the remaining influences of an ancient custom, which required that clerks should maintain the dress and appearance of a more highly paid class*".[48] Mill also observed the similarity between the treatment of personal servants and clerks. Noting that people paid personal servants and others they placed in positions of trust above the market value for their labour in return for an appearance and behaviour superior to that associated with a low standard of remuneration. Critically, Mill then suggested that "*Similar feelings operate in the minds of men of business in respect of their clerks*".[49] In other words, employers looked after their trusty clerks in a manner that reflected well on their own social standing.

America in the 1860s remained a largely agrarian economy with most communities producing food and goods for local consumption; there were very few ventures set up to service national markets – just a few banks and insurance businesses, merchants and shippers. The canals and railroads remained a long way off in terms of economic impact. Even the larger firms employed only modest numbers of clerks.

One of the more prevalent job roles was copyist, or scrivener, who would produce fair copies of draft legal documents using a quill pen and drying the wet ink with sand. Contracts would often require four fair copies to be made. Bookkeeping involved preparation of financial records, which had to be legible and accurate, and were often quite complex. At this time the bookkeeper covered the whole business rather than a particular area, or department. Many offices also had an 'office boy' or 'messenger' who would not only run messages to people within and beyond the business, but also undertake a wide range of menial tasks, as and when required. Clerks formed the largest group of office workers at this time, undertaking a wide variety of tasks.

Relationships between owners and clerks were similar to those in Europe. Owners controlled everything, issuing orders directly to their clerks; each had their own style and approach, which could be good news, or not, for the clerk. For example, the degree of training that took place would vary widely. There was generally a high level of trust, and interpersonal relationships were generally more patriarchal than master–servant in character. In the US, the owner–clerk relation was less impacted by the class backdrop found in the UK. US clerks would see their circumstances more in personal terms rather than as an outcome of social status. If relations soured a clerk would blame their employer's behavioural traits or admit his own failures rather than lay the situation at the door of class distinction.[50] However, again in common with Europe, trust was an essential feature of a good owner–clerk relationship.

Margery Davies made the point that one way of ensuring accuracy of work was to closely supervise, check and repeat tasks; but that such oversight involved a large investment of time. The only real way to avoid this waste was to build a relationship of trust. And when this was done it served to strengthen the tendency to personalise the relationship. Trust from the employer established a friendly rather than adversarial relationship; while trust from the clerk emerged from responding to his employer's goodwill.[51]

Davies also draws attention to the importance of upward mobility during this period. Many clerks were really apprentice owners or managers: they were analogous to today's business school student or management trainee. There were also those who never became owners, but who were trusted and competent office managers, with their grasp of office details and their ability to make decisions themselves, who became "*masters of the craft of office management*".[52]

A fair day's pay? In the early years, clerking was generally regarded as a well-paid job – especially when literacy and numeracy were in short supply, and when owners saw clerks as trusted colleagues, many of whom were on a journey towards independence. But as time lapsed, so did the individual skills of clerks and their working conditions, along with the relationship between owner and employer, and the level of satisfaction that clerks derived from their work, become enormously varied.

We saw earlier how Charles Lamb started work at East India House in 1800 with a 2-year probationary period, receiving a £30 gratuity. After 2 years this became a regular salary of £40 per year, which then increased steadily to £240 in 1815. At this point, as the result of an internal reorganisation, his salary doubled to £480, and in 1825, when Lamb requested retirement, his salary had grown to £730.[53] If nothing else the gradual rise in salary shows that it was possible to 'progress' in clerking and become more senior with, presumably, higher levels of responsibility. But how do the figures look in terms of contemporary averages?

Anderson makes the point that one of the characteristics of nineteenth century clerical work "*was the great variation in salary and status associated with different job functions*". He also contrasted different size offices. Thus in typical small offices there were few positions that carried any real status. For example, *"in Liverpool's cotton brokerage trade, most offices were run by several apprentices and only two or three full-time clerks, a cashier, a bookkeeper and a manager. The salaries of these clerks were low, an office manager earning no more than £120 to £150.*" By contrast, in larger commercial offices,

> *clerks earned better salaries and probably in better conditions. However, not only were they often far removed, physically and socially, from their employers, there was also a wide gulf between those holding senior positions and those engaged in the more routine clerical work.*

A senior clerk could earn £400 while none of those under him earned more than £150.[54] In 1865, Evans was able to report that most private bankers, who employed between 40 and 50 clerks each, paid an average salary £200 a year; while a "*post of confidence must be held before a maximum of £600 a year is gained*" adding that "*not more than two or three, if so many, of this character exist in a firm, as to give a chance for general competition*".[55] These examples seem to be broadly in line with Lamb's experience.

In many ways, the Bank of England was the apex of clerking. Evans reported that, mid-century, the number of clerks employed there was about 800, and the salaries ranged between £50 and £2,000. To attain the highest salaries required "*years of steady perseverance and great abilities for business*". The more general average salary at the Bank ranged from £170 to £300 a year.[56] Evans also demonstrated how highly structured was the salary regime at the Bank. No clerk could take office there until he was 17 years of age, when his salary commenced at £50 per year; and never after 25 years of age, when was paid £100 per annum. In the intermediate years up to the age of 21, the salary increased at the rate of 10% per annum, but after that period the salary rises were restricted to 8% per annum.[57]

According to Lockwood, by the middle of the nineteenth century two classes of clerk seem to differentiate themselves. On the one hand, banking, civil service, insurance and the more prosperous class of mercantile and business clerks were able to maintain a fairly respectable middle-class way of life without undue strain; on the other, the greater proportion of clerks whose salaries barely ran ahead of the wages of manual labour, were never really part of the middle class in an economic sense, but were always striving socially to identify themselves with it.

> *The range of remuneration in fields such as banking, insurance, government and commerce formed a pyramid-like structure in which the apprentice or routine clerks were found at the base earning anything between twenty and two hundred pounds a year, while at the apex the various grades of managing clerks earned salaries running into four figures.*[58]

Klingender shows a breakdown of the incomes of clerks at the Government's Customs Department and at Inland Revenue at mid-century (Table 5.1). The average salary in Customs is cited as £150, while that in Inland Revenues is £240. As can be seen, the range between 1st and 5th class clerks is quite wide.

Table 5.1 Customs Department and Inland Revenue salaries, mid-1800s

Role	*Customs*		*Inland Revenue*	
	No	*£ salary*	*No*	*£ salary*
Controllers, Chief Accountants	4	450–800	3	800
Principal clerks, etc.	13	320–450	–	–
1st class clerks	23	250–320	16	400–550
2nd class clerks	27	200–250	12	300–400
3rd class clerks	60	150–200	14	200–250
4th class clerks	42	105–150	16	150–200
5th class clerks	50	75–100	18	90–140

Source: Adapted from Klingender (1935) p11.

Orchard reported on salaries of clerks in Liverpool. To start, he counts 11 banks, employing about 200 clerks and 30 apprentices. Within these, bank managers and sub-managers average £2,500 and £1,000, respectively. Beneath them,

> *the secretary or chief accountant, as head clerk and responsible for the entire office work, draws about £700. Then come three cashiers, receiving £300 to £500, according to seniority; below them are the bookkeeper and the two corresponding clerks, at £300 each, and these are followed by a few apprentices and inferior clerks who perform the routine simple work for salaries ranging from £20 to £200.*[59]

Orchard notes that all of the staff are "*well-educated persons and (with the exception, perhaps, of the two at the top) give personal or guarantee security, not only that they will be honest, but also that they will not betray the secrets of the bank*".[60]

Orchard then counts 12 marine insurance offices (the least lucrative for clerks) employing 100 clerks and 20 apprentices. Here a manger will earn £300 to £1,500 and a staff of assistants from £60 to £400. In fire insurance, Orchard counted five head offices employing c220 clerks and 35 apprentices. In the larger establishments, with c100 staff or more,

> *the manager draws from £1,500 up to £4,000 and the sub-manager from £1,000 to £2,000. After them come one or two leading officials at £500, who are succeeded by five or six heads of departments at £150 to £350; and then troop the bulk of the clerks, the men receiving from £60 to £150, and the lads, as apprentices, 6s to 20s a week.*

In the ship owning sector, Orchard identified 134 firms employing 1,430 clerks, including 401 apprentices. Of these six or seven probably earned in excess of £750; with at least 120 earning £250 to £750. Of the remainder, around two-thirds earned over £80.

Irrespective of the business sector, it was increasingly the case that seniority was becoming more important to clerks as offices became larger. The longer a clerk worked for an office the more knowledge and insight he gained into how its different parts worked, while at the same time his value to the business increased. And as the number of clerks in any office generally increased, longevity also brought seniority. Inevitably, this led to more finely grained pay structures for clerks.

What do these various accounts of clerking salaries tell us about clerking in the mid-nineteenth century? The range of salaries shows more than anything else that clerking had moved on from the situation where a very small number of clerks worked for an owner–manager, to the era of the 'clerical factory'. Larger workforces, multilayered and departmentalised, were rapidly becoming the norm. Specialisation of tasks was bringing more routine and mundane roles.

An array of job titles suggests levels of seniority and respect which later dominated social relations in the corporate office. And, of course, the combinations of hierarchy and stratified reward brought with it a growing matrix of social relations, sitting between the 'labouring class' and the gentry and aristocracy – the middle class.

Class and behaviour

Many clerks saw their tenure as a stepping stone to management or even ownership. One of the better known examples of this is the founding of the Midland Bank, which

> *owes its origin to a bank clerk's discontent. Charles Geach, an enterprising young Bank of England clerk, who had been transferred in 1828 to the Bank's new branch at Birmingham, became dissatisfied with 'the slow course of promotion' and, as a result, while still in the Bank's service, launched in 1836, with the assistance of a number of local businessmen, a joint stock bank in Birmingham, the Birmingham and Midland, a 'puny provincial institution faced with the keenest competition', of which at the early age of twenty-eight he became the first manager.*[61]

The early-nineteenth century clerk was typically attired in black coat, suit, collar tie and hat; and the black coat was more than a sartorial statement: it indicated to everyone that while he was not an owner or gentleman, neither, far more importantly, was he part of the working class; he sat apart from the working class but rubbed shoulders with the upper class. By contrast, working "*with one's hands was associated with other attributes – lack of authority, illiteracy, lowly social origin, insecurity of livelihood – which together spelt social depreciation*".[62]

Just like everything else about clerks and clerking, their social cohesion defies generalisation. They emanated from all avenues of life. One curiosity, freely observable from the written testaments, is the relatively large number of writers and scholars who worked as clerks, largely as a means of ensuring a secure income. We have already met Charles Lamb at East India House and Cobbett, Dickens and Mill. There were yet others who simply filled large parts of their spare time with learning and other cultural and educational activities. Orchard describes clerks who were both highly proficient at clerking while being talented in other areas.

> *Some clerks who occupy positions where the highest clerkly qualities are needed so display these as to possess the entire confidence of their employers, yet are distinguished equally by skill in various accomplishments. Some are classical scholars, enjoying Plato and Horace; some are artists ... some chess players, some musicians, some theologians, some valued contributors to the press, some refined and eloquent speakers .... Another, but smaller section, have the advantages of a university education, of foreign travel, and of good society.*[63]

Orchard argues that this 'upper crust' of the clerking world has been assisted in their commercial careers because they "*have not neglected business proficiency*". However, he proceeds, there is a "*much more numerous and far less successful class of clerks*" who *having been imperfectly trained to business habits, or not allowing business requirements their due supremacy, become half-hearted dawdlers, neither hot nor cold, and consequently the reverse of what their employer seeks. For it cannot be too anxiously kept in mind that that 'well-educated clerks' are those, and only those, who can perform clerkly duties well.*[64]

Some eighteenth century clerks thus sought advancement not only through material improvement but also cultural awareness. Orchard suggested that employers

if just and far-sighted, must learn to see that the clerk who, by home study, manifests refined taste and varied aptitudes, is more worthy of regard than he whose spare time is spent in idleness, folly, or vice.[65]

They must recognise the truth that almost every clerk who has attained by diligence and capacity to a good position, and is admittedly 'up to the mark', has also studied in his leisure hours and become agreeably proficient in some intellectual art or pursuit. However, for some employers, broader learning and skills were a distraction for clerks. Orchard suggests that employers have a misplaced confidence in the clerking abilities of the scholars over the great mass:

He assumes ... that the man who has no faculties to comprehend anything nobler than the business routine will give his mind to his work more thoroughly than his intellectual and cultured fellow.[66]

Lockwood was forthright. He suggested that the wider achievements of clerks (including painting, music, history, literary composition), were appropriate to the status of an educated gentleman, but were definitely discounted by employers as of little significance in the tough world of commercial negotiation.[67]

Orchard was himself a clerk and served as the head of the Liverpool Clerks' Provident and Annuity Association. His detailed study provided a street-by-street breakdown of the numbers of firms and clerks in Liverpool. In so doing, he arrived at a total of 4,579 firms employing 16,158 commercial clerks, and an additional 4,000 office boys, or "*clerks in embryo*".[68]

Orchard painted an unprecedented picture of the city's clerks: their numbers and distribution, background and education, relationships with employers, and salaries. There were also chapters devoted to discontent, co-operation and apprenticeship. He begins his study by stating that while in London, Manchester or Liverpool clerks were themselves "*numerous enough to populate a good-sized city*" they were a largely little-known group of workers.

It is somewhat curious that ... one important class of society has been little noticed, though found chiefly in those places where literary men most do congregate. Its members are not hidden from sight, for literary men frequently meet with them.

Moreover, despite their aggregate number and their tendency to rub shoulders with 'literary men', as a class, they were largely ignored, or unappreciated:

... a body of men with strong distinctive traits, yet uncounted, uncared-for, misunderstood, or not understood at all, even by those who mix most closely with them – indeed, not even by themselves.[69]

Orchard not only used a quill for working, as one of his main passions in life was writing: he wrote and presented essays and eventually became a newspaper reporter, working on several papers before himself launching the *Liverpool Critic*.

Part of the reason for the fact that clerks were little understood as a class was that they never formed a cohesive professional group of workers. Anderson provides a number of reasons for this. First, they possessed "*no common knowledge base beyond the rudimentary skills of literacy and numeracy*". Secondly they lacked a "*clearly defined service orientation to society*". Thirdly they were "*not bound as most professional were by a code of ethics*".[70] A further important consideration is that clerks often worked very closely with their employers and considered themselves aspirant gentlemen. The clerks lacked any desire for working class militancy that would, in modern parlance, bite the hand that fed them.

Partly because of the close working relationship between owner and clerk, and partly because of the wider, early-Victorian social structure, clerking was characterised by a sense of morality. Under scrutiny was not only the competent completion of clerical tasks but also the very "*soul of individual employees*".[71] Indeed, some commentators of the era viewed a bank's influential role as extending far beyond the immediate confines of the counting house, to a role which acted upon the moral fabric of society as a whole.

> Banking exercises a powerful influence upon the morals of society. It tends to produce honesty and punctuality in pecuniary engagements. Bankers, for their own interest, always have a regard to the moral character of the party with whom they deal … Thus the establishment of the bank in any place immediately raises the pecuniary value of a good moral character.[72]

The paternal relationship between employer and clerk was often magnified through discipline. Even when recruitment relied on personal introduction and social relations, there was no guaranty that the clerk would always behave in a saintly fashion. In this respect, a junior clerk in the early-nineteenth century experienced a disciplinarian environment not dissimilar to school.

And just like school, discipline could often be strongly observed. For example, clerks could invite disciplinary action from their employers for any number of transgressions, including lack of punctuality. Similarly, in a well-defined office hierarchy, a clerk's respect for his seniors was taken for granted. Insubordination or even a breach of office protocol could provoke a hostile reaction from an employer. Employers showed a particular concern for two areas of personal conduct – indebtedness and intemperance.[73]

Writing about Hoare & Co., Jeacle described how staff who resided within the confines of the bank walls (approximately six) were "*more particularly exposed to the influence of the paternal hand*".

> *Their conduct was governed by a range of house rules as might be laid down by a legal guardian. Clerks had to be home by 11.30, any extensions to be granted only by special permission. Entertaining friends during the working week was simply out of the question. As a memo of 1893 declares: 'The practice of having anyone to dinner on a week day is strictly contrary to rules'.*[74]

However, the potential inflexibility of this dictatorial regime was matched by acts of great kindness on the part of the partners. The archive reveals that the bank was also an establishment where rewards were generous and kind acts of paternalism prevailed.

While owner–clerk relations could be described as paternal, as the years passed and as the clerking class grew (and became less select), so relations with owners took on a more contractual form. Orchard's extensive study of the clerking community in Liverpool observed that many clerks had come to have "*singularly overstrained conceptions of the part they perform in the great drama of commercial enterprise, and their just title to a proportion of the profits resulting therefrom*". Orchard recorded a widespread sense among employers that many clerks' hearts were not in their work and that they did as little as they could for their money; while citing one clerk who asked: 'why should we care for masters who grind our bones to make their bread?'

Little surprise then that Orchard observed that there "*is no cordiality between various large sections of Liverpool clerks and their employers. There is scarcely a pretence of friendliness*".

> *There are still firms whose partners invite their clerks to dinner once a year, and who give Christmas boxes, or make welcome presents when babies come to increase their humble friends' expenses ... . Nevertheless it is unfortunately true that, taken altogether, the masters' faces are hard and their tender mercies cruel.*
>
> *As a class, clerks are weary-souled and long-suffering. Their countenances are down-cast – they murmur rather than speak. Having no capital with which to invoke the law when oppressed, or to fall back on when suddenly left without employment, if disputes arise between master and servant, the servant finds himself practically a bondsman; and the power to act ungenerously being felt, is too often put in force.*
>
> *Quoting a merchant in Fenwick Street: My clerks do not provide me with capital, or assist me in a crisis, or arrange the extended combinations by which I make my money. They only do fixed clerkly work, for which they are paid, and which I could do at least as well were I disposed to assume their duties. If one leaves me, twenty others will compete for his post. To me my clerk on his stool is an instrument and nothing more.*[75]

There can be little doubt that by the 1870s, the personal relationship between employer and trusted clerk was changing as business itself was changing and becoming less personal. There remained enormous variability in the experiences of clerks across the country, and between sectors and companies; and Orchard's description of them being weary-souled and down-cast seems a little too general. But certainly the signs were present that the world of office work was on the edge of major change.

Summary. Simply in terms of its administrative and physical organisation, the East India Company was, in many ways, ahead of its time. While Charles Lamb loathed his time behind the heavy desk in a badly lit and ventilated room, in other respects the organisation was a forerunner of corporate structures of the later-nineteenth century. It was the side-effect of rapid growth in business that led to worsening working conditions towards the end of the eighteenth and into the nineteenth century.

The general trend was of a movement from smaller owner-managed businesses where clerking remained an aspirational occupation for gentlemen to a growth in larger firms with more job specialisation. The result for clerks was more mundane and repetitive work, and generally poor working conditions – uncomfortable furniture, poor lighting and ventilation, and overcrowding. Most buildings were not purpose designed; they were adapted for activities for which they were inappropriate.

Before hierarchies and management became commonplace, owner and clerk were, in many ways mutually dependent. The clerk knew everything about the business and was often in a position of trust. Owner and clerk were often working side-by-side, with a mutual interest in nurturing a cordial relationship. This relationship between owner and clerk helped to maintain the aspirational nature of clerking as a role. As we have seen here, the nineteenth century clerk typically wore a black coat, suit, collar tie and hat; an ensemble that sent out the message that while he was not an owner or gentleman, neither was he part of the working class. He sat neatly between the two and had the prospect of moving upwards; he sat apart from the working class and rubbed shoulders with the upper class.

Notes

1 Smith A (1776) *An Inquiry into the Nature and Causes of the Wealth of Nations* Reprint edited by Campbell RH and Skinner AS (1976) Two volumes Clarendon, Oxford pp109–110.
2 British Library https://blogs.bl.uk/asian-and-african/2017/01/east-india-company-headquarters-on-leadenhall-street.html Accessed 24 April 2020.
3 Lucas EV (1905) *The Life of Charles Lamb* Two volumes Methuen & Co, London Vol I p252.
4 Lucas EV (1905) *The Life of Charles Lamb* Two volumes Methuen & Co, London Vol I p253.
5 Lucas EV (1905) *The Life of Charles Lamb* Two volumes Methuen & Co, London Vol I p253.
6 Lucas EV (1905) *The Life of Charles Lamb* Two volumes Methuen & Co, London Vol I pp253–254.
7 Bowen HV (2008) *The Business of Empire: the East India Company and Imperial Britain, 1756–1833* Cambridge pp148–149.
8 Talfourd TN (1838) *The Works of Charles Lamb Volume 1* Harper & Brothers, New York p196.
9 Lamb C (1892) Essays of Elia: The Superannuated Man. In: *The Complete Works of Charles Lamb* London, Chatto & Windus pp163–168.
10 MacDonald W (ed) (1903) *The Works of Charles Lamb* JM Dent & Co, London p88.
11 Lucas EV (1905) *The Life of Charles Lamb* Two volumes Methuen & Co, London Vol I p140.
12 Lucas EV (1905) *The Life of Charles Lamb* Two volumes Methuen & Co, London Vol I p140.
13 Shepherd RH (1892) *The Complete Works of Charles Lamb in Prose and Verse* Chatto & Windus, London p163.
14 Smith E (1878) *William Cobbett: A Biography* Vol 1 Sampson Low, London p27.
15 Smith E (1878) *William Cobbett: A Biography* Vol 1 Sampson Low, London p27.
16 Attewell P (1989) The Clerk Deskilled: A Study in False Nostalgia *Journal of Historical Sociology* Vol 2 No 4 pp357–388 ISSN 0952-190.
17 Attewell P (1989) The Clerk Deskilled: A Study in False Nostalgia *Journal of Historical Sociology* Vol 2 No 4 pp357–388 ISSN 0952-190.
18 Zakim M (2006) The Business Clerk as Social Revolutionary; or a Labor History of the Nonproducing Classes *Journal of the Early Republic* Winter 2006 Vol 26 No 4 pp563–603.
19 Zakim M (2006) The Business Clerk as Social Revolutionary; or a Labor History of the Nonproducing Classes *Journal of the Early Republic* Winter 2006 Vol 26 No 4 pp563–603.
20 Anderson G (1976) *Victorian Clerks* Manchester University Press p18.
21 Zakim M (2006) The Business Clerk as Social Revolutionary; or a Labor History of the Nonproducing Classes *Journal of the Early Republic* Winter 2006 Vol 26 No 4 pp563–603.
22 Orchard BG (1871) *The Clerks of Liverpool: Being Ten Chapters on their Numbers, Distribution, Salaries, Grievances, Marriages, etc.* p4.
23 Lockwood D (1958) *The Blackcoated Worker: A Study in Class Consciousness* George Allen & Unwin, London pp23–24.
24 Cited in Sampson A (1995) *Company Man: the Rise and Fall of Corporate Life* Harper Collins, London p19.
25 Lockwood D (1958) *The Blackcoated Worker: A Study in Class Consciousness* George Allen & Unwin, London p53.
26 Cited in Klingender FD (1935) *The Condition of Clerical Labour in Britain* Martin Lawrence, London p3.
27 Orchard BG (1871) *The Clerks of Liverpool: Being Ten Chapters on their Numbers, Distribution, Salaries, Grievances, Marriages, etc.* p4.
28 Orchard BG (1871) *The Clerks of Liverpool: Being Ten Chapters on their Numbers, Distribution, Salaries, Grievances, Marriages, etc.* p26.

29 Orchard BG (1871) *The Clerks of Liverpool: Being Ten Chapters on their Numbers, Distribution, Salaries, Grievances, Marriages, etc.* p26.
30 Thackrah CT (1832) *The Effects of Arts, Trades, and Professions, and of Civic States and Habits of Living, on Health and Longevity* 2nd Edition Longman, London.
31 Thackrah CT (1832) *The Effects of Arts, Trades, and Professions, and of Civic States and Habits of Living, on Health and Longevity* 2nd Edition Longman, London pp176–177.
32 Thackrah CT (1832) *The Effects of Arts, Trades, and Professions, and of Civic States and Habits of Living, on Health and Longevity* 2nd Edition Longman, London p183.
33 Parsons CE (1876) *Clerks: Their Position and Advancement* Provost & Co, London p9.
34 Parsons CE (1876) *Clerks: Their Position and Advancement* Provost & Co, London p9.
35 Gilbart JW (1849) *Practical Treatise on Banking Volume 1* Longman, London pp194–195.
36 Locker FM (1984) *The Evolution of Victorian and Early Twentieth Century Office Buildings in Britain* University of Edinburgh PhD thesis p146.
37 Locker FM (1984) *The Evolution of Victorian and Early Twentieth Century Office Buildings in Britain* University of Edinburgh PhD thesis p147.
38 Locker FM (1984) *The Evolution of Victorian and Early Twentieth Century Office Buildings in Britain* University of Edinburgh PhD thesis p148–149.
39 Locker FM (1984) *The Evolution of Victorian and Early Twentieth Century Office Buildings in Britain* University of Edinburgh PhD thesis p153.
40 Lockwood D (1958) *The Blackcoated Worker: A Study in Class Consciousness* George Allen & Unwin, London p20.
41 Anderson G (1976) *Victorian Clerks* Manchester University Press p31.
42 McKinlay A (2002) 'Dead selves': the Birth of the Modern Career *Organization* Vol 9 No 4 pp595–614.
43 Galassi F & Newton L (2001) My Word is My Bond: Reputation as Collateral in Nineteenth Century English Provincial Banking *Warwick Economic Research Papers No 599*, Department of Economics, University of Warwick. Cited in Jeacle I (2010) The Bank Clerk in Victorian Society: the Case of Hoare and Company *Journal of Management History (Archive)* Vol 16 pp312–326.
44 Anderson G (1976) *Victorian Clerks* Manchester University Press p14.
45 MacDonald (ed), *Works of Charles Lamb* p187.
46 Anderson G (1976) *Victorian Clerks* Manchester University Press p13.
47 Mill JS (1848) *Principles of Political Economy* Volume 1 p469.
48 Mill JS (1848) *Principles of Political Economy* Volume 1 p469.
49 Mill JS (1848) *Principles of Political Economy* Volume 1 p482.
50 Davies MW (1982) *Women's Place at the Typewriter* Temple University Press, Philadelphia p23.
51 Davies MW (1982) *Women's Place at the Typewriter* Temple University Press, Philadelphia p23.
52 Davies MW (1982) *Women's Place at the Typewriter* Temple University Press, Philadelphia p26.
53 Lucas EV (1905) *The Life of Charles Lamb* Two volumes Methuen & Co, London Vol I p140.
54 Anderson G (1976) *Victorian Clerks* Manchester University Press pp20–21.
55 Evans DM (1865) *The City, or The Physiology of London Business with Sketches on 'Change, and at the Coffee Houses* Baily Brothers, London p8.
56 Evans DM (1865) *The City, or The Physiology of London Business with Sketches on 'Change, and at the Coffee Houses* Baily Brothers, London p6.
57 Evans DM (1865) *The City, or The Physiology of London Business with Sketches on 'Change, and at the Coffee Houses* Baily Brothers, London p7.
58 Lockwood D (1958) *The Blackcoated Worker: A Study in Class Consciousness* George Allen & Unwin, London pp22–23.
59 Orchard BG (1871) *The Clerks of Liverpool: Being Ten Chapters on their Numbers, Distribution, Salaries, Grievances, Marriages, etc.* p28.
60 Orchard BG (1871) *The Clerks of Liverpool: Being Ten Chapters on their Numbers, Distribution, Salaries, Grievances, Marriages, etc.* p28.
61 Richards RD (1937) The History of the Midland, the Westminster and the National Provincial Banks Economic History Vol 3 No 12 pp430–440.

62 Lockwood D (1958) *The Blackcoated Worker: A Study in Class Consciousness* George Allen & Unwin, London p99.
63 Orchard BG (1871) *The Clerks of Liverpool: Being Ten Chapters on their Numbers, Distribution, Salaries, Grievances, Marriages, etc.* pp14–15.
64 Orchard BG (1871) *The Clerks of Liverpool: Being Ten Chapters on their Numbers, Distribution, Salaries, Grievances, Marriages, etc.* p15.
65 Orchard BG (1871) *The Clerks of Liverpool: Being Ten Chapters on their Numbers, Distribution, Salaries, Grievances, Marriages, etc.* p16.
66 Orchard BG (1871) *The Clerks of Liverpool: Being Ten Chapters on their Numbers, Distribution, Salaries, Grievances, Marriages, etc.* p15.
67 Lockwood D (1958) *The Blackcoated Worker: A Study in Class Consciousness* George Allen & Unwin, London p21.
68 Orchard BG (1871) *The Clerks of Liverpool: Being Ten Chapters on their Numbers, Distribution, Salaries, Grievances, Marriages, etc.* p10.
69 Orchard BG (1871) *The Clerks of Liverpool: Being Ten Chapters on their Numbers, Distribution, Salaries, Grievances, Marriages, etc.* pp3–4.
70 Anderson G (1976) *Victorian Clerks* Manchester University Press p132.
71 McKinlay A (2002) 'Dead selves': the Birth of the Modern Career *Organization* Vol 9 No 4 p601.
72 The Banker's Clerk, 1843, p158 Cited in: Jeacle I (2010) The Bank Clerk in Victorian Society: the Case of Hoare and Company *Journal of Management History (Archive)* Vol 16 pp312–326.
73 Anderson G (1976) *Victorian Clerks* Manchester University Press pp32–33.
74 Jeacle I (2010) The Bank Clerk in Victorian Society: the Case of Hoare and Company *Journal of Management History (Archive)* Vol 16 pp312–326.
75 Orchard BG (1871) *The Clerks of Liverpool: Being Ten Chapters on their Numbers, Distribution, Salaries, Grievances, Marriages, etc.* p4 pp37–38.

6 Offices take shape

The Rebuilding of London Act 1666 sought to make the City more fire proof, introducing some basic measures such as regulating street widths, the heights of houses and construction materials – timber formed buildings were banned outright, as were projections and jetties over streets. The Act directed that henceforth there should be only four 'Sorts' of buildings: houses fronting by-lanes; houses fronting streets and lanes of note, houses fronting high and principal streets, and mansion houses, "*for citizens, or other persons of extraordinary quality*", not fronting either of the three cited ways. The Act was limited in ambition and limited in effect. Enforcement was a key weakness as vested interests and complicated ownerships frustrated comprehensive redevelopment.

The Building Act 1774 (formerly known as the Fires Prevention (Metropolis) Act 1774) was given royal assent in order to reinforce the earlier legislation and to regulate the design and construction of new buildings in London. The provisions of the Act also standardised the construction template of buildings and sought fireproof building exteriors (for example, timber ornamentation was banned).

The same Act also replaced the four 'Sorts' of buildings detailed in the 1666 Act with seven classes or 'rates' of buildings. Four related to houses, while the others referred to commercial buildings including mills and workshops. The four rates were defined by size and value (thereby being linked to the financial standing of their occupants), with a code of structural requirements for the foundations and external and party walls for each of the rates. Each rate had to conform to its own structural code for foundations, thicknesses of external and party walls, and the positions of windows in outside walls. For all rates, the 1774 Act stipulated that all external window joinery was hidden behind the outer skin of masonry.

It mandated inspection of new buildings by building surveyors to ensure that buildings were compliant with the regulations. The Act also brought into being the first legislation relating to human life and escape, rather than just building safety, and made parishes responsible for permanent provision of working firefighting equipment. The four rates are shown below.

- **First Rate.** Valued at over £850 and occupying a plot of over nine squares of building (900 sq ft (84 sq m)). These houses were typically for nobility and gentry. The occupants would frequently not own the house, but would rent and use it as their town house as a temporary alternative to their larger country house.
- **Second Rate.** Valued at between £300 and £850 and occupying a plot of five to nine squares of building (500–900 sq ft (46–84 sq m)). These houses were typically for professional men and merchants, and might face notable streets or the Thames.

DOI: 10.1201/9781003582205-8

- **Third Rate.** Valued at £150 to £300 and occupying a plot of three-and-a-half and nine squares of building (350–500 sq ft (33–46 sq m)). These houses were typically for clerks and faced principal streets.
- **Fourth Rate.** Valued at less than £150 and occupying a plot of less than three-and-a-half squares of building (350 sq ft (33 sq m)). These houses were typically for mechanics or artisans, and would be found in minor streets.

It was this legislation that led to the rapid spread of a uniform appearance of so many streets and squares across Georgian London, much of which survives to this day (for example, Bedford Square). Olsen noted this point, as well as highlighting the versatility of such structures, including their adaptability as offices.

> *Eighteenth century London, apart from its churches and an occasional public building, consisted of a collection of one standard pattern of house, in one of four sizes determined by the Building Acts. It could be converted into a hotel or lodging house by putting a brass plate on the front door, and letting the rooms separately. Different furniture and a different plate on the door turned it into an office building. A display window on the front, ground-floor parlour transformed it into a shop.*[1]

The new style and uniformity was not to everyone's taste, least of all to that of Sir George Laurence Gomme FSA (1853–1916) who was a public servant and leading British folklorist. He helped found both the Victorian Society and the Folklore Society. In his later career, Gomme worked for the London County Council, from 1893 as statistical officer, and from 1900 as Clerk to the Council. From this later position, Gomme used his influence to persuade the LCC to set up the Blue Plaque commemorative scheme. In 1898, Gomme gave vent to his despair at the replacement of the picturesque houses of Tudor and early Stuart London by "*the hideous monstrosities of Harley Street, Gower Street*", going on to claim that the "*Georgian spirit of architecture was against art, and declared for so-called utilitarianism, as if utility could exist without the element of art*".[2]

The Lombard Street tradition

Lombard Street was the epicentre of the birth of modern banking; indeed, the name has in many ways been synonymous with banking activity. It was, of course, immortalised in Walter Bagehot's 1873 *Lombard Street,* where he described the street thus

> *The briefest and truest way of describing Lombard Street is to say that it is by far the greatest concentration of economical power and economical delicacy that the world has ever seen.*[3]

The origins of Lombard Street stretch back to the fourteenth century, when it was named after the Lombards, merchants from Genoa, Lucca, Florence and Venice (most of Italy was once part of the Lombard Kingdom). The Italian merchants began arriving in London around 1290, and the street that bears their name was first recorded in around 1320. The Lombards' legacy lives on in everyday language, bequeathing words, including bankrupt, cash, creditor, debtor, diary, journal and ledger as well, of course, as "£. s. d" (Lire, Soldi, Denari, or pounds, shillings and pence), the standard monetary system across much of Europe until decimalisation in the eighteenth and nineteenth centuries, and in the United Kingdom until 1972.

As already noted, it was from the largely domestic premises in and around Lombard Street that many of the early banks and insurers emerged, often combining office, or shop, on the ground floor, with domestic accommodation above. And as we saw in Chapter 3, by mid-century, a few of the more established banks began to experiment with purpose-built buildings – albeit retaining the dual commercial-domestic functionality and the Georgian town house architecture. Private banks Charles Asgill (Lombard Street) and Drummonds (Charing Cross) were cited as examples.

This Lombard Street tradition[4] continued through the eighteenth century and one such building was that built for Child & Co at Temple Bar at the turn of the century. The building went through several phases of redevelopment and Figure 6.1 shows the building as it was in 1850. The ground floor is clearly of a domestic scale, and similarly the

> *upper floors showed a domestic simplicity, reflecting their use as residential space for partners and unmarried clerks. The dormer attic windows further underlined that this was a place of residence as much as a place of business. What is striking, even in comparison to new private banking-houses further east in the City, was the simplicity of design.*[5]

This hybrid house-bank approach perpetuated for most private banks built both in the City and West End, through to the 1820s. In both markets, the interior spaces were similar in their simplicity and utility.

> *Everyday business, such as paying in bills or cashing, would be done at a long mahogany counter dividing the floor of the shop. More personal discussions and financial arrangements, especially those involving an aristocratic clientele, took place in the parlour, a private space usually modelled on a Georgian drawing room.*

The emphasis on privacy and discretion in financial matters for the aristocracy provided a common cultural context for the banker: the success of his business "*rested upon his private wealth, symbolically and materially embodied within the private banker himself*".[6]

Another building in the Lombard Street tradition was the building commissioned by banker William Praed (1747–1833) at 189 Fleet Street in 1802 from architect John Soane. William Praed had been a partner in the West End house of Cocks, Biddulph & Co., of Charing Cross until 1792, before deciding to establish his own bank 10 years later. Along the way, in 1778, Praed married Elizabeth Tyringham daughter of banker and MP Barnaby Blackwell. This family connection led to a close relationship with, of course, Sir John Soane, to whom he turned to build a new banking house, at 189 Fleet Street.[7] Soane started his commission in June 1801, with negotiations over some vacant ground to be incorporated in the new banking-house and then presented initial plans to Praed and his partner Mr Babbage within a month. Booker describes the features of Soane's work:

> *he extended an architectural treatment to the whole facade, masking the interior division between bank and living area by unifying external decoration and the use of round-headed windows in the shop, first floor and attic. The panelled and fluted pilasters harked back to Soane's earlier work in the consols office and stock office of the Bank of England, while the fret-work frieze and decorative features of the cornice-balustrade found echoes in Soane's later rebuilding of the Bank of England's perimeter wall.*[8]

CHILD'S BANKING-HOUSE, FLEET STREET, NEXT TEMPLE BAR, 1850.
From a Drawing by Findley, 1855, in Mr. Crace's Collection. (*See page* 36.)

Figure 6.1 Child & Co Bank, Fleet Street, c1850.

Another important innovation was the placing of the entrance to the right, rather than in the centre, which had been common practice. This according to Booker meant an improved quality of light in the shop,

> *as the clerks were no longer placed between the counter and the back wall. ... . customers would enter a lobby from the street and then turn left into the 'Front Shop', where the counter would run along the east wall ... The rest of the ground floor was relatively simple, with a strong room in the centre and a 'Back Shop' to the rear, whilst a further door from the entrance lobby led to a staircase and the upper floors where partners and clerks would reside.*[9]

Praed's Bank was finished towards the end of 1801 and occupied in early-1802, at a cost of just over £10,000, including purchase of the ground, construction, fit-out and architect's fees. William Praed himself referred to the building as 'the most elegant and convenient' in the City, an assessment supported by none less than *The Times* newspaper.[10]

Detailed records of how banks used their premises in the eighteenth century are scant, but what evidence exists suggests that, even at this early stage, they were serially changing and adapting their premises to suit changing business circumstances. For example, Booker describes how the rapid growth of many businesses meant that sites were frequently outgrown and over staffed: "*Banks which 50 years earlier had two or three principals and as many clerks, now had a payroll of perhaps thirty.*" He goes on to cite a couple of interesting examples.

> *Coutts & Co. had reached 59 The Strand in 1739. No sooner, it seems, had Paine remodelled the bank in 1770-71 than it outgrew itself, and the partners were forced to lease property around William Street and John Street. James Paine again made alterations in 1780-83 but within a few years another house had been added and two more were leased in 1799. The early 19th century saw the addition of part of a Baptist chapel, an adjacent house (58 The Strand) and property at the west corner of John street.*
>
> *An even better example is Glyn's bank. Within thirty years their new building in Birchin Lane, more a dwelling-house than a bank, was too small. Worse, they had lost their position in Lombard street and it was expensive to attempt the return. In 1788 they moved across the road to 11 and 12 Birchin Lane, and merged the two buildings. Other property nearby was leased in 1790 and thrown-in with the rest in 1801.*
>
> *In the next few years five more nearby houses were added and, at last, in 1821, they got back to Lombard Street, buying nos. 66 and 67 which they rebuilt as one. In 1824 they moved in, relinquishing 11-13 Birchin Lane.*[11]

As well as the banks, insurance companies were also busy adapting domestic premises to commercial usage. Mention was made in Chapter 3 of Charles Asgill's bank building at 70 Lombard Street, which, a decade after the eponymous owners' death and following the demise of the business in 1796, was occupied by insurer Pelican in 1797.

Pelican had been set up in 1797 by Phoenix Assurance, as the Pelican Life Office. The new occupier stamped its identity on the building with an allegorical sculptural group on the building's otherwise plain façade. The group was designed by Lady Diana Beauclerk and sculpted by John de Veere of the Coade Stone Factory. While the building was demolished in 1915, the sculpture survives as an exhibit in the Horniman Museum.

After 110 years of business, in 1907 Phoenix reabsorbed its own offspring, Pelican Life Assurance, at that time known as the Pelican and British Empire Life Office, and thus began its career as a composite insurer. Sun Alliance and London acquired Phoenix Assurance in 1984.

Similar stories of eighteenth and nineteenth century premises' adaptation and expansion are told for Barclay, Bevan & Co. in Lombard Street; Messrs. Smith & Co. also in Lombard Street; Child's Bank at Temple Bar, and Gosling's Bank. Sir John Soane, who built his reputation for work on the Bank of England, remodelled 56, Pall Mall for Ransom, Morland & Hammersley in 1791 and 62, Threadneedle Street for Grote, Prescott & Grote in 1818. The elder Cockerell designed a bank parlour for Cocks, Biddulph & Co., at Charing Cross in about 1800. The number of such alterations and refrontings which passed unrecorded is probably large.[12]

Through adaptation and remodelling, in the late-eighteenth and early-nineteenth centuries, buildings were beginning to transition from homes being used also as bank premises to purpose-built bank buildings – albeit often continuing the tradition of providing domestic accommodation on upper floors. This was when offices ceased being referred to as 'shops'. Purpose-built office buildings appeared tentatively and then proliferated in response to the growing and evolving role of the City and in response to the changing nature of the organisations occupying them.

These examples typify a period when banks were small and lacked capital for investment in property on a large scale. Their continuous adaptation of property through remodelling interiors and adding neighbouring properties was a practical response that allowed them to expand and grow their businesses in a piecemeal fashion. Booker suggests that a reluctance to surrender living accommodation for administrative purposes to add, say, a Discount Office, Country Office, Stock Office, Town Ledger Office or Transfer Office to the original shop and parlour, led to the compromise solution of remodelling or refronting, perhaps disguising a medley of little offices behind a unifying facade.[13]

However, during this period 'large' companies became more common, and black-coated workers grew in number, each occupying increasingly specialised roles in increasingly sub-divided companies. Nowhere was this more the case than in London and in the banking and insurance sectors, to provide credit and cover for the expanding trade passing through the Port of London. And London's skyline was continuously changing as it grew into one of the most vibrant and cosmopolitan cities in the world. It would not be long before a definitive break with the past would be made, and purpose-built, single-use office buildings would emerge.

In America, from the outset, offices were purpose-built rather than adapted from residential use. Perhaps the earliest example is one that can still be seen today: Schermerhorn Row in New York City (see Figure 6.2). On the south side of Fulton Street, from Front Street to South Street, this block was one of the biggest structures in New York when it was built between 1811 and 1812 when the City was expanding rapidly. Situated very close to the docks where trading vessels docked, the block was developed by Peter John Schermerhorn (1749–1826) a captain, merchant, investor and real-estate developer. The buildings were counting houses, early offices, for the shipping companies powering the city's growth. The block was saved from demolition in the 1960s, and Schermerhorn Row is now a feature of the South Street Seaport and home to the South Street Seaport Museum.

Bank houses to office buildings: the first office boom

The later-eighteenth and early-nineteenth century saw rapid growth and change in the banking and insurance sectors. The result was that many businesses were outgrowing their converted residential premises and looking for larger buildings to cope with growing numbers of clerks and more footfall from clients. As noted by Booker, buildings "*were outgrown and overstaffed.*

Figure 6.2 Schermerhorn Row, New York.

Banks which 50 years earlier had two or three principals and as many clerks, now had a payroll of perhaps 30".[14] Booker illustrates the point with reference to Glyn's Bank.

> *In 1788 they moved across the road to 11 and 12 Birchin Lane, and merged two buildings. Other property nearby was leased in 1790 and thrown in with the rest in 1801. In the next few years five more nearby houses were added and, at last, in 1821, they got back to Lombard Street, buying Nos 66 and 67 which they rebuilt as one.*[15]

Booker notes that the same pattern of growth and property acquisition and rebuilding was seen at Coutts & Co, Child's Bank and Gosling's. Particularly after Waterloo, purpose-built offices began to spread quickly. One of their notable features was a sudden change in scale. The new office buildings, and indeed other new buildings such as warehouses and exchanges, were much larger than the predominantly residential stock that typified the City.

> *The warehouses in the City, the banks and other offices, are in many instances towering above their former height ... . Deadened as our sensibilities are by the still more monstrous erections of our own time, it is hard to recapture the sense of awe and wonder that the new office buildings, blocks of flats, and hotels evoked when they first appeared, towering above their Georgian neighbours.*[16]

The relationship between scale of company and scale of building came to be understood as an interdependency. This while converted town houses continued to provide satisfactory quarters for small businesses,

> *the increasing scale of activity of the nineteenth century insurance, banking and other businesses came to demand larger and more impressive buildings, both to house their various operations and to serve as architectural expressions of their wealth, respectability, and reliability.*[17]

Gradually at first, a new building form spread through the City and beyond, sometimes built for specific and single occupiers and sometimes as office blocks designed wholly for sub-leasing. The cumulative impact was to presage the emergence of a more uniform, single function City to replace the more diverse one that had grown up on a wide range of commerce and trade as well as homes.

The nature of office buildings changed greatly, mainly in response to the more formal and more structured form of working brought on by growing company size and breadth of business being undertaken. Long gone were the days – at least for many of the larger business – when a single clerk could know the detail across a form's whole business. The outcome around 1830 was a new building form, quite distinct from before around 1830. Locker describes the banking hall as "*one of the more memorable architectural accomplishments in the Victorian office building evolution*". He goes on to suggest that the "*consistency in size and grandeur in these halls all across the country is perhaps as remarkable in itself as are some of the finer individual examples*".[18] Banking halls were

> *... intended to be efficient and healthy working places for the numerous clerks required in an industry dependent on ledger books and hand calculations ... The requirement that tellers be close to clerks , and that clerks all be under the eye of the head clerk resulted in the common banking hall arrangement of a contiguous block of desks covering most of the available floor space, and placed under the best available natural lights. It helped make banks ideal working places.*[19]

The growing similarity in overall approach to the interior design and layout of banks can to some extent be acknowledged as from the influence of on individual, namely James William Gilbart (1794–1863). A banker and the first general manager of the London and Westminster Bank (one of the first JSBs to emerge in 1834 following the Bank Charter Act of the previous year), Gilbart wrote some of the earliest and most influential manuals on banking, including in 1827 his two volume *A Practical Treatise on Banking*.[20]

Volume 1, Section 11 is dedicated to 'The Administration of the Office', which considers 'The Arrangement of the Office'. This is a fascinating piece of writing and merits quoting at length. It is, in effect, a design brief for a bank building; perhaps the earliest attempt to formalise the main design parameters of an office building. The following discourse is summarised from pages 192–196 of Gilbart's original work.

The section begins by stating that a bank should be located in the most respectable part of a town and that it should "*a handsome building*", and that such expenditure as required for this purpose "*is no sin against economy, but an outlay to be repaid through profits*". He then gets into the layout of the building, which should contain a private residence for the manager (note the JSB language of 'manager' rather than clerk), to be separated from the office by a single, locked door. The domestic quarters should be constructed so that no activity in the domestic area can be heard in the bank.

Gilbart then goes on to write in some detail about three key design features – space, light and ventilation.

> ***Space*** *A banker should ensure that his clerks have adequate room to do their work comfortably. Every accountant knows that he can often work faster if he can have two or more books open at the same time; but if his space is so confined that he must shut up one book, and put it away, before he can use another, he will get on more slowly. The cashiers, too, will be much impeded if they are obliged to stand too close to each other; and the public will be huddled together, and will often count incorrectly the money given to them.*
>
> *Want of space will necessarily occasion errors, from the confusion it produces, and from one clerk being liable to interruption from the noise or vicinity of the others. A banker should therefore take care that his office is large enough for his business; and that it will admit of being enlarged in case his business should increase. Ample space is also conducive to the health of the clerks, as there will be more air to breathe, and the atmosphere is less likely to become polluted by the burning of lamps and candles.*
>
> ***Light*** *It is well known in every London bank, that fewer mistakes are made by the clerks in summer than in winter. Abundance of light prevents mistakes, and saves all the time that would be employed in the discovery of errors. Light is also of great importance to the cashiers in detecting forged signatures and bad or counterfeit money.*
>
> ***Ventilation*** *The air in a close office is not only rendered impure by the number of people that breathe it, and by the burning of gas, but it also contains very frequently particles of dust arising from the floor, through the number of people constantly walking in and out. It is almost impossible for persons so circumstanced to enjoy for a length of time even moderate health. A portion of this evil may be mitigated by a good system of ventilation. To obtain this should be regarded as an object of the first importance.*

Following these environmental issues, Gilbart then goes on to advise on the layout of space. In what was perhaps one of the very earliest examples of space planning. He goes into great detail about how different functional areas should be arranged, suggesting that the counter, desks and other furniture should be arranged so as to "*enable any given number of clerks to discharge their duties with the greatest efficiency, and so as best to promote the public convenience*". He states that the counter should be readily accessible for the visiting public (customers) and of "*sufficient length to meet the requirements of the business*", and that some banks will have two counters, one for paying, and the other for receiving. He then moves on to the cashiers desks, which should be "*sufficiently wide apart for the public to be promptly served, and to stand without jostling one another*".

> *At other banks the cashier does not enter the credits, but merely agrees the amount with the customer, and then passes them to a clerk, who enters them in the Waste Book. In the same way, when a cheque is presented for payment, he gives it to a clerk behind him, who enters it, and hands the notes to the cashier, who pays out the gold and silver.*

Gilbart then remarkably describes an element of space use amounting to what we refer to today as 'hot desking':

> *When the business is large, extra or supernumerary cashiers are appointed, who take the place of the regular cashiers when they are absent at dinner or otherwise, so that during the whole of the day all the cashiers' desks are occupied.*

He then touches upon the dynamics of space use rather than simply the layout, by addressing how queueing can be minimised. His solution is both simple and effective

> *To relieve the counter, the payment of bills that have been presented in the morning and not paid, is usually received at a separate desk or office. All these are expedients that should be adopted when necessary, to save the time of the public. There are few things that try a man's temper more than to be kept waiting a long time at a banker's counter; and he will be very apt to give vent to his impatience by quarrelling with the clerks, or reproaching the establishment.*

Next on the space planning menu is optimising communications. Gilbart recommends the importance of grouping together those clerks whose duties require them to have frequent communication with each other. If this rule is not observed, he says, then "*the clerks will lose much time in the course of the day in passing from one part of the office to the other; and the work will not be so expeditiously performed*". It is especially desirable, he continues, "*that the ledger keepers should be placed close behind the cashiers; so that if a doubtful cheque be presented for payment, the cashier may be able to show it to the ledger keeper, and be informed if he may pay it, without being observed by the party presenting it*".

Finally, Gilbart addresses the issue of supervision, or oversight and enforcing discipline. He starts this section by recommending that the desk of the chief or head clerk should be placed in such a position that allows him to see all over the office. Such a position will allow him, for example, to monitor the counter and intervene when it is crowded by allocating additional clerks to assist the cashiers. Similarly, should any "disputes take place between the clerks, or between the cashiers and the public, he will *come forward and settle the matter before the dispute is carried to high words*"; while he will also be able to observe "*customers who come frequently to the counter, and from their transactions he will often draw conclusions respecting their circumstances which will be serviceable to the bank*".

The clerks should not have their faces or their backs towards the window, rather the light should fall on them sideways. These matters may appear trifling, but they will not be deemed unimportant to those who are entrusted with the practical administration of an office. It is only by attention to minute things that the business of an office can be well conducted.

So there you have it: space planning for nineteenth century banks! Gilbart's work is unique in its prescriptions for interior space layout being specifically organised for business efficiency. His work was widely adopted in both JSBs and among the private banks. Gilbart was made a Fellow of the Royal Society in 1846 and devoted much of his life to banking, writing and various societies. He died at his home at Brompton Crescent and is buried in a Gothic Revival Grade II-listed mausoleum at West Norwood Cemetery, accredited to the architect Sir William Tite.

Locker describes some of the impacts of Gilbart's advice. For example, the requirement that tellers should be positioned close to clerks, and that clerks should be under the eye of the head clerk resulting in the common banking hall arrangement of a contiguous block of desks placed under that best available light. Again with respect to lighting the general ease of lighting spaces from the street front made the banking hall with a side entrance and a counter perpendicular to the street the most common patter for Victorian banks. Finally, that adequate lighting required office spaces to be very shallow, quickly distinguishing them from houses and establishing the offices as a separate building type. A 4.5m (15 ft) depth was considerably shallower than the 7.3m (24 ft) typical of the Georgian House, and closer to the dimension of the third rate houses common in the City at the time.[21]

As the nineteenth century approached its middle age, the banking and insurance companies were maturing into the kinds of businesses that we would recognise today, and as they did so, the buildings they occupied also took on their modern form. Among the clearest traits were the emergence of larger buildings, with grander banking halls and Italianate architecture – all moves to convince customers (and competitors) of each bank's strength and solidity. Olsen suggests that this had something to do with the growth in professional builders, in so far as architects were having to justify themselves and their fees to sceptical clients by producing increasingly extravagant designs, leaning towards the overstated, the individual, the idiosyncratic and on originality at all costs.[22]

Another factor for the increasing grandeur of office buildings might also have been the growing availability of capital, most notably via the joint stock banks. Indeed, Summerson is clear "*the monumentalising of bank architecture only came with the intrusion of the joint-stock banks into London*".[23] As noted earlier, 1833 legislation permitted joint stock banks in London for the first time (albeit for deposit only) and then from 1844 they were granted full rights to operate. The legislation providing a critical impetus to the creation and growth of well-capitalised businesses, and so the number of joint stock banks grew rapidly.

The impact was as profound, comparatively, as Big Bang in the City in 1986. Being able to raise share capital, banks were able to expand their businesses on an unprecedented scale. This in turn led to larger organisations, requiring larger buildings. Which is where a secondary impact of the new legislation came into play: companies were able to find the necessary capital to invest in real estate. And they did so on a grand scale.

> *The joint stock banks ... attempted to differentiate themselves in appearance from their predecessors, the private banks ... (the) banking-houses of the London private bankers were exactly that: banking-houses. The buildings that they operated from were decidedly domestic in appearance. Joint stock banks moved vigorously away from this style.*[24]

It was not only the joint stock banks that were busy replacing their old houses with office buildings; the insurance companies were also busy. Among the earlier, more impressive examples were several by Charles Robert Cockerell (1788–1863). These included his 1831 design for Westminster, Life and British Fire Office (1831); London and Westminster Bank (1838) and the Sun Fire Office (1841), all in London; and for Bank of England Buildings in Plymouth (1835), Bristol (1847), Manchester (1845) and Liverpool (1848).

Sun Insurance occupied a new, purpose-built headquarters with the style of a fashionable mansion in 1849. The Royal Insurance in Lombard Street commissioned successive large headquarters from John Belcher (1841–1913). The first was built on a site bought by the Royal Insurance in Lombard Street in 1857. Summerson notes that the new 1857 building "*became at once the dominant building in the street*". But the building proved a short-term solution: within just 6 years the site and building was disposed of and a new building – again designed by Belcher – was built nearby opposite the corner of Clement's Lane. Summerson links that fact that "*the Royal had discovered and tested the benefits of publicity and was unique among the offices in spending £20,000 or £30,000 a year on advertising*" with the form of the 1863 building, which

> *with its four panelled and plastered stories and fulsome display of allegorical sculpture, was sheer advertisement. The Royal of 1863 marks the point where insurance companies*

> *began to use architecture less as a symbol of quality and integrity than as an effusion of conspicuous waste.*[25]

Suddenly the office economy was gaining enormous ground and a boom in office building was gathering steam led by the banks and insurance companies. The level of construction between the 1850s and the 1880s was, in comparative terms, similar to the post-war rebuilding of the 1950s and the 'Big Bang' boom of the 1980s. The 1850s "*saw the erection of at least ten insurance buildings in the City, important enough to receive comment in the press*".[26] By the 1880s, London's position at the epicentre of international trade and finance was firmly established. By the time Bagehot published *Lombard Street* in 1873, "*the street itself was almost entirely lined with buildings built or rebuilt since 1857, all ... in lavish style and with the assistance of some notable architects*".[27]

Over the course of the later-nineteenth century the City was rebuilt. More particularly, it was repurposed. The physical environment was transformed to cope with new, dynamic, expanding businesses. Between 1855 and the turn of the century, around 80% of the City's physical fabric was rebuilt. One such new building in Lombard Street was that for the Royal Insurance Company, built in 1859 (Figure 6.3).

Lombard Street was more or less completely rebuilt, with a ripple effect along adjacent and nearby streets including Bishopsgate, Cheapside, Cornhill, Leadenhall Street, Lothbury and Threadneedle Street. The Georgian domestic architecture was demolished and replaced as quickly as it had originally spread. Suddenly houses adapted for use by banks and insurance companies were being replaced by purpose-built office buildings. This commercialisation of the City was reflected in a rapidly shrinking residential population: the City of London saw its population fall from 129,000 in 1851, to just 51,000 in 1881. Indeed, in 1871, the *Architectural Magazine* stated that

> *There are few spaces in Europe of equal size which can compare for modern architectural interest with the mile radius round the Exchange. The edifices in which the larger city merchants are housed by day are palaces compared with the cold and monotonous dwellings which even our most aristocratic families inhabit at the West End.*[28]

The time was, in many respects, revolutionary. Indeed, what happened in this relatively small window from the 1850s to the 1880s formed the basis upon which countless office buildings and workplaces would be built and occupied over the following century.

The newly capitalised banks and insurance companies were fundamentally changing not only the physical fabric of the buildings within which they undertook their business, but also their scale of operation, business processes and public image. New buildings during this period included those for London and Westminster in Lothbury (1838), City Bank on the corner of Threadneedle Street, built in an Italianate palazzo style (1856); National Discount Company's palazzo-style building in Cornhill (1858), the General Credit and Finance Company's 'Venetian Gothic' offices on Lothbury (1868) and National Provincial's head office on Bishopsgate (1866) were all typical of this era and approach.

Summerson argues that the rapid and widescale rebuilding had in view three objects: "*commodity, prestige and investment*". Commodity was provided by "*the generous reception area on the ground and a handsome boardroom on the first floors, strong rooms, boiler rooms and clerks' lunch rooms in the basement, and a caretakers'; flat at the top*". The second element, prestige, was afforded "*by a highly enriched masonry façade towards the street*"; and the third element, investment, "*by the allocation of all the remaining space to lettable offices, flexibly*

NEW OFFICES OF THE ROYAL INSURANCE COMPANY, LOMBARD-STREET.

Figure 6.3 Royal Insurance Company, Lombard Street, 1859. © Illustrated London News Ltd/Mary Evans.

partitioned between the iron columns which behind the solid stone façade, formed the main structure".

This was the new pattern of the fifties and in principle it has hardly changed in 120 years. A city still almost uniformly Georgian and domestic was being broken up into a city of competing commercial blocks. A city that lit itself by candles and oil lamps was moving into the world of gas sun-burners.[29]

Towards grandiose and ostentatious

As we saw earlier, private banks expanded quickly during the later-eighteenth century, but that their approach to building remained typically one of incremental change – conversions and refurbishments rather than new buildings. The approach taken by joint stock banks was very different: their answer "*to the problem of rapid growth was to build a head office of such monumental proportions that there was anticipated room for internal expansion over many years*".[30]

In the period immediately following the 1833 Act, most joint stock banks continued to purchase and lease existing premises. Noting that this applied both to main offices in the City and branches in the West End, Booker suggests that "*there were more than enough suitable buildings for adaptation or conversion*" and that to have done otherwise "*might have been considered provocative or aggressive by other banking interests*". Many of the existing premises came available from failed banks and "*the lure of a ready-made strong room appears to have countered any misgivings from superstition*".[31] Another reason was the fact that joint stock banks were severely constrained in their activities in London.

As already stated, the 1833 Act allowed joint stock banks only to deposit in London, they were not allowed to issue notes within a 65 mile radius of London. This naturally limited their growth in the capital and thereby, tamed their premises' ambitions. However, when the Joint Stock Companies Act of 1844 abolished the 65 mile rule and allowed incorporation without specific legislation, joint stock banks were emboldened.

Whether it was architectural ambition or financial abandon, there can be little doubt that office buildings very quickly became grander and more ostentatious, and compared to the Georgian City of London the Victorian cityscape, employed larger-scaled ornament and more stone and glass.[32] The buildings themselves were being used to reflect the status and power of their occupiers. The message was aimed at their actual and potential customers, suggesting that the bank was strong, reliable, a safe place and well capitalised.

More and more banks, insurance companies and related businesses were seeking offices with a prestigious address and a decorative façade. From the mid-1850s to the late-1860s, most joint stock banks moved to new premises that were, by comparison to the norm up to this point were grand and ornamented.

Kynaston cited City Bank in 1856, on a corner of Threadneedle Street, with its "*imposing Italianate palazzo style*", and the new limited liability discount houses which, "*likewise put a premium on appearance*". He also referred to the National Discount Company's "*palazzo-style building in Cornhill, completed in 1858 with distinctive cinquecento motifs and detailing*"; and the General Credit and Finance Company's "*stunning effort*" at 7 Lothbury in 1868 (Figure 6.4).[33]

The organisations most prominent in creating buildings on a far grander scale and with far more ornament were the joint stock banks. They used their buildings more than the private bankers ever thought possible directly to amplify their presence in the city, to attract new customers and to convey a sense of solidity. The Georgian houses of the past were left standing.

Figure 6.4 General Credit & Discount, Lothbury. © Erik Brown, 2020.

> *It is not only in the interior appearance that the business of the joint stock banks differs from that of the private bankers, but also in the exterior appearance of the establishments themselves. A joint stock bank cannot be carried on without a fine building, like those of the London and Westminster, the Union of London, and the others. Height of walls, extent of plate-glass windows, liveried porters at the entrance, and large chandeliers, seem to constitute the security of these concerns; and not sober, business-looking establishments, with anciently-fashioned railed counters, as in the case of the majority of our private banks.*[34]

Black describes one particular architectural feature of the new bank buildings that helped redefine the relationships between bankers and their customers: the banking hall. Thus while the private bankers and their largely wealthy customers engaged in face-to-face contact "*within the intimacy of small and personalized office spaces*" to reinforce "*the bonds of trust so important to validate money transactions*", the new banks took a different path. They replaced the private banks' receptions and parlours with large new banking halls

> *to accommodate a larger, more impersonal, and more socially diverse customer base. The attention devoted to ever more imposing banking halls reflected this bureaucratization of banking. For the new clientele, trust in the fidelity of the joint-stock banker relied more on the scale and richness of the bank's architecture to compensate for the loss of personal contact.*

Black goes on to describe how another building feature, the entrance, was

> *With the London and County Bank the facade had a much stronger central emphasis, whereas at the City Bank the entrance started to break out, with an arch and overhanging ornament. Such designs indicated the growing status of the new institutions within the widening public sphere for money transactions.*

Another fine building from this time was London & County Bank's headquarters in Lombard Street. Completed in 1862 to a design by CO Parnell. Booker describes the ground floor which introduced three-quarter columns with rusticated bands and windows with rounded corners and carved lintels; and then cites an article from *Building News*, which described the building thus

> *The Lombard Street front is beautifully proportioned; but it is largely indebted for its magnificence to the great care with which every detail has been studied, and to the absence of any weakness of design or trifling eccentricity. It is rich without being overladen with ornament.*[35]

Black describes National Provincial's new building on Bishopsgate, where the entrance was recessed "and the bay's cornice breaks back for emphasis, framed by more elaborate statue groups".

> The openness to a new banking public was here unmistakable. Overall, the centrality and verticality these arrangements gave to the banks' facades, disrupted the general horizontal unity of the traditional streetscape of the Georgian City, further reinforcing in architectural form the challenge of the new banking companies.[36]

Not everyone was in favour of the grandeur and ostentation increasingly common in the new buildings of banks and insurance businesses. Kynaston cites the *Bankers' Magazine* which, in 1863, suggested that they were "*succumbing to a love of show, leading eventually to excessive expenditure and sometimes to embarrassment.*"[37] Similarly, WJ Loftie was the author of an imposing volume called *London City*, published by the Leadenhall Press in 1891. In the final pages he was even more forthright suggesting that his brief survey had left him with

> *an impression of sadness on the mind. London has been practically rebuilt since the beginning of the present reign, yet how little is there of good architecture to be seen anywhere.*

And then really lets rip, suggesting that some of the new buildings

> *would be a disgrace to any city. The designs of costly structures, covered with coloured marbles and polished granite, are often unworthy of a gin palace ... Everywhere money has been lavished, brains have been spared. The maximum of ornament is associated with the*

minimum of design, and the chief modern buildings of the City are much more remarkable for their cost than for their beauty.[38]

The following paragraphs describe four grandiose and ostentatious buildings from this period: London & Westminster Bank; Atlas Assurance; City Bank and National Provincial Bank.

London and Westminster Bank. The London and Westminster Bank was the first joint stock bank in London, opening in 1834 at premises at 38 Throgmorton Street, with a branch at 9 Waterloo Place in the West End. The bank was well capitalised and it led the way for grander and more ostentatious bank buildings. Having outgrown its initial premises, the bank assembled a large freehold plot at 41 Lothbury, facing the Bank of England, and appointed Charles Cockerell and William Tite (1798–1873) as joint architects for the new building. Cockerell was commissioned to design the façade, while Tite was architect of the rear and interior. The new building opened in Lothbury on 26 December 1838.

Neither the site nor the choice of architects were without some meaning. London and Westminster was the first bank to challenge the dominant role of the Bank of England. It instigated a Bill in the House of Commons to challenge the 65 mile rule, which was passed only to be beaten in the House of Lords, no doubt following vigorous lobbying from friends of the Bank. Relations between the establishment and the newcomers deteriorated, not helped by the decision by London private banks to refuse access to the clearing house. Choosing a site overlooking the Bank could be seen as a belligerent gesture, while employing its architect could have been interpreted as defiance.

Black describes the main features of the building. The front elevation in Portland Stone was 24.4m (80 ft) in length, and the building 27.4m (90 ft) deep. The façade was in one plane, broken only by piers at each end. While the front displayed "*solidity, purpose and strength ... qualities to be admired in the office of a capital-rich banking institution providing a home for the savings of the growing metropolitan middle class*", there was also ornamentation: "*the first floor windows possessed side panels between the rusticated piers available for embellishment. Here caducei and fasces were sculptured in alternate disposition*". At the extremities of the front were

bold piers, each displaying a statue of a seated female figure. These figures were emblematic of the commercial interests of the London and Westminster Bank, with the statue at the eastern end carrying a shield decorated with the arms of the City of London, whilst that at the western end displayed the arms of the City of Westminster.

William Tite's interior was very grand. The new banking hall provided an air of confidence to the setting of daily formalized money exchanges, where the separation of clerks and customers underscored the mixing of public and private space so important for reputable money transactions. It was precisely these qualities of internal space that caught the attention of contemporary critics.[39]

The *Atlas* noted how the principal banking hall

is not only by far the most considerable in the building, but is unequalled in importance by anything of the kind in London, except in the case of some offices of the Bank of England, and in altitude it exceeds even them.

On the east and west sides of this spacious banking hall were aisles, surmounted by balustraded galleries, and divided from the centre by arcades of three arches springing from Doric columns.[40]

The central space was topped by a dome pierced by a large circular skylight, providing a solution to the pervasive problem in the City of ensuring adequate light, where many lofty buildings crowded in on an essentially medieval street plan. Additional light was obtained, according to the *Atlas*,

> *by triple windows, occupying the flat semi-circular spaces left by the pendentives of the dome, on the three sides which are exposed to the view of a person entering, in the manner of the imperial Roman baths ... smaller semi-circular windows are introduced likewise in the three arches on the north side ... so that a very sufficient light has been contrived throughout this vast apartment.*[41]

The bank was more grand than ostentatious, certainly compared to many of those that followed. Thus Booker recognised the importance of the building in terms of "*a new attention to functionalism which dictated to some extent the exterior appearance*", For example, "*the problem of light was uppermost in the architects' minds, and the central doorway, the bays of tall, tiered windows and the domed banking hall were working towards the same end*". Furthermore, the banking hall grew in size to cover an area "*too large to be lit satisfactorily from windows on the outside wall*". The decision was then taken to pierce through the upper floors to create a light-giving dome, which meant that "*the main entrance could return to the centre of the front elevation, in grand manner*" and that the resulting interior columns offered "*opportunities for decoration and embellishment*".

Atlas Assurance. In the early-nineteenth century, the office economy was dominated by insurance and banking companies, and it was these two sectors that put an abrupt end to the Georgian architecture of the Lombard tradition and swept in the commercial architecture of the Victorian era together with its global ambitions.

The insurance companies were quick off the mark with a number of grand buildings, elaborately detailed and often in the Italianate style. One splendid example that has remained fairly close to its original design through to the present day, stands at the corner of Cheapside and King Street in the City of London. This office building is Atlas House, commissioned by the Atlas Assurance Company (originally founded by a group of merchants and bankers in 1807 at Will's Coffee House). Following offices at Bush Lane (near Cannon Street station), and then Coleman Street (near Guildhall), Atlas acquired the site on the corner of Cheapside and King Street in the mid-1830s. At the time, Cheapside was one of the city's busiest roads and a hub of business and trade.

The company's directors hoped that a new head office on the site would convey a sense of prestige to potential customers. The board appointed one of the country's top architects, Thomas Hopper (1776–1856) to design the new headquarters (see Figure 6.5). The 1836 design took inspiration from classical Italian architecture. The ground floor is made of Cornish grey granite, while the upper storeys are Portland stone. The upper storeys' windows feature alternate curved and triangular pediments, with single, Corinthian pilasters between. The top of the building was framed with a balustrade. Business was going so well for the company, the directors splashed out on the new technology of internal gas lighting. They also seem not to have been concerned about paying window tax as the building has many windows.

In 1857, the building was extended northward, and again in 1893–1894. Celebrated English architect Alfred Waterhouse (1830–1905) designed the late-nineteenth century alterations, including a grand entrance at 1 King Street, topped by a sculpture of Atlas. The piece was carved from Portland Stone by Thomas Tyrrell for Farmer & Brindley.

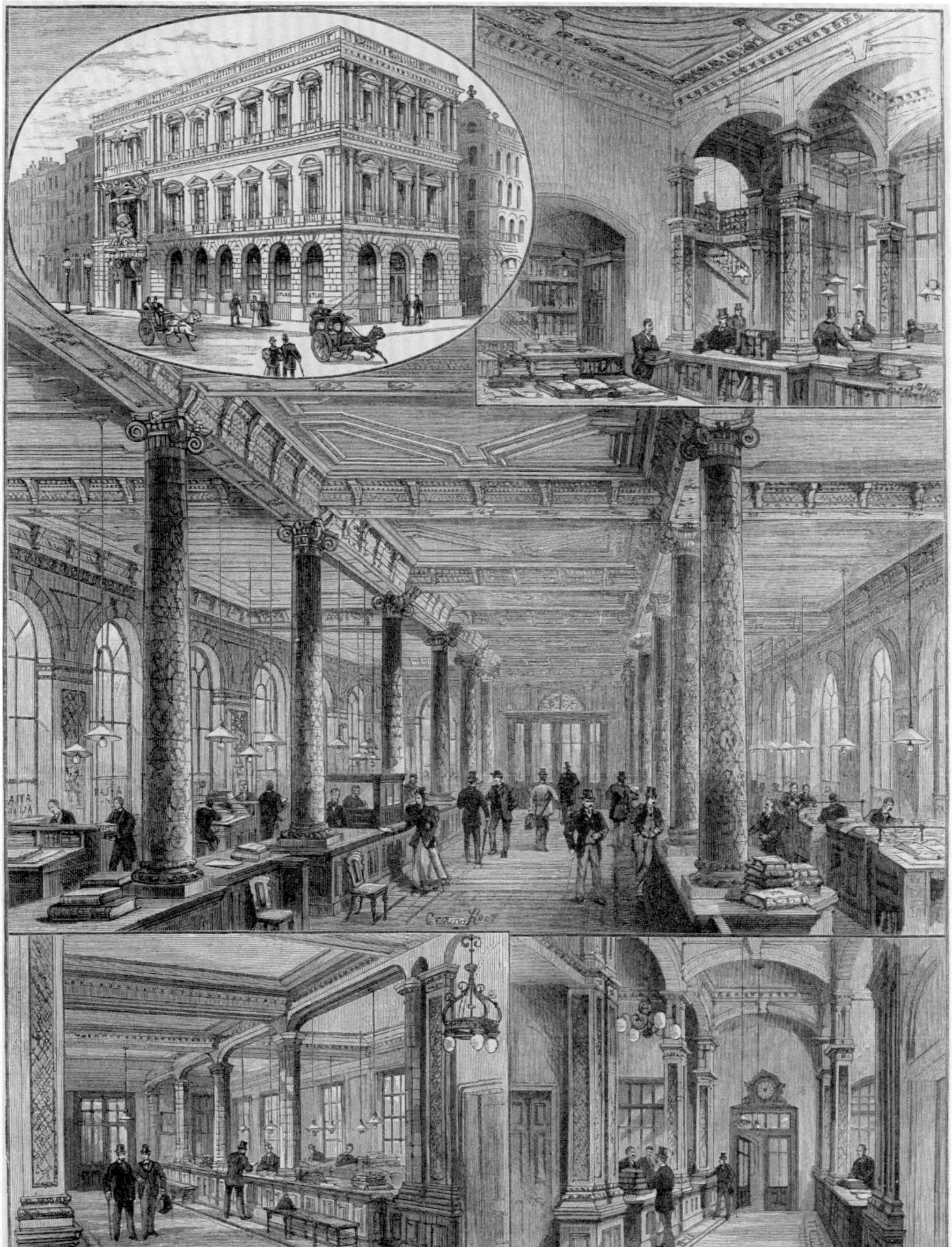

Figure 6.5 Atlas House, Cheapside, 1894. © Illustrated London News Ltd/Mary Evans.

After 123 years on Cheapside, it was the end of an era when Atlas Assurance was acquired by the Royal Exchange Assurance in 1959. Although Atlas Assurance no longer exists, its name lives on in the building, which was Grade II listed in 1972. It was later designated as part of the Guildhall Conservation Area in 1981. Today, the building includes a retail bank branch on the ground floor, with rented office space on the upper floors. The grandiose nature of the Atlas building was repeated for Westminster Life and British Fire Office on the Strand (1831) and the Sun Fire Office on Threadneedle Street (1841), both designed by Charles Robert Cockerell.

City Bank. Founded in 1855 the City Bank formed a part of the second wave of grand buildings by joint stock companies. By the 1860s, the bank had established a network of West End and suburban bank branches. The bank initially established its City foothold in a temporary office at 2 Royal Exchange Buildings (in the style that modern companies establish their presence through a serviced office).

Black describes how the bank's directors "*moved with remarkable speed to assemble a new site in Threadneedle Street*" to build a purpose-built headquarters on a prestigious site. Site assembly was complex and involved the "*purchase of two leasehold plots, together with a rental agreement with another party before the prestigious corner site facing on to Threadneedle Street was secured*".[42] Architect William Moseley (1799–1880) was commissioned to design the new office, which was completed in 1856, and remains to this day (Figure 6.6).

Figure 6.6 City Bank, Threadneedle Street.

Black describes how, at a cost of £31,145 including purchase of leaseholds, building, fixtures and fittings, Mosely produced

> *an imposing Italianate palazzo design, distinguished by its rounded corner entrance with a heavily rusticated ground floor and a deeply projecting cornice at roof level … . The palazzo style allowed richness of detail and a conscious expression of symbolic capital without the loss of propriety necessary to convey the safe conduct of banking business.*
>
> *Contemporary commentators also drew attention to the refined taste and craftsmanship displayed in the cabinetwork and ornamental fittings which, when combined with the feature dome and slender granite columns, gave a real richness of effect to the interior.*

Black continues to explain that in common with many other new joint stock banking companies, The City Bank was

> *involved in the production of new symbolic spaces and sought to create a distinct identity* for *a new and more public form of the money economy. Palatial style and grandeur offered a distinct break with the Georgian restraint of the City's private bankers, signalling to the new metropolitan middle class an attractive combination of both innovation and security in their money transactions.*[43]

National Provincial Bank. Perhaps the grandest and the most ostentatious bank building from this period was the head office of the National Provincial Bank (later National Westminster) on Bishopsgate.

The bank was founded in 1833 and had grown quickly into an extensive regional branch system, with over 120 branches scattered across the country by the time it built its new headquarters. It was well-capitalised: when building commenced in 1862, the paid-up capital of the bank increased from £450,000 in 1853 to £1.08m in 1866.[44] Net profits stood at £395,709 in 1863, £523,252 in 1864, and £483,612 in 1865. The business was profitable and expanding.[45]

As noted earlier the Bank of England held a monopoly on issuing bank notes within a 65 mile radius of London, which meant that banks could operate in London so long as they did not issue notes. Against this backdrop, National Provincial had built up an impressive network of regional offices but was not active in London. Its head office was in London, but this performed administrative services alone. However, the lure of the rapidly expanding London market proved too great, and so the bank "*ceased issuing new notes in 1862 in order to move its financial services into London and the lucrative markets that could be accessed from the capital city, as well as the wealthy areas that surrounded it*".[46]

In 1840, the bank acquired a Georgian mansion, Salvador House, on Bishopsgate and converted the residential premises into an office. According to Locker, the bank reconfigured the ground floor, knocking down dividing walls, to accommodate 40 clerks in contiguous seating, with accounting and customer waiting area comprising the remainder of the ground floor. On the first floor, Watson created rooms for principals and a clerk's residence.[47] This building was, however, just the start of the bank's rise to grandeur, with the realisation of its ambitions in its building on Bishopsgate (Figure 6.9), designed by John Gibson (1817–1892).

In 1835, Gibson joined the office of Charles Barry, who he assisted by working on the drawings for the new Palace of Westminster. Then in 1844, Gibson won the commission to design the National Bank of Scotland in Glasgow, which was completed in 1847. Alongside this, he also designed the Imperial Assurance Building on Threadneedle Street in 1846. But his biggest break came with the National Provincial Bank.

Gibson's commission of National Provincial's 1866 building provided a "*magnificent classical banking hall in Bishopsgate ... resplendent with Corinthian columns and marble pillars*" which could "*have eased the doubts of even the most neurotic depositor*".[48] According to Black, the new headquarters was a magnificent architectural statement.

> *Gibson's design was for a single-storey building, bolder and grander than the Bank of England itself ... with Corinthian columns running the full height to the cornice. Gibson broke with the established pattern of multi-storey building, signalling by the single-storey structure that this important and wealthy institution, new among the established metropolitan banks, had no need of upper stories for speculative letting space. That the facade was given over entirely to the expression of the banking hall further underlined the innovative nature of the building.*[49]

Figure 6.7 shows that bank as it is today – almost exactly as it was in 1866. The image emphasises height and grandeur; solidity and trust, and it was a successful attempt to establish what we might call today the bank's 'competitive position': it was at least as good as its rivals. Victoria Barnes and Lucy Newton provide a wonderful insight into the genesis of the design of the building, exploring the symbols and messages that the leaders of the bank wished to convey

Figure 6.7 National Provincial Bank, Bishopsgate.

to the market. They note a *Bankers' Magazine* article, reporting that the "*magnificent building*" was "*an important addition and ornamentation to the architecture of the city*", highlighting the impressive statues and panels of the exterior, and the "*banking room, commonly called 'the shop' [which] is 118 feet in length*".[50] Ward-Jackson summarized the building thus: "*the most extravagant of the City's Victorian joint stock banks, makes it mark principally through the profusion of its sculptural adornments*".[51]

No less a figure than John Summerson described the building as "*by far the most exciting, and perhaps really the queen of all the City banks since Soane*":

> *a single-story building of noble scale, bolder and grander than the Bank of England itself, with Corinthian columns running full height to a cornice ranging with the tops of the tall houses next door and even then pointing further into the sky with groups and statues. Within is a stately Corinthian hall lit from three domes.*[52]

By the 1870s, the office economy was maturing rapidly. As we saw in Chapter 4, companies were growing in size and complexity and consequently their buildings became larger and more complex (and more ornate!). During this time, living accommodation on upper floors came to be replaced by lettable floor space, and it was not long before buildings consisting entirely of lettable floor space were to be built. The combination of these factors encouraged the emergence of a mature property development sector, producing a range of purpose-built and speculative buildings. In particular, they encouraged the emergence of the professional, speculative developer.

Speculation and the modern office

In the nineteenth century, a growing number of banks let out the upper floors of their grandiose buildings in what we would call today 'office suites' – self-contained spaces for smaller firms looking to gain a foothold in the City. And by the 1850s and 1860s, most new office buildings contained a proportion of lettable space. But just as owner-occupier buildings with sub-let space were taking shape another trend was emerging, one that would become the normal pattern of development and occupation through to the current day. This was the development of speculative buildings.

A speculative building is one that is designed, funded and developed for leasing in its entirety on the open market, for any occupier. It is speculative because the occupier and the financial return carries a good deal of uncertainty. The spread of speculative and other purpose-built buildings was, by mid-century, starting to give the City of London a look and feel more akin to its image today than of the eighteenth century. As described in *The Builder* in 1868

> *The City of London ... is becoming more and more the office of the world .... Stately buildings replace the ugly and cramped houses of the Georgian era, and these buildings are almost entirely parcelled out in offices.*[53]

And again

> *The semi-public structures ... which, as banks, insurance companies, offices, chambers, or however else designated, have been there erected within the last year or two, have done*

> *more for its architectural character than had been done by its professedly monumental edifices.*[54]

Quite quickly, the development industry evolved – the coalescing of architect, developer, constructor, funder and advisor. Development companies began to proliferate with the single end of buying up sites and covering them "*rapidly with ostentatious buildings containing nothing but office space*".[55]

In Chapter 2 we met Nicholas Barbon, who established a template, or modus operandi, for large-scale, speculative developers, albeit of residential districts. While it is with some justification that he could be charged with helping to establish the reputation of developers as a rapacious and single-minded class, sometimes with questionable ethics, he did nevertheless also help take development approaches into a new era.

Bowley writes that "*until at least about the middle of eighteenth century the building owner himself or his architect, or agent, employed each craft directly*".[56] Before Barbon – and indeed for a long period after – most buildings were custom built for individuals on their own land, using their own labour and materials, whether residential or commercial. The buildings were relatively simple in form and straightforward to construct. Commercial buildings were typically general purpose in nature, occupied by small enterprises who did not make particular demands on building form. The notion of speculative development was a novel one, and independent developers who acted between landowners and occupiers of buildings were relatively rare. Barbon changed all of this.

Barbon the developer was a response to the increasing scale and complexity of development. He was paving the way for a new role: the speculative developer. The developer emerged to assemble land and capital, to organise the labour, to oversee the delivery of the physical building and to earn a financial gain upon completion. The developer became an intermediary between the landowner and the occupier of the building: he assumed a level of risk and responsibility, and was rewarded accordingly. As the template became established, developers began to purchase land and develop in their own right. But whether as an intermediary or as a direct player, as projects became larger and more complex, so the need for working capital and access to credit became paramount. In the early days, wealthy individuals and building materials firms were active in this sense, but increasingly there emerged a critical role for banks and insurance companies.

The first such building in the City, and possibly the world, dates from around 1823, and was described by architect and surveyor Edward l'Anson (1812–1888). Born in the City of London, he was the eldest son of Edward P. I'Anson (1775–1853), also a surveyor and architect. He was educated at the merchant Taylors' School, and articled to his father and subsequently entered the office of John Wallen, in the City. At the close of his indentures I'Anson travelled for 2 years. On returning in 1837, he entered into practice, both as assistant to his father and as an independent architect. From 1850 their firm, Edward I'Anson and Sons, operated from a building next to the family home, at 9 St Laurence Poutney Lane.

His first important building in the City, commissioned in the same year he returned from travels and entered practice, was the Royal Exchange Buildings, designed for Sir Francis Graham Moon in 1837. This was an innovative scheme because l'Anson used concrete for the internal works, previously used only in minor projects. The scheme also established his reputation as a leading architect in the City. I'Anson developed a career as a pioneering designer of purpose-built commercial buildings in the City, many in the Italian style, including 146 Queen Victoria Street in 1868 (Figure 6.8). He also designed the Corn Exchange in Mark Lane (1875),

Figure 6.8 146 Queen Victoria Street.

a Portland stone building in the classical style; and the neo-Gothic school of the Merchant Taylors' Company, charterhouse.

He delivered many papers to the RIBA, including in 1864 *Some Notice of Office Building in the City of London*, in which he made particular reference to purpose-built office buildings. Pevsner relates how he told his audience that within his recollection *merchants dwelt in the City over their counting houses and next to their warehouses', but that thirty years ago, i.e., in the 1830s, 'certain houses [were] let out in separate floors and used as offices ... . 'The first building which I remember to have been erected for that special purpose was a stack of office buildings in Clement's Lane at the end nearest to Lombard Street.' This dated from about 1823 and was by an architect called Voysey.*[57]

From this we can determine that the first speculative office building in London was built in the City of London in 1823. The process found favour and before long, speculative buildings became the norm. In a later paper, l'Anson elaborated on the nature of speculative buildings.

When I first began to build on the new London Bridge approaches, previous to 1840, City offices as now constructed were not thought of; the houses were built as shops and dwellings, or as warehouses, and it was the same in Moorgate Street. Since that time, however, a distinct type of construction has been evolved, which is now, perhaps, nearly perfect ... . In these buildings the greatest attention is paid to the size of the rooms may be obtained in the smallest space, and that they be abundantly lighted: convenience of access, the position of doors, stairs, fireplaces and water and gas supply are all carefully studied.[58]

I'Anson was elected a Fellow of the RIBA in 1840 was chosen president in 1886. In the same year he was elected president of the Surveyors' Institution.

T Roger Smith and WH White, in a paper read to the Society of Arts in 1876, summarised the rise of the specialised office building:

The huge blocks of offices for merchants and professional men now to be found in the City ... are comparative novelties ... Little more than fifty years ago a great many merchants resided east of Temple Bar often over their counting houses. Even at that time, a few houses were let in separate floors, and used as offices; but buildings erected expressly for offices were unknown, and one of the first houses planned and constructed for that purpose was a stack of offices in Clements-lane, built about 1823, During the last twenty years the growth of this class of building has been enormous.[59]

Kynaston refers to the burgeoning specialist offices from mid-century that "*started to mushroom almost everywhere, many in the vicinity of the Stock Exchange in order to house stockbroking and jobbing firms*".[60] One, built in 1854, was 5 Throgmorton Street. Owned by the two leading brokers James Capel and John Norbury, this housed not only their own firm but also several others. Another, constructed by the mid-1860s, was at Warnford Court just off Throgmorton Street, where one of the early tenants was the new firm of de Zoete & Gorton.

If Capel and Norbury were only part-time property developers, by the 1860s there was at least a handful of full-time operators based in the City, including the City of London Real Property Company, the Registered Land Company and the City Offices Company. "*It could be a chancy business, but with interest rates low and the formation of new financial companies all the rage, there was big money to be made as City rents soared*".[61] Speculative development soon became commonplace.

The City of London Real Property Company was founded in 1864 by John Innes (1829–1904) and his older brother James. John had an early career as a wine merchant but then went on to become a property developer. He purchased land in Merton and created the Merton Park garden suburb in Surrey. Accumulating significant wealth in the process, John became a philanthropist. In his will he left part of the Merton estate for the establishment of a horticultural

Figure 6.9 59–61 Mark Lane.

institute, the John Innes Centre, now located in Norfolk and the compost that bears his name is still being sold today.

Soon after establishing their company, the Innes brothers commissioned George Aitchison (1825–1910) to design what is now 59–62 Mark Lane (Figure 6.9) as a speculative block of offices in 1864. The four-storey Byzantine-influenced building, which remains today, is dressed in Portland stone. The upper levels are *arcaded with carved caps to columns, with intricate patterning, including abstract plant motifs, around and between the arches. The structure comprises wrought-iron beams on cast-iron columns with fireproof brick-arched floor.*

Summerson compares the building with the "only slightly less remarkable" 103 Cannon Street. Built 2 years later by the Registered Land Company and designed by architect Fred Jameson, the building

> *is more positively Romanesque and owes a great deal to Crown Life, but both buildings show how close the City medievalism of the sixties ... could come to a reasonable and sympathetic modernity". Summerson describes the building's fenestration as "simple, rhythmical, and luminous, the modelling bold in relief.*[62]

In 1866, The City Offices Company, bold development business, embarked on two very large schemes in successive years: a £70,000 block in Lombard Street in 1866 and an £80,000 block on the site of the old Bull Inn, running through from Bishopsgate to Old Broad Street, in 1867 – both huge sums for the time. Summerson describes projects on such scale as

> *new and highly significant of what was going to happen to the City in the next hundred years. Architecturally, they aimed to overtake the banks and insurance offices in external show. Good address and an extravagant facade were justified by the sort of rents that would pay for the site and the building within a very few years.*[63]

Funding arrangements, construction contracting and intermediation all evolved quickly, and the West End and the City began to take on their modern shape in bubbles of property development. Some of these bubbles created whole new districts, and were distinctive not only in a temporal sense, but also in their architectural footprint. The emergence of Regent Street, with the architecture of John Nash and James Burton, and the patronage of George, the Prince Regent (later George IV), is perhaps one of the most distinctive. John Blashfield built Kensington Palace Gardens in the mid-1840s, and John Duncan and Jacob Connop transformed the area that became Ladbroke Grove.

Summary. Bank and insurance office buildings emerged in domestic premises in and around Lombard Street, often combining office, or shop, on the ground floor, with domestic accommodation above. By mid-century, a few of the more established banks began to experiment with purpose built buildings. Thus the 'Lombard Street tradition' arrived.

Even at this early stage, banks and insurance firms were serially changing and adapting their premises to suit changing business circumstances and over time, buildings transitioned from homes being used also as bank premises to purpose-built bank buildings. This was when offices ceased being referred to as 'shops'. Purpose-built office buildings appeared tentatively and then proliferated in response to the growing and evolving role of the City and in response to the changing nature of the organisations occupying them.

The nature of office buildings changed greatly, mainly in response to the more formal and more structured form of working brought on by growing company size and breadth of business being undertaken. Another factor for the increasing grandeur of office buildings was the growing availability of capital, most notably via the joint stock banks. Mid-century legislation granted full rights to joint stock banks to operate in the City, providing a critical impetus to the creation and growth of well-capitalised businesses.

The impact was as profound, comparatively, as Big Bang in the City in 1986. Being able to raise share capital, banks were able to expand their businesses on an unprecedented scale. This in turn led to larger organisations, requiring larger buildings that were grandiose and ostentatious compared to any earlier buildings. The early-nineteenth century also saw the arrival of speculative buildings, and the first such building in the City dates from around 1823.

By around 1870, the office economy, offices and office work were on the brink of a new era of radical change. The final few decades of the nineteenth century witnessed the transformation of the American economy, the mechanisation of office work and the feminisation of the office workforce. The period also saw a divergence in office workplaces between Europe and North America as the latter adopted the skyscraper. Working conditions improved significantly as purpose-built buildings suitable for large corporate organisations spread and new approaches to office management emerged. This was the start of the era the defined the term 'white-collar factory'.

Notes

1 Olsen DJ (1976) *The Growth of Victorian London* Penguin Books, London p24.
2 Cited in Olsen DJ (1976) *The Growth of Victorian London* Penguin Books, London p37.
3 Bagehot W (2009) *Lombard Street: A Description of the Money Market* Seven Treasures Publications (Original work published 1873) p6.
4 Booker JML (1984) A Study of the Design of British Banks from the 18th Century to Modern Times Vol One D.Phil. Thesis, University of York p10.
5 Black IS (2003) Private Banking in London's West End, 1750–1830 *The London Journal* Vol 28 No 1 pp29–59.
6 Black IS (2000) Spaces of Capital: Bank Office Building in the City of London, 1830–1870 *Journal of Historical Geography* Vol 26 No 3 pp351–375.
7 Black IS (2003) Private Banking in London's West End, 1750–1830 *The London Journal* Vol 28 No 1 pp29–59.
8 Booker J (1990) *Temples of Mammon: the Architecture of Banking* Edinburgh University Press p13.
9 Booker J (1990) *Temples of Mammon: the Architecture of Banking* Edinburgh University Press p8.
10 Booker J (1990) *Temples of Mammon: the Architecture of Banking* Edinburgh University Press p13.
11 Booker JML (1984) A Study of the Design of British Banks from the 18th Century to Modern Times Vol One D.Phil. Thesis, University of York p13.
12 Booker JML (1984) A Study of the Design of British Banks from the 18th Century to Modern Times Vol One D.Phil. Thesis, University of York p13.
13 Booker JML (1984) A Study of the Design of British Banks from the 18th Century to Modern Times Vol One D.Phil. Thesis, University of York p15.
14 Booker J (1990) *Temples of Mammon: the Architecture of Banking* Edinburgh University Press p9.
15 Booker J (1990) *Temples of Mammon: the Architecture of Banking* Edinburgh University Press p10.
16 Cited in Olsen DJ (1976) *The Growth of Victorian London* Penguin Books, London p64.
17 Olsen DJ (1976) *The Growth of Victorian London* Penguin Books, London p120.
18 Locker FM (1984) *The Evolution of Victorian and Early Twentieth Century Office Buildings in Britain* University of Edinburgh PhD thesis p53.
19 Locker FM (1984) *The Evolution of Victorian and Early Twentieth Century Office Buildings in Britain* University of Edinburgh PhD thesis p55.
20 Gilbart JW (1827) *A Practical Treatise on Banking* (Two vols) Bell & Daldy, London.
21 Locker FM (1984) *The Evolution of Victorian and Early Twentieth Century Office Buildings in Britain* University of Edinburgh PhD thesis pp55–62.
22 Olsen DJ (1976) *The Growth of Victorian London* Penguin Books, London p72.
23 Summerson J (1977) The Victorian Rebuilding of the City of London *The London Journal* Vol 3 No 2 pp163–185.
24 Barnes V & Newton L (2019) Symbolism in Bank Marketing and Architecture: the Headquarters of National Provincial Bank of England *Management & Organizational History* Vol 14 No 3 p220.

25 Summerson J (1977) The Victorian Rebuilding of the City of London *The London Journal* Vol 3 No 2 pp163–185.
26 Summerson J (1977) The Victorian Rebuilding of the City of London *The London Journal* Vol 3 No 2 pp163–185.
27 Summerson J (1977) The Victorian Rebuilding of the City of London *The London Journal* Vol 3 No 2 pp163–185.
28 Cited in Olsen DJ (1976) *The Growth of Victorian London* Penguin Books, London pp121–2.
29 Summerson J (1977) The Victorian Rebuilding of the City of London *The London Journal* Vol 3 No 2 pp163–185.
30 Booker J (1990) *Temples of Mammon: the Architecture of Banking* Edinburgh University Press p11.
31 Booker J (1990) *Temples of Mammon: the Architecture of Banking* Edinburgh University Press p45.
32 Abramson DM (2005) *Building the Bank of England: Money, Architecture, Society 1694–1942* Yale University Press.
33 Kynaston D (1994) *The City of London Volume One: A World of its Own 1815–1890* Chatto & Windus, London p245.
34 Evans DM (1865) *The City, or The Physiology of London Business with Sketches on 'Change, and at the Coffee Houses* Baily Brothers, London p16.
35 Booker J (1990) *Temples of Mammon: the Architecture of Banking* Edinburgh University Press p129.
36 Black IS (2000) Spaces of Capital: Bank Office Building in the City of London, 1830–1870 *Journal of Historical Geography* Vol 26 No 3 p371.
37 Kynaston D (1994) *The City of London Volume One: A World of its Own 1815–1890* Chatto & Windus, London p245.
38 Loftie WJ (1891) *London City* Leadenhall Press, London pp279–280 Cited in Kynaston D (1995) *The City of London Volume Two: Golden Years 1890–1914* Chatto & Windus, London p28.
39 Black IS (2000) Spaces of Capital: Bank Office Building in the City of London, 1830–1870 *Journal of Historical Geography* Vol 26 No 3 p361.
40 *The Atlas Newspaper*, December 29, 1838. Cited in Black IS (2000) Spaces of Capital: Bank Office Building in the City of London, 1830–1870 *Journal of Historical Geography* Vol 26 No 3 p360.
41 *The Atlas Newspaper*, December 29, 1838. Cited in Black IS (2000) Spaces of Capital: Bank Office Building in the City of London, 1830–1870 *Journal of Historical Geography* Vol 26 No 3 p360.
42 Black IS (2000) Spaces of Capital: Bank Office Building in the City of London, 1830–1870 *Journal of Historical Geography* Vol 26 No 3 p362.
43 Black IS (2000) Spaces of Capital: Bank Office Building in the City of London, 1830–1870 *Journal of Historical Geography* Vol 26 No 3 p362.
44 Orbell J & Turton A (2001) *British Banking: A Guide to Historical Records* Routledge, London pp403–404.
45 Barnes V & Newton L (2019) Symbolism in Bank Marketing and Architecture: the Headquarters of National Provincial Bank of England *Management & Organizational History* Vol 14 No 3 p219.
46 Barnes V & Newton L (2019) Symbolism in Bank Marketing and Architecture: the Headquarters of National Provincial Bank of England *Management & Organizational History* Vol 14 No 3 p219.
47 Locker FM (1984) *The Evolution of Victorian and Early Twentieth Century Office Buildings in Britain* University of Edinburgh PhD thesis p36.
48 Kynaston D (1994) *The City of London Volume One: A World of its Own 1815–1890* Chatto & Windus, London p245. Locker FM (1984) *The Evolution of Victorian and Early Twentieth Century Office Buildings in Britain* University of Edinburgh PhD thesis
49 Black IS (2000) Spaces of Capital: Bank Office Building in the City of London, 1830–1870 *Journal of Historical Geography* Vol 26 No 3 p368.
50 Barnes V & Newton L (2019) Symbolism in Bank Marketing and Architecture: the Headquarters of National Provincial Bank of England *Management & Organizational History* Vol 14 No 3 p220.
51 Ward-Jackson P (2003) *Public Sculpture of the City of London, Public Sculpture of Britain*. Liverpool, Liverpool University Press p35.

52 Summerson J (1977) The Victorian Rebuilding of the City of London *The London Journal* Vol 3 No 2 pp163–185.

53 *Building* (1868) Vol 26 p501 Cited in Olsen DJ (1976) *The Growth of Victorian London* Penguin Books, London p89.

54 *Building News* (1857) Vol 3 p1 Cited in Olsen DJ (1976) *The Growth of Victorian London* Penguin Books, London p121.

55 Summerson J (1977) The Victorian Rebuilding of the City of London *The London Journal* Vol 3 No 2 pp163–185.

56 Bowley M (1966) *The British Building Industry: Four Studies in Response and Resistance to Change* CUP, p334.

57 I'Anson E (1864) Some Notices of Office Buildings in the City of London *Transactions of the RIBA* Vol 15 1864–65 pp25–36 As described by Pevsner NA (1976) *A History of Building Types* Thames & Hudson, London p214.

58 From a paper by Mr I'Anson, read at the RIBA, December 2nd, 1872, reported in *The Builder* 4th January 1873 pp6–7.

59 Smith TR & White WH, in a paper read to the Society of Arts in 1876 *Building* Vol XXXIV 1876 p305 Cited in Olsen DJ (1976) *The Growth of Victorian London* Penguin Books, London p122.

60 Kynaston D (1994) *The City of London Volume One: A World of its Own 1815–1890* Chatto & Windus, London p246.

61 Kynaston D (1994) *The City of London Volume One: A World of its Own 1815–1890* Chatto & Windus, London p246.

62 Summerson J (1977) The Victorian Rebuilding of the City of London *The London Journal* Vol 3 No 2 pp163–185.

63 Summerson J (1977) The Victorian Rebuilding of the City of London *The London Journal* Vol 3 No 2 pp163–185.

Part III

1870–1945

The white-collar factory

Resulta BS7 Adding Machine (1930s)

Office Era	Time period	Work & Place Themes	Office Technology
Market	**Part I: 1670-1770**	**Trade**. Goldsmiths as bankers. Central banking. Coffee Houses. Joint Stock Companies. Exchanges. Insurance.	Coal fires. Candles. Copying. Ledgers. Paper. Bills of Exchange. Quill pen. Cash book. Shop counter.
	Part II: 1770-1870	**Organisation.** Stock Exchange. Clearing House. Private and merchant bankers. Joint Stock Banks. Office buildings. Speculative development.	Gas lighting. Copying. Ink Pen. Ledgers. Paper. Stenography. Polygraph. Correspondence book.
Factory	**Part III: 1870-1945**	**Production**. Rise of corporation. Division of labour. Work: fixed, manual, repetitive and mechanical. Female workforce. Rise of professions.	Mechanisation. Adding machine. Electricity. Telegraph. Telephone. Typewriter. Copiers. Elevator. Air conditioning. Fluorescent light.
Corporate	**Part IV: 1945-2000**	**Process**. Globalisation. Scale. Bureaucracy. Command & Control. 'Corporate man'. Work landscapes. Birth of digital office.	Photocopier. Mainframe. System Furniture. Golf Ball Typewriter. PC. MS Office. Mobile phone. Internet. Email.
Digital	**Part V: 2000 onward**	**Knowledge**. Agile work. Data. Power.	Smart phone. The Cloud. Social Media. Augmented Reality. Automation. Artificial Intelligence. Robotics. Virtual Reality.

7 A new organisation of society

For the first half of the nineteenth century the office economy continued to be dominated by small, largely family run businesses, and remained one which was comprised mostly of financial and insurance interests, both of which were closed worlds, little understood by very few beyond those with a direct interest. Particularly in banking, family dynasties held very privileged positions – economically and socially – and it was no surprise that they viewed joint stock banks with some disdain as they looked out from their domestic quarters in their bank houses over increasingly grand and imposing bank buildings. But, as serial financial crises showed, the 'system' of small banks was fragile.

The later decades of the nineteenth century were a period of transformational economic and social change. It would be no exaggeration to say that it marked a transition between eras, as the agrarian economy yielded to the industrial economy and a new organisation of society. Urbanisation and population growth saw a huge concentration of economy and society in cities, where factories and workshops corralled large numbers of people around mechanised processes. Property was transformed from the simplicity of land and cash to financial services and shareholding. Universal education, advances in medicine and political enfranchisement helped revolutionise society. And the range of goods and services consumed by society grew exponentially as the office economy and commercial activity generally spawned a growing white-collar workforce, one with more leisure time and with more disposable income.

Shops stocked with comparative and exotic goods, rather than just the needs of day-to-day existence became commonplace, and in London shopping areas emerged in places like Cheapside, Mayfair and Piccadilly, along with that great Victorian innovation, the department store. Allders (Croydon, 1862); Army & Navy (Victoria Street, 1871); Bentall's (Kingston, 1867); Fenwick's (Bond Street, 1882); John Barker (Kensington, 1870); John Lewis Partnership (Oxford Street, 1864) and Whiteleys (Westbourne Grove, 1867) all emerged during this time. London's suburban tentacles spread ever outwards as the central area hosted rapidly growing commercial activity.

The expansion and growing complexity of commerce had to be supported by increasingly sophisticated trading. By 1875, Britain was supplying not only between one-third and one-half of the world's trade in manufactures, "*but also the facilities that made possible most of the rest of trade and investment*".[1]

> *City banks advanced large sums of short-term credit to importers and exporters and arranged the finance of long-term investments overseas. The Stock Exchange was the centre of the world's capital market; and London's insurance companies had no serious international competitors.*[2]

DOI: 10.1201/9781003582205-10

London was the world's principal money market, and it further consolidated its position through the expansion of foreign issues. For example, by 1890, some 30% of the capital stock of American railways, and an even larger percentage of the Argentine and Indian railways, was held in Britain.

This period saw the transition from an economy built primarily on small family owned businesses into one dominated by the manager class working in corporate organisations. In common with the USA, the hectic growth of the railways in Britain called for the rapid growth of new organisations, with a sense of punctuality and discipline, which helped set the template for subsequent, more complex organisational structures.

The later decades of the nineteenth century were blighted by the Great Depression. Triggered by a financial crisis in the USA in 1873, and spreading to Britain and Europe. Stock markets had been riding high for a prolonged period, but fears of a financial bubble reached fever pitch in Vienna, leading to the collapse of the Vienna Stock Exchange in the same year. Depending upon which measures are used, the Great Depression is widely reported as lasting through to the 1890s. There was a tangible impact on clerking as employers lost business confidence and as profits were squeezed; there followed an almost inevitable downward pressure on salaries and upward pressure on competition for jobs.

This was also the era that saw the office economy evolve from one comprising almost entirely banking and insurance interests into a broader financial and business services sector, including accounting, advertising, finance and investment, legal, marketing, real estate and surveying, as well as the head offices of manufacturing firms and a growing public sector. The office economy grew sixfold in the latter half of the century.

Within this maelstrom of change, London was practically rebuilt. As noted by Powell, "*Business meant money and numbers and these meant counting houses*". The result was the emergence of novel building forms – modern offices: "*Banks, insurance offices and allied business increasingly built new premises in the City ... many of these manifestations of Victorian commercial spirit were imposing and costly, expressing power, prestige and dependability*".[3] For example, there was Mansion House Chambers, built in 1872 on the south side of Queen Victoria Street, which contained some 500 rooms, and "*epitomised the dawning age of the office block*".[4]

London had modernised. The city had accepted its declining role as 'the world's port' and was now embracing its role as the office of the world. London was no longer *doing* the trade, it was *organising* the trade, *funding* the trade, *insuring* the trade. Telecommunications meant that goods could be directed rather than received and then re-exported. From an office in the City, merchants could direct their operations across the globe.[5] As Michie points out, while Britain's share of the world's shipping fleet fell sharply, the City of London

> *dominated the movement of international shipping ... it was the only centre with all the information all the time, it was important to be represented there if cargoes were to be picked up.*[6]

By 1910, around 30% of all quoted securities in the world were listed on the London Stock Exchange[7] and by 1914, 80% of all issues on the London market were for overseas borrowers.[8] Improvements in communications reinforced London's role as the leading centre in an increasingly integrated world securities market. And its growth as a commercial and international trading centre, together with its position as the hub of world communications, led inevitably to it becoming the world's foremost shipping market. By 1914, the British merchant fleet accounted for around half the world's tonnage, and one-third of the fleet was then engaged in trade that never touched Britain's shores at all.[9] Key to London's emerging role was technology because, as Michie notes,

> *The result of this communications revolution was that it became possible to conduct a global trading business from an office in the City, maintaining constant contact supplemented by rapid visits and the receipt and despatch of samples and catalogues.*[10]

Increasingly, merchants, brokers and agents from both Britain and abroad gravitated to the City, organising and controlling world trade. Not only trade itself, but shipping, finance and a whole host of ancillary services fundamental to world trade such as insurance, law and accountancy located themselves in the Square Mile.[11] But in some respects, this period also proved to be the high-water mark, particularly in the City of London.

> *In 1908 it was estimated that at least a million people a day flowed into and out of the City at a time when the telephone was only slowly becoming an accepted means of conducting business.*[12]

One grand whole: the business ecosystem

The workforce of the typical counting house in the middle decades of the nineteenth century, was small. BG Orchard, writing in 1871, estimated that Liverpool's 4,590 firms employed 17,398 clerks, less than four each on average. Orchard notes that there were an additional "*4,000 office boys, clerks in embryo*". But despite the average, some firms were much larger: Orchard cites "*Four employers alone could array 500 clerks*".[13] Banking and insurance were both restructuring, which meant numerous mergers to create ever-larger businesses with growing head offices. In the banking sector, Lloyds Bank employed 2,880 staff by 1909; London County and Westminster, 3,250 by 1914 and Midland Bank, 5,000 by 1914.[14]

Although small by present-day standards, life-insurance companies at this time were large compared with many Victorian employers. The Clerical Medical, Eagle, Edinburgh and Economic – all mid-sized insurers processing from 300 to 600 policies yearly – employed between five and eight clerks in the 1850s. But the Prudential's head office, which already numbered 245 staff in 1873, increased to 700 by 1892 and to 2,600 by 1920; while the staff at the Royal Exchange's head office went from 76 to 149 between 1890 and 1900, then doubled again by 1913, to 314.[15] The scale of the Prudential Assurance Company's undertaking was amplified by Black and Muddiman. By 1891, the company controlled nine million life policies.

> *Premiums were collected weekly, which translated into around 470 million transactions per year, each of which had to be recorded both locally, by the collector, and centrally. Collection work was undertaken by an army of over 11,000 field agents ... . Work in the field was supported by another army: swathes of clerical workers, in the company's London head office, which by 1905 numbered 1,500.*[16]

As the authors observe "*a miracle of data processing in an age without computers*".

Similarly, Anderson points to the larger firms among Liverpool's steamship owners, suggesting that by 1871, the city's 54 steamship companies employed over a thousand clerks and apprentices. Likewise 12 Liverpool firms dealing with marine insurance employed 450 clerks and apprentices, an average of nearly 40 clerks per firm; while in 1875, the Mersey Docks and Harbour Board employed 300 clerks and apprentices.[17]

And other large employers, including manufacturers, the railways, utilities and government were growing in scale and complexity and employing growing armies of clerical workers. Heller records two of the larger organisations: Railway Clearing House in Seymour Street, 2,500 by 1914 and London County Council in Trafalgar Square, 1,267 by 1909.[18]

As the Industrial Revolution got into full swing, the rapidly expanding rail and canal companies, the extractive and manufacturing industries, food processing, shipping, textiles and utilities, all had voracious appetites for money; they needed it in growing quantities and in ever more complex arrangements, in an increasingly international context. In short, industry needed money to pay wages, to invest, to expand and to modernise. In this context, it is necessary to recall that as late as the First World War, office activity remained a minority activity in London. In the pre-war City of London, for example, manufacturing and commodity businesses accounted for 46% of jobs compared to just 9% in banking and finance.[19]

Office activity as we understand it today was very limited: much of it was ancillary to commercial and trading enterprises. Such 'factory offices' first employed a timekeeper and a foreman's clerk, whose task was to assist the foreman by keeping track of the work in progress and its stages of completion, while keeping the records of workers, materials and tasks. "*Out of these rudimentary functions grew the modem cost, planning and scheduling, purchasing, and engineering and design sections*".[20] In 1865, journalist David Evans described the City as

> *a world within itself. Centred in the heart of the metropolis, with its innumerable capacities for commercial pursuits, it presents at first sight, to a stranger, a most mysterious and unfathomable labyrinth of lanes and alleys, streets and courts, thronged with a bustling multitude, whose various occupations, though uniting in one grand whole, seem to have no direct association with each other.*[21]

Evans was describing both the physical environment and the increasingly complex relationships that were "*uniting in one grand whole*" – the business ecosystem.

Braverman recognised that most offices at this stage were subsidiary and complementary to the productive labour processes carried on elsewhere within the same corporation. "*But with the development of monopoly capitalism came the extraordinary enlargement of those types of enterprises which, entirely separated from the process of production, carry on their activities either chiefly or entirely through clerical labour*".[22] He went on to show how, in purely clerical industries, such as banks and credit agencies, there was "*only one mode of labour, the clerical, and below the managerial level the labour employed consists almost entirely of clerks who work in offices and service workers who clean the offices*". The only thing that prevents this from being the case with brokerage and investment houses and insurance companies is the need for a large number of salespeople.[23] But the situation was changing, rapidly.

> *The office function was clearly due to grow, and to grow quickly. And in addition it would need to separate off from other facets of urban activity, because of the growing complexity and intricacy of organisation needed to finance, administer and support the industrial revolution.*[24]

The key points here are the *growing complexity and intricacy* of activities needed to *finance, administer and support* wider economic change. This is critical because it hints at the roots of the hugely complex web of activities that form the office economy ecosystem as we know it today. The scale of activities was becoming such that single, often family owned companies could no longer import/export, ship, wholesale/retail, finance and underwrite their activities. Specialist functions were emerging beyond the insurers and bankers in the later decades of the nineteenth century: accounting, advertising, brokerage, engineering, legal, surveying, and related professions began their inexorable growth and evolution into the twentieth century office economy.

From early days, banks were the apex businesses: they controlled the money and everything flowed from that. Merchant banks emerged out of commodity trading and money lending, with most retaining a "*mercantile legacy additional to their major functions of guaranteeing trade credits ('accepting') and organizing capital through share and bond issues*".[25] In 1845, there were 60 private banks in the City, comprising 217 partners.[26]

There then followed a period of enormous growth and change, with companies emerging, growing, consolidating and diversifying; all the time becoming larger and more complex. Between 1865 and 1885 alone, private banks were absorbed by joint stock banks at the rate of 2.25 per annum. Private-to-private amalgamations took this average up to nearly five. At the end of the period, there were (nationally) 250 private banks and 120 joint stock banks. The combined 370 banks had 2,500–2,700 bank offices, including head offices.[27]

In 1891, the President of the Institute of Bankers said the business "*has changed greatly from its old, easy, comfortable character: the private banker has the feeling that his day is past. His great joint-stock neighbour over shadows and disturbs him.*" And foreseeing the inevitable, he said many more would be driven "*into the haven of amalgamation*".[28]

In the same year, there were 24 amalgamations of private banks, either with one another or with joint stock banks. Clapham commented in 1938 that "*So many have never been arranged in a single year before or since*". In the 11 years from 1891 to 1902 there were 114 bank amalgamations. By the outbreak of the First World War, the number of private banks had shrunk from 250 to around 12: "*for the country as a whole, private deposit banking was extinct ... . It came to nearly every one of them within the next ten years*".[29]

By the turn of the century, the bankers had grown in number and scale. Table 7.1 lists the main merchant banks in existence by 1900, and the families associated with them. The banks varied in size and scale, but perhaps the largest, most powerful and best connected globally was NM Rothschild and Sons.

This rapid consolidation of the banking sector around the turn of the century reflected the growth of joint stock companies among the bankers' clients – extractive industries, factories, food processing, printing, shipping, textiles and utilities. All were growing in scale, often operating internationally and becoming more complex organisations, their demands on banks similarly changing – greater scale, more international and more complex. Banker and client were co-evolving, with the former becoming ever more concentrated in the City of London, where head offices grew in scale and complexity.

And in the City they met a particular class of people. As Cassis suggested, there "*has never been any real doubt about the position of the banking community at the top of the social hierarchy. The social origins of bankers are a first indication of this*". Data presented by Cassis show that between 1841 and 1861, 52% of bankers and bank directors went to public school and Oxbridge; this rose to 72% in 1861–1880.[30]

In the early days, insurance was relatively expensive and limited to the wealthy – the new capitalists rather than traditional merchants – but the market quickly matured so that, by mid-century, with economic expansion and the attendant growth in wealth, the sector reached the wider population. By 1850, there were some 180 life insurers providing £150m of life cover in the UK, insuring the growing professional classes – doctors, lawyers, clergymen and teachers. Victorian society embraced the values of thrift and prudence and used life insurance as a form of savings and as security for business loans. Life insurance demand rose most among the professional classes, with new insurers specialising in insuring doctors, lawyers, teachers, farmers and clergymen.

Table 7.1 English merchant banks and their owners, 1900

Merchant Bank	*Main shareholders (family)*	*Foundation*
Arbuthnot Latham	Arbuthnot	1833
Baring Brothers	Baring	1772
Robert Benson & Co	Benson	1810
Brown Shipley	Brown	1880
Robert Fleming & Co	Fleming	1880
Fruhling & Goschen	Goschen	1832
Anthony Gibbs & Sons	Gibbs	1808
JK Gilliat	Gilliat	Pre-1850
Glyn, Mills & Currie	Glyn, Mills, Lawrence	1753
Guinness Mahon	Guinness	1836
CJ Hambro & Sons	Hambro	1839
Frederick Huth & Co	Huth, Meinertzhagen	Pre-1850
Jardine Matheson	Keswick	1848
Kleinwort	Kleinwort	1830
Lazards	Lazard, David-Weill	1877
Samuel Montagu	Montagu, Samuel, Franklin	1853
JS Morgan & Co	Morgan	1838
Ogilvy Gillander	Gladstone	1840
Ralli Bros.	Ralli	1820
Robarts, Lubbock & Co	Robarts, Lubbock	Pre-1800
PP Rodocanachi	Rodocanachi, Sechiari	1860
NM Rothschild & Sons	Rothschild	1804
M Samuel & Co	Samuel	1878
ED Sassoon & Co	Sassoon	1840
J Henry Schroder	Schroder, Tiarks	1804
Stern Bros	Stern	1856

Source: Lisle Williams (1984).

The sector was highly dynamic, with widespread bankruptcies and mergers, but growing rapidly. Premiums trebled between 1870 and 1914; and by 1914 there were 94 life insurers insuring lives for a value of £870m. Growth was driven mostly by demand for investment products like endowment insurance, another UK insurance innovation that guarantees a payment at the end of a fixed term, or earlier if the insured dies. By 1913, endowments accounted for 62% of all life-insurance policies.[31]

Well into the 1900s life-insurance premiums remained unaffordable for most British workers. However, in the mid-1800s a new type of life insurance emerged to serve the needs of the better-off working classes. Industrial life insurance used a weekly or monthly door-to-door collection of small premiums to cover a lower sum insured – industrial insurance was intended to cover the cost of burial, rather than provide for dependents. The sector was to become dominated by Prudential Mutual Insurance Investment and Loan Association, later known as Prudential, which had issued 25m policies by 1905.

By 1914, there were 39m industrial assurance policies in the UK. However, the average amount insured under an industrial life-insurance policy was just £10 compared with £345 under ordinary insurance policies in 1905.

As the financial structure of businesses became more complex, especially with the growth of large joint-stock companies like railways, the need for experts skilled in handling the

> *companies' affairs arose. Whereas clerks and book-keepers could be left with maintaining the accounts reflecting everyday business, events such as financing, risk assessment, acquisition, merger or public flotation all required specialist advice and expertise.*[32]

Arrival of the professionals. Cue the arrival of accountants,

> *whose business it was to provide that type of expertise. In particular, it was necessary to protect the interests of the shareholder in a public company from fraudulent management, and so increasingly public companies instituted audits from independent firms for this purpose.*

As *"early as 1880, there were some 500 accountants and 50 auditors in the City*". Deloitte began in London in 1845 to audit the accounts of the spate of new joint stock companies being formed; and they were followed by Cooper Brothers, William Peat, George Touche and John and Alexander Young, all of whom established eponymous firms in the late nineteenth century.[33]

The City's legal firms rose in much the same way as the accountants, but developed at a much slower pace and on a much smaller scale. Both Linklaters and Paines appeared in the mid-1840s, at the time of the railway mania. By 1893, Linklaters were employing 44 clerks and generating profits of £5,000 per annum.[34] And just as the accountants and lawyers coalesced around the bankers and industrialists so too, in time, did supporting professionals including architects, engineers and surveyors.

The preceding few paragraphs, in an extremely simplified form, show how the office economy business ecosystem evolved, meeting a rapidly growing demand for a wider breath of services, from its core of banking and insurance into a plethora of businesses required for their professional, independent and expert advice. In the mid-nineteenth century, the core functions of banking, insurance and professional services began to form a strong cluster of inter-related business services in the City of London, leading eventually to today's Global Financial Centre.

The growth of lawyers and accountants foreshadowed the wider rise of the 'professional class'. Among the earliest professions to organise were the medical and legal professions. Solicitors, barristers, surgeons and general practitioners all had national organisations by 1832. They were followed by architects (1834), actuaries (1848), surveyors (1868), accountants (1870), insurers (1897), structural engineers (1908) and town planners (1914).

From the very beginning, 'professionalisation' was imbued with an aspirational quality; it was deemed to reflect the social standing of those who referred to themselves as having a profession: something different to, or more crudely, better than, a run-of-the-mill job. In many countries, professional status endowed upon the individual higher social status, more secure employment and even political power, or at least, influence. But there was another, equally important factor: depending upon your point of view: professions either helped to maintain a certain level of standard in the service they provided, or they were a closed shop for those with a vested interest in maintaining prices and reducing competition.

No doubt professionalisation reflected a blend of these factors, but it is undeniable that they carried social and cultural importance as well as functional identity. Neal and Morgan observed a staged process to the establishment of the early professions:[35]

- the occupation became full-time rather than practiced as part of another activity;
- a training regimen was established, formalised by the articles system, in which articled assistants were trained by established professionals;
- a professional association was formed in order to obtain greater status;
- qualifying exams were introduced to stress high levels of competence;

- a Royal Charter was sought to secure legal protection of the profession, protecting members from competition from non-members, and
- routes to academic qualifications were formed to enhance standards and status.

The specific origins and early development of each profession varied widely, and the number of professions multiplied as the demand for technical knowledge and skills proliferated. But they were clearly territorial in nature: placing boundaries around certain activities that could not be undertaken by unqualified tradesmen. This was different to the growth of *managerialism* and the management class.

The emergence of London's business ecosystem was an early example of agglomeration economics. Agglomeration is the invisible glue that holds the office economy together in dense clusters. Whether in accounting, advertising, banking, consulting, finance, legal and media; or in coding, developing, graphic designing, programming and software engineering; or in architecture, construction, design, engineering, planning and surveying, the office economy is bound together in enormously complex webs of relationships, involving partnering, collaborating, networking and socialising, and which drive innovation and higher output. At their most basic, they form vast job markets. At their most sophisticated they form sustainable and competitive business ecosystems. These strengths are not replicable in a fully dispersed world.[36]

Increasing numbers of studies drew attention to the concentration of office services in large metropolitan areas, and empirically verified the 'glue' that held the companies so tightly together, in terms of the location of markets (clients and suppliers), highly skilled labour markets and the agglomeration benefits associated with minimising transaction costs.[37] Other studies showed how office services created competitive advantage in urban areas including quality, creativity and innovation, spatial proximity, flexibility, timeliness of delivery, scope of services offered and adaptability to client needs – all a long way from the simpler explanations based on classical economics of earlier times.[38]

As the new century unfolded, of course, the world was plunged into the tragedy that was the First World War, a decade after the completion of which began the Second World War. Almost three decades lapsed covering the wars and the period between.

Rise of the American corporation

In the mid-nineteenth century European and American economies operated on a largely owner-managed, entrepreneurial, small-scale business model, characterised by many layers of production and distribution. Owners managed all aspects of their businesses, and most firms operated from the town or city where they were based; from a single building or complex of facilities, and were largely serving local markets. They employed one or two clerks to assist with paperwork; keep the accounts up to date and copy correspondence to customers and suppliers. Merchants effected sales and distribution for wholesaling and retailing.

In America, all of this changed fundamentally, and at speed, with the spread of the railroad across the entire continent. The railroads foreshortened distance, bringing producers and merchants alike to much bigger markets, while at the same time opening up vast new territories. So it was that the railroad companies became the template for the modern white-collar organisation: the railroads became vast new enterprises in their own right.

> *No other business enterprises in the 1850s had as large initial costs, operating expenses, payrolls, and required so many, so varied, and so technically difficult decisions as did the new large railroads.*[39]

The role of the railroad and government. The railroad system required financing to build new track, stations and stock, and it employed rapidly growing numbers of people in roles, including driving, conducting, ticket sales, station masters, construction, engineering, surveying and, yes, a growing army of clerical workers involved in administration, bookkeeping and payroll to ensure control over the vast tentacles of the business. Complex communication systems were established, and timetables developed.

> *One of the first and most significant challenges raised by the coming of the large railroad was its administration. There were few precedents to guide those groping to devise rational ways to supervise, coordinate, and plan for the use of far more men, money, and equipment than any other private enterprise had hitherto had to administer.*[40]

Very quickly railroad companies were employing a new management class – people whose role it was to ensure that other people fulfilled their roles. The managers were also tasked, of course, in the context of vast geography and primitive telecommunications, with making decisions on behalf of the business. Whereas previously foremen, overseers, senior craftsmen and supervisors had been trusted employees, it was never within their gift to make business decisions *on behalf of* the business.

According to Chandler, the first railroad to systematise its operations was the Baltimore & Ohio line. In 1847, when its President, Louis McLane and its Chief Engineer, Benjamin H. Latrobe, began to make plans to complete its construction, they agreed that the road needed a new system of management.

> *By the 1847 reorganization, Latrobe and McLane had set up one of the very first functionally departmentalized, administrative structures for an American business enterprise. Of the two great departments, the first, headed by the Treasurer, supervised finances. The second, run by the General Superintendent, administered operations and transportation, and was further subdivided along functional lines. A smaller, third department in the central office, headed by an Attorney, handled a wide variety of legal matters involved in the operations of the railroad ranging from the obtaining or revising of charters to the settling of claims for damages and losses.*[41]

Further key developments included the creation of a central office manned by general executives, including the President, Vice Presidents, and a General Manager, who concentrated less on day-to-day operation and more on long-term problems of cost determination, competitive rate setting and strategic expansion.[42]

Although railroad company structures would evolve and become more complex as the railroad system grew and integrated, the McLane and Latrobe model became the template on which the whole US railroad system would be administered. The sheer scale of the endeavour ensured that management came to resemble a distinct profession, with the associated social recognition.

Local, regional and national managers and operational and administrative managers were all bringing new systems and processes to the running of the railroad. In doing so, they created administrative functions, or departments, which themselves grew rapidly, creating head offices, regional offices and so on. The new offices corralled specialist and lower-level clerk functions alongside higher-level generalist activities; all were overseen by an increasingly complex web of middle managers, senior managers and even vice presidents. The highest level of management reported to the Board.

Figure 7.1 State, War and Navy Building, Washington DC.

Alongside the railroads, the US government was also ballooning in size, no less so than in the military. The State, War and Navy Building was built between 1871 and 1888 (Figure 7.1). Reputedly the largest building in the world at the time, the building's external detailing is extraordinary. With five storeys, it ranged over four hectares (ten acres), a "*fit symbol of the sprawling bureaucratic growth of the Federal Government*".[43] Today, the building is known as the Eisenhower Executive Office Building and forms part of the White House estate.

Stephens and Lubar observe that early US government office buildings "*mirrored the increase in size and complexity of the work that went on within them. In their monumental simplicity, Washington's first federal office buildings were deliberately designed to demonstrate stability and permanence*".

Another interesting Government building is the Pension Building (1885) by federal architect Montgomery C Meigs, one of the few architects "*who expressed any concern about the men and women who worked inside the office buildings*". He observed that Federal buildings typically suffered from poor lighting and ventilation in their long corridors, and sought instead to counter this with a single row of rooms on each of the four floors surrounding a great and lofty hall. "*His design reflected the needs of the workers and the ideology of government work*". The latter point related to rooms being open to view, with business being conducted in a transparent nature.[44] The Pension Bureau employed 600 clerks in the building. Despite a range of shortcomings, the

Pension Building and contemporary structures reflected rapidly evolving corporate thinking as much as any factory floor.

The impact of the organisation of railroad and government on the emergent corporate office organisations was noted by Sampson. The new buildings and corporations looked inhuman and anonymous, "*but their first character was established by individual entrepreneurs who understood how to build up new kinds of organisations*" which themselves were modelled on railroads and the US Army. Within these organisations they "*constructed hierarchies of engineers, middle-managers and accountants to connect up the complex systems, allocating resources and co-ordinating distribution and selling.*"[45]

The impact of private corporations. In the early-nineteenth century, the United States was largely agricultural and mercantile and overall remained smaller than the leading European economies. All this changed rapidly in the latter half with growth spreading rapidly across the continent, spurred on by the increasingly dense railroad network. Chandler notes that before the mid-nineteenth century "*there were no multifunctional, multinational, and multiproduct enterprises administered through managerial hierarchies*", and that these first "*appeared during the 1850s and 1860s to co-ordinate the movements of trains and flow of goods over the new railroad networks and messages over the new telegraph system*" before being used "*to manage the new mass retailing establishments – the department stores, mail order houses, and chains or multiple shops-whose existence the railroad and the telegraph made possible*".[46]

The rise of American corporations saw its physical manifestation in the rise of tall buildings, and then the iconic skyscraper. Figure 7.2 illustrates one of the earliest of the tall buildings, the Equitable Life Assurance Building at 120 Broadway, opened in 1870. The contrast with anything built in Europe by this time – in terms of height and bulk – is stark.

The 1880s and 1890s saw the transformation of the US economy: it came to be dominated by large, highly capitalised, complex 'vertically and horizontally integrated' businesses, which gained huge economies of scale by stripping out the add-on costs of intermediaries, brokers, merchants and distributors. Davies notes that the small and highly competitive firms that had dominated production in the antebellum United States gave way to giant corporations integrated vertically and horizontally in the merger movement that swept through industry during the 1890s.[47]

By the onset of the First World War: "*it had the largest industrial sector in the world … . The economy in toto produced more than the economies of the three chief belligerents in World War I taken together*". By this time it also had among the highest levels of per capita income in the world.[48]

The shift from owner-managed to corporate structure was rapid, profound and transformative. "*By the end of the century the American business world of fifty years earlier was almost unrecognisable.*"

> *Before long the masters of these new forces were not so much the individual financiers and inventors who had first unleashed them, as the emerging company men who knew how to operate inside them, and to impose their own will over the new structures.*[49]

Lowe expanded upon this description of the role of offices at the time.

> *… the formalised procedures, hierarchical chain-of command, and specialized division of labour of bureaucracies were supposed to inject efficiency into organizational life*" and that

Figure 7.2 Equitable Life Assurance Building, New York, 1870.

> they did this with an increased focus on internal operations. As firms increased in size and complexity, "*office clerks processed a rising tide of data on production, costs, personnel, and internal communications*".[50]

Vertical integration brought under single control the various stages of the supply chains (production, sales and distribution); while horizontal integration involved diversification into related areas business. This led to the spread of large, integrated industrial corporations that sought control of as much of the production and distribution process as possible in order to impose greater efficiency.

The resulting large, complex, multilayered, multidepartmental organisations would come to dominate the twentieth century office economy. Most importantly perhaps, for the first time, office functions separated from production and factories and occupied buildings dedicated to management and administration. Firms engaged in food, oil and steel were the first to transform, followed by electricity distribution, chemicals and motor manufacturers. Soon, these early pioneers of office buildings were to be joined by the professional services firms described above.

This was the dawn of faceless bureaucracy, the corporation. It was here that the template was laid down for hierarchy, departmentalisation, communication systems and internal processes – essential features not visible on a more impactful sepia photo of a typing pool. Various levels of managers were established, clear lines of authority were devised, and formal rules were created to govern the company's operations. The managerial revolution helped to create a 'new' middle class. Unlike the older middle class, which consisted of farmers, shopkeepers, and independent professionals, the new middle class was made up of white-collar managers in the corporations.

It was from this firmament of change and upheaval that the modern corporation literally took shape, with its complex structures, hierarchies, internal systems and processes. But the rise of the corporation carried more than economic and organisational significance: its impact on wider society was also profound.

> *The growth of corporate capitalism and the dynamic organizational revolution that swept the American industrial belt in the few decades preceding the First World War profoundly and irreversibly altered such areas as work culture, education, family life, and consumption.*[51]

The number of large bureaucratic organisations mushroomed as increasing specialisation in work led to the rapid expansion of workers to manage, supervise, co-ordinate and administer the new routines of colleagues. Mechanisation and bureaucratisation went hand in hand, with the former allowing firms to reach new heights of scale, which, in turn, led to ever-greater organisational distance from owner to customer – a gap filled by bureaucrats.

A central factor of bureaucratisation of corporations was their sub-division and organisation into departments. As previously noted, this process was necessitated by the growing size and complexity of businesses, when it was no longer possible for one capitalist, or even a small group, to make all the decisions.[52] Davies observed that before the Civil War there had been four basic clerical jobs in the office: copyist, bookkeeper, messenger or office boy, and clerk. This relatively simple range was expanded and elaborated following the war, with the division of labour most pronounced in the largest offices. File clerks, shipping clerks, billing clerks and other 'semi-skilled' workers began to appear.[53]

The workplace rapidly ossified into a rigid (social) hierarchy, graded according to the authority granted to each level to plan and execute their own work while planning and supervising the work of others who did not plan work or make decisions, but delivered assigned work. The template survived for over a century. In 1997, Frank Duffy observed that the office in its present form

> exists directly as a consequence of the managerial *changes that occurred in the western world at the end of the nineteenth century when it became both possible and necessary to exercise control over manufacturing and distribution through accumulating and manipulating large amounts of information. More and more people became necessary to carry out what were inherently very dull and repetitive tasks, largely paper-based … . The top-down style of management associated with organising such routine tasks was never associated with sensitivity or imagination.*[54]

The growth in office workers during the late-nineteenth century was accompanied by a step change in investment in machinery to improve efficiency and productivity. Innovations such as typewriters, bookkeeping machines and adding machines all spread with remarkable speed. All of which enabled the expansion of businesses and corporatism.

From the perspective of the twenty-first century it is difficult to appreciate the socio-cultural impact of the emergence of large, faceless organisations that seemed to operate to an alien set of rules. The economy was being radically reorganised, and Woodrow Wilson captured the profundity of the situation memorably: "*We are in the presence of a new organization of society … . We have changed our economic conditions, absolutely, from top to bottom; and, with our economic society, the organization of our life.*" Furthermore, he argued, the truth is

> *we are all caught in a great economic system which is heartless. The modern corporation is not engaged in business as an individual. When we deal with it, we deal with an impersonal element, an immaterial piece of society. A modern corporation is a means of co-operation in the conduct of an enterprise which is so big that no one man can conduct it, and which the resources of no one man are sufficient to finance.*[55]

Clerks as parasites. In terms of social impact, the emergence of white-collar workers was regarded with more suspicion in the US than it was in Europe. While in the latter clerking grew gradually as a respectable occupation for gentlemen, before bringing cleaner and safer work not only to men but also women, the American experience was much more sudden and altogether different. Saval has recently suggested that clerical work failed to conform to how most Americans thought about themselves and their culture.

> *Clerks didn't work the land, lay railroad tracks, make ammunitions in factories, let alone hide away in a cabin by a small pond to raise beans and live deep. Unlike farming or factory work, office work didn't produce anything. At best, it seemed to reproduce things. Clerks copied endlessly bookkeepers added up numbers to create more numbers, and insurance men literally made more paper … . He was a parasite on the work of others, who literally did the heavy lifting.*[56]

In the early stages of the American office, many office functions were the separated admin, marketing and support functions of manufacturers, railroad companies and so on rather than 'pure offices', such as bankers and accountants. In this sense, their growth and spread raised questions about social structures:

> *offices became highly ambiguous spaces in the fast-developing world of American capitalism. Were clerks part of the growing industrial working class, replacing the artisans and small farmers of the old-world economy? Or were they merely stopping points on the way to becoming part of the 'ruling class'? The answer appears to be that they were somewhere uncomfortably in between: not 'middle class' exactly, or not yet – the phrase was never used, and the concept hadn't yet sprung up among nineteenth century Americans.*[57]

The reaction to the growth of clerks in America was beautifully captured by Zakim[58] who described the tens of thousands who got their living in one way or another by penmanship as being

> *the most overt expression of the age's new emphasis on buying and selling. Whether at work in counting rooms, credit agencies, trust companies, law offices, banks, insurance brokerages,*

auction firms, import houses, or commission businesses ... it was the clerk who effectively administered all the new markets in the new market society.[59]

He went on to argue that the clerk had become a "*visceral means for Americans to talk about the revolution that was overtaking – and undermining – the nation's traditional ethic of industry*", and that the clerk

played an instrumental role in the social redefinition of an economy in which property ceased to function as the static foundation of agrarian life and became the fungible object of commodity exchange. What once anchored a family competence, in other words, was now the stuff of business initiative.[60]

Zakim then went on to describe how the "*clerk's new industrial labor violated another common truth, for as hard as he worked he did not produce anything. The clerk had no actual product to show for all his long hours on the job*". Furthermore, his "*strenuous idleness*" placed the clerk "*outside an Anglo-American tradition that identified labor as the source of value. The clerk was, in this respect, quite worthless. And yet, strangely enough, his labor proved to be essential to industrial growth*".[61]

As we have seen, the early stages of the development of corporations involved a move away from mainly owner-managed businesses, in which an entrepreneur, sometimes a partnership, oversaw the entirety of a business' operations supported by a small group of clerks. This then yielded to larger businesses with departmentalised structures and multiple layers of responsibility. The previously simple structure of clerk, copyist, bookkeeper and office boy gave way to a far more elaborate array of roles and tasks. Different kinds of clerks (billing, shipping, filing) and different job requirements (skilled, semi-skilled, unskilled) were overseen by a variety of managers (general, operational, admin, etc.). Use of office technology could also designate roles: typists, copyists, stenographers and so on.

The office economy was experiencing – as had the factories earlier – a growing division of labour and the organisation of the work itself into a series of discrete work flows. Individual workers were active within a smaller and smaller slice of their organisations' operations as they became more specialised. The increasing division of labour reinforced the usefulness and necessity of hierarchical structures. A clerical worker who executed and understood only a small section of a firm's operations would not have the knowledge necessary to make a decision about problems that were not in his or her immediate purview.[62]

The age of the professional manager had arrived. Managers typically had little or no ownership of the business, but they took *de facto* control of the business by reducing the role and influence of the owner.

As part of this process the new companies established offices from which to centrally manage, direct and administer operational functions. The template laid down by the railroad companies was duplicated, modified and developed by countless banks, insurance companies, manufacturers and retailers. This was the dawn of 'corporate' offices, with their armies of similarly dressed workers, undertaking largely repetitive, processed-based work in highly divisionalised, layered companies, that were to dominate the twentieth century office economy. The new businesses practiced centralised command and control management, and there was a consequential rise in white-collar work as the emerging behemoths generated a demand for process management, administration, purchasing, marketing, sales, accounting and so on.[63]

It is often stated that offices inherited their early layout from factories, with serried rows of typists often shown arranged on a floor akin to a production line. But the truth is more nuanced than this simple interpretation, because the template for the US office was established before mass factory production lines; and both were more influenced by the structure of railroad companies and the federal government.

The speed and scale of change in America was in stark contrast to the European experience. Payne points out, for example, that comparisons between the extent of merger activity in terms of the number of consolidations, the capitalisations of the amalgamated firms, and the degree of market control achieved, show that merger activity was on a much greater scale in the United States. In seeking to account for this, Payne concedes that the reasons are complex, but postulates that the answer "*is bound up with Britain's early start, the size and character of the market for British goods*", which seem logical; but he then adds, significantly, the resource and technological bases of British industry (which were lower); a greater pride in self-sufficiency of the British industrialist (which could be read as a nod to tradition) and the fact that British entrepreneurs could utilise methods of market control closed to their American counterparts (means of practicing closed shops).[64] In other words, there were distinct cultural differences in approach.

The European experience

Thomas Burke (1886–1945) was an author who wrote extensively – both fiction and non-fiction – about London. Following a tough upbringing, when Burke was 16, he started work as an office boy – a job he is widely reported to have detested. His first successful publication was *Limehouse Nights* (1916), a collection of stories centered around life in the poverty-stricken Limehouse district of London. His *London in My Time* (1934) begins with a reflection on the celebrations for Queen Victoria's Diamond Anniversary in 1897.

> *The roofs and windows of London are rippling with red-white-and-blue; even the poorest dwelling shows its three-hap'ny flag. Every street-organ is playing and every boy whistling, Leslie Stuart's Soldiers of the Queen … . London is celebrating the Record Reign and sixty years of what it thought was Progress, never guessing that more progress was to be packed into the next thirty years than the whole previous hundred years could show.*

Despite the 60 years of 'progress', it was a London

> *that still held many of the fixtures and much of the atmosphere of what has come to be known as the Dickens' London. A London of horse-trams with halfpenny fares, and of hansom cabs; of crystalline bells and spattering hoofs. A London with winters of slush and fog of a richer sort than any known to-day, and summers of dust and clam; the slush and dust being its heritage from the horse-traffic.*[65]

And it was a London where

> *There were still numbers of dens below ground in which ill-paid clerks worked for small firms under all-day artificial light. There were dusty garrets up four pairs of stairs in which six or seven people worked together … . The general impression left with me, after the short time I spent there, was of fustiness, superannuated gloom, and cramp; plus, of course, the tight-lipped commercial spirit which took no account of these things so long as it could do business.*

Wherever one looked the scene was colourless. The buildings had a tone peculiar to the City, for which there is no name. It wasn't black or brown or dun or drab. It was not so definite as the tone of mud or the tone of cobwebs or manure. It was such a tone as you might get from wet smoke mixed with army blanket fluff and engine smuts.[66]

While everything seems to change, on another level things change little. Two examples serve to show this blending of change and continuity that so characterised the European experience: Child & Co and NM Rothschild.

In 1973, Philip Clarke published *The First House in the City,* to mark the 300th anniversary of banking by Child & Co at 1 Fleet Street (Figure 7.3).[67] In its tracing of the bank's business

Figure 7.3 Child & Co, 1 Fleet Street.

ups and downs and of the family at the helm of the business, the book also touches upon the premises occupied by the bank.

Sir Francis Child (1642–1713) oversaw "*the crucial period of change from goldsmithing to pure banking, a change and an expansion of techniques which took place, for those days, over a relatively short space of time*".[68] Between 1673 and 1879, the bank prospered and its premises expanded incrementally, with

> *programmes of alteration and rebuilding following in the wake of purchases. Thus today Child's occupies land upon which once stood a tavern called the Sugar Loaf and Green Lettuce and the Cock Alehouse; the Olde Devil Tavern, home of the Apollo club.*[69]

When its Temple Bar building (shown earlier in Figure 6.1) was compromised by nearby excavation the opportunity was taken to build new premises. Work began in November 1879 and the new premises on Fleet Street opened for business on 23 August 1880, since when

> *the front of the building has remained unchanged although the interior was redesigned in 1960. Upper rooms, formerly the bedrooms of the clerks and servants, were converted into offices* ... .".
>
> "*The clerks' old dining room is now the manager's dining room. At the front of the house the Long Room on the first floor, which, with its windows looking onto Fleet Street, was once used for sedate board meetings, has been given over to the hum and clatter of data processing machines.*"

Burke describes the underlying continuity of London despite the superficial change; and Clarke does the same for a single bank and building. Both serve as introductions to the powerful influence of culture and tradition in our understanding of the built environment and the workplace. The following brief synopsis of Rothschild Bank's changing premises underscores this cultural influence.

Nathan Mayer Rothschild (1777–1836) moved to Manchester from his native Frankfurt in 1799. He established a business, Rothschild Brothers, as a commission agent on his father's account, dealing primarily in textiles. Over the following decade, he revolutionised the burgeoning textile trade through his use of ready cash to buy up large quantities of materials and his reliance on a network of family and friends to ship Manchester goods throughout Europe. He built a strong reputation as a successful merchant operating on the principle of high turnover and modest profits, taking advantage of the opportunity to bypass the middle men and purchasing textiles direct from the manufacturers, quickly turning the £20,000 with which he arrived into £60,000.

He was soon trading on his own account. Items in the Manchester warehouse, listed in his Stock Book of 1805 include arungos, baftaes, bandanas, books (probably of material), beads (mock pearl), cambrics, corals (red), cane table mats, cornelian beads, dimities, emeralds, white flush cambric handkerchiefs, jacconets, ink (china), lappett sprigs, muslins, plain buff nankeens, pullicates, pearls, pockets, shawls, thicksetts, velveteens, vandykes, wine.

David Lockwood argued that even as late as 1958, the majority of English offices, had not been greatly affected by the rationalising trends in office management. He suggested that it is "*extremely easy, and very common, to overestimate the degree to which this process has occurred*", and that in most cases, "*the size of the office and the nature of clerical work set narrow limits to the application of production line methods*". Further, only government departments and central statistical offices of large private firms could afford the introduction of complicated statistical and accounting machinery, such as for example the Hollerith punch card machine (see Chapter 9).

As we shall see in Chapter 12, post-war European offices experimented with different types of layout, but none of these could be described as Taylorist (or Fordist as some prefer) white-collar production lines. Sampson found parallels with the US railroad in Britain. He observed that the hectic growth of the railways in Britain called for rapid and complex new organisations with a sense of punctuality and discipline, which were quite foreign to most of the country.[70] However, and tellingly, he also suggested that the precedent for the railway companies on this side of the Atlantic was the army, and so the railways "*naturally devised uniforms for their porters and other workers ... . For their managers they turned to retired army officers and majors and captains ran many of the chief companies*".[71]

Consequently many European organisations were structured like the army, built-up from the bottom, comprising ever greater units of command: section, platoon, company, regiment, battalion, brigade, each run by an appropriate rank, with deference and discipline holding the whole together. The development of office organisational structures on this side of the Atlantic was thus influenced far more by culture and social structure than in the US. Here we sought to shoehorn age-old cultural structures into the rapidly changing economic structures and fairly unchanging physical structures: classic European muddling through. Not so in America. The integration of the US economy happened with very little historical baggage: new organisations and new buildings for a new age.

We have already covered many aspects of the cultural influence on the European office. The rise of private banking, catering for the needs of the aristocracy and wealthy merchants, in converted houses where clerks and managers lived 'above the shop', together with the faltering start of joint stock banks that were viewed with much suspicion. Sampson captured the essence of the situation: while the "*earlier social structures – the country estates, the regiment or the cathedral close – were beginning to fade ... the old patterns were re-emerging in new disguises within company headquarters*", where

> *the British style of country-house life was recreated in the new company buildings which were often designed to look like the mansions of the aristocracy. The great hall became the front office, the grand staircase became the directors' lift, the footmen became chauffeurs; while the sub-classes of housekeepers, nannies, governesses, companions or housemaids were recreated in the new layers of office managers, personnel officers, private secretaries or tea ladies.*[72]

The fact that many mid-nineteenth century European offices shared characteristics with homes (as we saw in Chapter 6) was little surprise because many were in fact homes – converted for dual use as domestic quarters and business shop. Either the owner and family or manager lived above the shop or, later, the clerks did so. And the office spaces themselves were decorated with tables, chairs, carpets, lighting and curtains that would all be familiar in an affluent town house.

Even after they passed through the owner-manager (or aristocrat-artisan) phase, many European firms continued to operate as family businesses, with owner-directors overseeing salaried staff within corporate structures (comprising officers, non-commissioned officers and other ranks). Alternatively, a partnership structure was established but with the same effect: the owners of the business, the partners, oversaw all matters of import, supported by salaried staff. In both cases – companies and partnerships – the socio-cultural background of the owners transferred to the management culture and layout of the offices themselves.

Apprentices, junior clerks, senior clerks and principal clerks described layers of skill and expertise while discipline extended to life outside of the workplace as well as within. Young clerks were treated like raw recruits – often with suspicion and a premeditated sense that they needed to be controlled. This aspect was captured beautifully by de Fraine in his memoirs of life at the Bank of England – starting there as a 16 year old in 1886. He describes his Principal as having "*cultivated all the qualities of a Squeers*" (Wackford *Squeers was a* cruel, one-eyed, Yorkshire schoolmaster and head of Dotheboys Hall boarding school for unwanted children in Dickens' Nicholas Nickleby).

> *To begin with, he would allow no talking, and most certainly no laughing during work; and he timed visits to the WC. `Unlike other Principals ... he preferred to pussy-foot about the office all day long, watching and listening.*
>
> *I can hardly exaggerate the nervousness with which one sensed his soundless approach, or the relief when one saw his extraordinarily shining bald head bobbing up and down in a distant part of the office. Any lapse would bring down a steam-hammer harangue to crush what was usually a very small nut.*

In just this one example, one can read so much into British social history: layers or hierarchies; self and imposed discipline; formality and impersonal relationships; the primacy of the group over the individual. De Fraine captures deep social structures, which ran straight through the emerging office. Centuries of social stratification and deference; long cultural history and deeply embedded aristocratic and military traditions meant that Europeans were less ready than the Americans to accept the values and styles of the new corporate businessmen. European socio-cultural history led to a slower, more guarded acceptance, and meant that handing responsibility for running large commercial enterprises over to a novel professional class of manager would always be more problematic.

Cultural and rational traditions

The fact that European offices had been evolving for over a century before American offices really got going, led to European and American offices evolving in very different ways. The result was that, by the dawn of the new century, the European office – its physical features and corporate structures, as well as the relationships between those who worked in these early modern office buildings – was a world away from the burgeoning office blocks in the US. The differences would characterise European and American buildings through to the late-twentieth century. One way to capture these differences is to describe the European model as the cultural office (the outcome of deep social and cultural influences), and the American model as the rational office (the outcome of modern industrial thinking). This contrast in styles is summarised in Table 7.2 and discussed further in Chapters 10 and 12.

Table7.2 summarises aspects of the economy, urban geography, the real estate supply process, business structure and buildings, and how these influenced the development of the cultural office in Europe and the rational office in the US. An important point to make at the outset is that Europe was not homogenous in terms of the office economy and London in particular became something of a 'cultural-rational hybrid'. Nevertheless, the model is a helpful tool in highlighting certain key traits which help explain differences in the design, occupation and management of office workplaces. For a start, the model suggests that scientific management (see Chapter 10) was not the single most important factor determining modern office workplace designs and cultures that is generally accepted.

Table 7.2 The cultural and rational traditions

Cultural office (Europe)	*Rational office (North America)*
Economy	
Colonial commercialism	Frontier entrepreneurialism
Localised markets	Enormous scale
Slow, historic evolution	Rapid vertical integration
Customised products	Standardised products
Stakeholder capitalism	Shareholder capitalism
Urban setting	
Medieval street pattern	Grid iron layout
Anti-tall building	Pro-tall building
Small plot size; complex ownership	Large plot size; simple ownership
Mixed-use centres	Central Business District
Limited, short commute	Widespread, long commute
Supply market	
Tight planning & building regulation	Laissez faire planning
More occupier-driven (except London)	More developer-driven
Traditional	Innovative
One-off design/construction	Repetitive design/construction
Business	
Typically small scale	Typically huge scale
Paternal/individualistic culture	Corporate/group culture
Worker involvement	C-suite control
More social/class status	More bureaucracy
Deference culture	Conformity culture
Buildings	
Low rise	High rise
Small, irregular floor plan	Large, contiguous floor plan
Low density; inefficient layout	High density; efficient layout
Preference for sub-divided space	Preference for open plan
Domestic in character	Factory in character

The changes that took place in the office around the turn of the century were perhaps more momentous, more transformative than any stage in the four centuries covered by this book. They saw the change from 'old' to 'modern'; the emergence of the corporation; the arrival of the office as a distinct facet of urban growth; the transformation of work from skilled craft to office work; the extinction of the office clerk and the arrival of the office worker; the mechanisation of the office and the feminisation of the office workforce.

Office work prior to 1900 involved generally small businesses employing just a few clerks. In a well-staffed office there might be a clerk and an office boy, together with a copyist and a bookkeeper. The clerk oversaw everything and had a detailed grasp of the workings of the whole business. Clerking was a job for gentlemen, relatively well paid, and many clerks went on to become directors or even owners of businesses. The workforce was entirely male. The working conditions were generally poor (though by no means like factories). Employer–employee relations were typically close, sometimes paternal. But the environment was a respectful one, although sometimes the owner's personal traits and temperament made life less pleasant for the clerk.

After around 1900, there was a dramatic increase both in the volume of work and in the demand for clerical labour. As the capitalist economy grew and evolved, so businesses became larger, more complex and unwieldy. The workforce was feminised and the workplace mechanised. The path was one towards multilayered, heavily departmentalised corporations, where work was broken down into ever smaller areas of responsibility. In the early years, this simply meant a bifurcation of clerical work into skilled clerking – management, bookkeeping and secretarial – and lower-level tasks such as stenography, copying, printing and so on. The first group retained an element of control over their work and were granted the right to some decision-making; the second group, as Braverman might have said, were not expected to participate in the conception of their work.

There was a clear shift in the nature of office work around the turn of the century: the Edwardian office and the work within it, had a quite different feel to that of its Victorian predecessor. In the next chapter, we discuss the metamorphosis of the all-round clerk into general office worker and the feminisation of the workforce, and in Chapter 9 we address the mechanisation of the office. But before that, the office was also changing in terms of how its utility, or functionality, was perceived: new business structures, new work processes and new approaches to business management led directly to new thinking on how offices should be designed, laid out and managed. Greater density of people and the encroachment of machines led to new approaches to ventilation, heating and cooling. Similarly larger floorplates led to innovation in lighting. All of these issues fed through to an evolving building design.

Locker observes that with the introduction of lighting and heating "*the modern era for office services began in the years leading up to the First World War. Such major change in so short a period is extraordinary.*" He goes on to suggest that the changes were encouraged "*by the overwhelming rationale of scientific office management, which in itself was producing instrumental changes in office building planning*". With the concurrent changes in building structure and the fundamental changes in style, the years leading up to the war were almost revolutionary.[73]

As we shall see in Chapter 10, in the early years of the twentieth century, 'scientific management' became popular in the USA as well as a growth in business-focussed publications seeking to offer advice on management matters. Furniture and equipment providers also multiplied, along with a plethora of magazines marketing the latest fashions and innovations in each. One of the earliest and most comprehensive office management books, *Office Organisation and Management*, was published in 1906 by Lawrence Dicksee and Herbert Blaine. The book highlighted the transition from Victorian to Edwardian thinking. Noting the commercial invasion of Great Britain by 'our American cousins' it argued

> *The day has gone past never to return when great commercial undertakings could afford to have their offices conducted on the slow, if steady and reliable, lines of old, and it is now realised that the office is to a business what the main spring is to a watch.*[74]

Locker describes office records as having been regarded primarily as legal documents "*providing rule of thumb guides for business decisions*" and as a consequence, "*clerks were often regarded as necessary, but non-productive; their roles within the office were often poorly defined. By the Edwardian years major changes in office work were underway*". The new approach to the office – whether scientific or not need not detain us here – changed the very essence of the office function,

> *from mere record keeping to accounting, overseeing, predicting production needs and market demand, and providing data for business decisions. Office tasks were no longer 'non-productive', but instead were recognised as integral components of business.*[75]

Dicksee and Blaine's description of an ideal office became a prescription for the early-twentieth century office. For ease of supervision, they suggested one large room for clerks, especially in larger organisations. Separate rooms would be necessary only for the manager and the chief clerk. The latter's office was to be raised above the general office, with glass walls for easy supervision of the clerks. Natural light, proper ventilation and sanitation were stressed. Telephones were to be in a separate room or box to avoid messages 'leaking out'. Typewriters, for acoustical reasons were required to have their own isolated room. The general office, whilst one space, was to be sub-divided into departments by railings, counters, or low partitions, with books, documents and 'spheres of influence' remaining within each. They decreed that office furniture should be simple and dust free, co-ordinated with the office supplies and equipment then gaining general use. Desks were no longer to be pigeon-holed roll top, but sloped-top and spindle-legged, with few drawers. Supplies were kept in cupboards, not desks. The traditional ledger system was replaced by loose leaf books and card files.[76]

The large open office remained the single most important characteristic of early twentieth century office management techniques. The traditional pattern of small rooms, containing very small groups of workers, amongst whom there was some rapport and sense of teamwork gave way, especially in America, to vast, cavernous spaces where perhaps hundreds of people were organised in serried rows of desks, rather like a production line in a factory. The central open area was certainly the favoured layout solution, due to its efficiency, which was seen to diminish once space began to be sub-divided due the enhanced need for additional floorspace, ventilation, lighting and heating, not to mention reduced communication. The factory layout, or school room layout in smaller offices, reinforced the diminution in the status of office workers compared to their clerking forebears. They had become 'units of production' rather than respected and trusted employees.

Just as the counting house of the eighteenth century yielded to the grandiose and ornate buildings preferred by the brash new joint stock companies of the nineteenth, so these were giving way to the efficient buildings of the twentieth. Organised like factories, with large open plan areas, separation of 'doing and management' and strict time keeping, offices came to resemble white-collar factories. Management became a dominant force in the organisation of resources and physical space in office buildings. Indeed, the growth and spread of management from the early years of the new century set a template for a further century. In his classic, mid-century tome, *The Practice of Management*, management guru Peter Drucker was effusive in his appraisal of the importance of management:

> *The emergence of management as an essential, a distinct and a leading institution is a pivotal event in social history. Rarely, if ever, has a new basic institution, a new leading group, emerged as fast as has management since the turn of the century. Rarely in human history has a new institution proven indispensable so quickly…*[77]

Summary. In the later years of the nineteenth century, the office building emerged in its modern form: large, single-use structures, dedicated to management and administration, reaching its apogee from the 1870s with the skyscrapers of the USA. The period also saw a fundamental shift in the office economy with the rise of the corporation. While joint stock companies introduced corporate structures, including shareholdings and professional managers, the rise of corporations took this to a new level of complexity. The massive economic expansion of the Victorian period culminated in the establishment of the office economy as a distinctive feature of the urban landscape.

It was of course, in the United States more than anywhere else that the modern global corporation evolved and alongside it, of course, the enormous filing case in the sky – the skyscraper. Ever larger clerical workforces gathered daily in these mighty structures to benefit from cheap and reliable electricity, elevators, lighting and heating.

And in both Europe and America, burgeoning clerical workforces experienced rapidly developing workplace equipment and technologies. Desks, chairs, table lamps, meeting tables and filing cabinets all evolved; adding machines, copying machines and typewriters provided a mechanical dimension; telegraphs and telephones added a sense of the exotic. And then there was stationery: ball point pens, carbon paper, card indexes, envelopes, loose clip files, paper, paperclips, pencils, notebooks and waste baskets – all manner of equipment and stationery to transform office work.

This was the dawn of the faceless bureaucracy, the corporation, when the template was laid down for hierarchy, departmentalisation, communication systems and internal processes. The workplace rapidly ossified into a rigid (social) hierarchy, graded according to the authority granted to each level to plan and execute work. The managerial revolution helped to create a 'new' middle class. Unlike the older middle class, which consisted of farmers, shopkeepers and independent professionals, the new middle class was made up of white-collar managers in the corporations.

The fact that European offices had been evolving for over a century before American offices really got going, led to European and American offices evolving in very different ways. The result was that, by the dawn of the new century, the European office – its physical features and corporate structures, as well as the relationships between those who worked in these early modern office buildings – was a world away from the burgeoning office blocks in the US. The differences would characterise European and American buildings through to the late-twentieth century. One way to capture these differences is to describe the European model as the cultural office (the outcome of deep social and cultural influences), and the American model as the rational office (the outcome of modern industrial thinking).

Notes

1 Dunning JH & Morgan EV (1971) *An Economic Study of the City of London* Economists Advisory Group George Allen & Unwin, London p33.
2 Dunning JH & Morgan EV (1971) *An Economic Study of the City of London* Economists Advisory Group George Allen & Unwin, London p33.
3 Powell C (1996) *The British Building Industry Since 1800: An Economic History* (2nd Edition) E&FN Spon, London p14.
4 Kynaston D (1994) *The City of London Volume One: A World of its Own 1815–1890* Chatto & Windus, London p288.
5 Harris R (2021) *From Clerical Factory to Digital Hub: London's Global Office Economy* Routledge, London p46.
6 Michie RC (1992) *The City of London: Continuity and Change, 1850–1990* Macmillan, London p39.
7 Rose H (1994) *London as an International Financial Centre: A Narrative History* The City Research Project, Corporation of London p21.
8 Rose H (1994) *London as an International Financial Centre: A Narrative History* The City Research Project, Corporation of London p23.
9 Rose H (1994) *London as an International Financial Centre: A Narrative History* The City Research Project, Corporation of London p22.
10 Michie RC (1992) *The City of London: Continuity and Change, 1850–1990* Macmillan, London p39.
11 Harris R (2021) *From Clerical Factory to Digital Hub: London's Global Office Economy* Routledge, London p155.

12 Michie RC (1992) *The City of London: Continuity and Change, 1850–1990* Macmillan, London p13.
13 Orchard BG (1871) *The Clerks of Liverpool: Being Ten Chapters on their Numbers, Distribution, Salaries, Grievances, Marriages, etc.* pp9–10.
14 Heller M (2003) *London Clerical Workers 1880–1914: the Search for Stability* PHD Thesis University College London pp52.
15 Alborn T (2008) Quill-driving: British Life-Insurance Clerks and Occupational Mobility, 1890–1914 *British History Review* Vol 82 Spring 2008 pp31–58.
16 Black A; Muddiman D & Plant H (2007) *The Early Information Society: Information Management in Britain before the Computer* Ashgate, Aldershot pp136–137.
17 Anderson G (1976) *Victorian Clerks* Manchester University Press p10.
18 Heller M (2003) *London Clerical Workers 1880–1914: the Search for Stability* PHD Thesis University College London pp53.
19 Rose H (1994) *London as an International Financial Centre: A Narrative History* The City Research Project, Corporation of London p6.
20 Braverman H (1974) *Labor and Monopoly Capital: the Degradation of Work in the Twentieth Century* Twenty Fifth Anniversary Edition, Monthly Review Press, New York p207.
21 Evans DM (1865) *The City, or The Physiology of London Business with Sketches on 'Change, and at the Coffee Houses* Baily Brothers, London pV.
22 Braverman H (1974) *Labor and Monopoly Capital: the Degradation of Work in the Twentieth Century* Twenty Fifth Anniversary Edition, Monthly Review Press, New York p204.
23 Braverman H (1974) *Labor and Monopoly Capital: the Degradation of Work in the Twentieth Century* Twenty Fifth Anniversary Edition, Monthly Review Press, New York p208.
24 Cowan P; Fine D; Ireland J; Jordan C; Mercer D & Sears A (1969) *The Office: A Facet of Urban Growth* Heinemann Educational Books, London p28.
25 Lisle Williams M (1984) Merchant Banking Dynasties in the English Class Structure: Ownership, Solidarity and Kinship in the City of London, 1850–1960 *The British Journal of Sociology* Vol 35 No 3 September 1984 pp333–362.
26 Evans DM (1865) *The City, or The Physiology of London Business with Sketches on 'Change, and at the Coffee Houses* Baily Brothers, London p8.
27 Clapham JH (1938) *An Economic History of Modern Britain: Machines and National Rivalries 1887–1914* Cambridge University Press p278.
28 Thomas Salt (1891) *Journal of the Inst of Bankers* Dec 1891 Cited in Clapham JH (1938) *An Economic History of Modern Britain: Machines and National Rivalries 1887–1914* p280.
29 Clapham JH (1938) *An Economic History of Modern Britain: Machines and National Rivalries 1887–1914* pp279–282.
30 Cassis Y (1985) Bankers in English Society in the Late Nineteenth Century *Economic History Review* Vol 38 No 2 pp210–229.
31 Haueter NV (2017) *Swiss Re A History of UK Insurance* Private publication.
32 Michie RC (1992) *The City of London: Continuity and Change, 1850–1990* Macmillan, London p176.
33 Michie RC (1992) *The City of London: Continuity and Change, 1850–1990* Macmillan, London p177.
34 Michie RC (1992) *The City of London: Continuity and Change, 1850–1990* Macmillan, London p182.
35 Neal M & Morgan J (2000) The Professionalization of Everyone? A Comparative Study of the Development of the Professions in the United Kingdom and Germany *European Sociological Review* Vol 16 No1 pp9–26.
36 Harris R (2021) *From Clerical Factory to Digital Hub: London's Global Office Economy* Routledge, London p325.
37 Coffey WJ (2000) The Geographies of Producer Services *Urban Studies* Vol 21 No2 pp170–183.
38 Lindahl DP & Beyers WB (1999) The Creation of Competitive Advantage by Producer Services Establishments *Economic Geography* Vol 75 pp1–20.
39 Chandler AD (1965) The Railroads: Pioneers in Modern Corporate Management *The Business History Review* Vol 39 No 1 pp16–40 Special Transportation Issue p17.
40 Chandler AD (1965) The Railroads: Pioneers in Modern Corporate Management *The Business History Review* Vol 39 No 1 pp16–40 Special Transportation Issue p17.

41 Chandler AD (1965) The Railroads: Pioneers in Modern Corporate Management *The Business History Review* Vol 39 No 1 pp16–40 Special Transportation Issue p17.
42 Chandler AD (1965) The Railroads: Pioneers in Modern Corporate Management *The Business History Review* Vol 39 No 1 pp16–40 Special Transportation Issue p17.
43 Stephens C & Lubar S (1986) A Place for Public Business: the Material Culture of the Nineteenth Century Federal Office *Business & Economic History* Second Series Vol 15 pp165–179.
44 Stephens C & Lubar S (1986) A Place for Public Business: the Material Culture of the Nineteenth Century Federal Office *Business & Economic History* Second Series Vol 15.
45 Sampson A (1995) *Company Man: the Rise and Fall of Corporate Life* HarperCollins, London p39.
46 Chandler AD (1986) The Beginnings of the Modern Industrial Corporation *Proceedings of the American Philosophical Society* December 1986 pp382–389.
47 Davies MW (1982) *Women's Place at the Typewriter* Temple University Press, Philadelphia p28.
48 Gallman R (2000) Economic Growth and Structural Change in the Long Nineteenth Century In: Engerman SL & Gallman R (2000) *The Cambridge Economic History of the United States* Cambridge University Press pp1–56.
49 Sampson A (1995) *Company Man: the Rise and Fall of Corporate Life* HarperCollins, London p31.
50 Lowe GS (1984) The Enormous File: the Evolution of the Modern Office in Early Twentieth-Century Canada *Archivaria* 19th January pp137–51.
51 Zunz O (1987) The Collar Line: Clerical Workers in America at the Turn of the Century In: Schröder WH & Jarausch KH (Eds.) *Quantitative History of Society and Economy: some International Studies* (pp79–93) St Katharinen: Scripta Mercaturae Verl.
52 Davies MW (1982) *Women's Place at the Typewriter* Temple University Press, Philadelphia p29.
53 Davies MW (1982) *Women's Place at the Typewriter* Temple University Press, Philadelphia p30.
54 Duffy F (1997) *The New Office* Conran Octopus Limited, London p14.
55 Wilson W (1913) *The New Freedom: A Call for Emancipation of the Generous Energies of a People* Doubleday, Page & Co., New York p10.
56 Saval N (2015) *Cubed: a Secret History of the Workplace* First Anchor Books, New York p14.
57 Saval N (2015) *Cubed: a Secret History of the Workplace* First Anchor Books, New York p27.
58 Zakim M (2006) The Business Clerk as Social Revolutionary; or a Labor History of the Nonproducing Classes *Journal of the Early Republic* Winter 2006 Vol 26 No 4 pp563–603.
59 Zakim M (2006) The Business Clerk as Social Revolutionary; or a Labor History of the Nonproducing Classes *Journal of the Early Republic* Winter 2006 Vol 26 No 4 pp563–603.
60 Zakim M (2006) The Business Clerk as Social Revolutionary; or a Labor History of the Nonproducing Classes *Journal of the Early Republic* Winter 2006 Vol 26 No 4 pp563–603.
61 Zakim M (2006) The Business Clerk as Social Revolutionary; or a Labor History of the Nonproducing Classes *Journal of the Early Republic* Winter 2006 Vol 26 No 4 pp563–603.
62 Davies MW (1982) *Women's Place at the Typewriter* Temple University Press, Philadelphia p39.
63 Shlaes J & Weiss MA (1993) Evolution of the Office Building In: White JR (1993) *The Office Building: From Concept to Investment Reality* Counsellors of Real Estate, Chicago p9.
64 Payne PL (1967) The Emergence of the Large-scale Company in Great Britain, 1870–1914 *The Economic History Review New Series* Vol 20 No 3 pp 519–542 December 1967.
65 Burke T (1934) *London in My Time* Rich & Cowan, London 1934 p1.
66 Burke T (1934) *London in My Time* Rich & Cowan, London 1934, pp231–232.
67 Clarke P (1973) *The First House in the City: an Excursion into the History of Child & Co to Mark Its 300th Year of Banking at the Same Address* Perivan/Williams Lea Group, Southend-on-Sea.
68 Clarke P (1973) *The First House in the City: an Excursion into the History of Child & Co to Mark Its 300th Year of Banking at the Same Address* Perivan/Williams Lea Group, Southend-on-Sea p27.
69 Clarke P (1973) *The First House in the City: an Excursion into the History of Child & Co to Mark Its 300th Year of Banking at the Same Address* Perivan/Williams Lea Group, Southend-on-Sea pp15–16.
70 Sampson A (1995) *Company Man: the Rise and Fall of Corporate Life* HarperCollins, London p24.
71 Sampson A (1995) *Company Man: the Rise and Fall of Corporate Life* HarperCollins, London p24.
72 Sampson A (1995) *Company Man: the Rise and Fall of Corporate Life* HarperCollins, London p62.

73 Locker FM (1984) *The Evolution of Victorian and Early Twentieth Century Office Buildings in Britain* University of Edinburgh PhD thesis p322.
74 Cited in Locker FM (1984) *The Evolution of Victorian and Early Twentieth Century Office Buildings in Britain* University of Edinburgh PhD thesis p227.
75 Locker FM (1984) *The Evolution of Victorian and Early Twentieth Century Office Buildings in Britain* University of Edinburgh PhD thesis p226.
76 Cited in Locker FM (1984) *The Evolution of Victorian and Early Twentieth Century Office Buildings in Britain* University of Edinburgh PhD thesis pp227–228.
77 Drucker PF (1954) *The Practice of Management: a Study of the Most Important Function in American Society* New York, Harper & Brothers p3.

8 Gentleman all-rounder to office worker

In later-nineteenth century Europe, white-collar work remained a minority activity. There were of course the capital-owning (landed and business) and professional middle classes (doctors, clergy, teachers), but salaried middle and lower middle class, white-collar workers were few in number. In 1851, there were just 95,000 clerks in Great Britain (fewer than 2% of whom were female).

But things were beginning to change. Industrialisation and rapid expansion in trade was leading to an unprecedented demand for experts in accounting, engineering, finance, insurance, law and surveying, as well as administrators to manage orders, sales, receipts, inventories, accounts and so on. Paper work was on the march, and with it the need for systems and processes for storing information; the transformation that led to the dominance of the office economy in the twentieth century was gathering pace. The number of clerks grew to 613,000 at the turn of the century, before doubling again to 1.3m by 1921 (39% of whom were female).[1] Women were also increasingly employed in commercial occupations; the social barriers to respectable female employment were weakening and the range of suitable jobs widening.[2]

The fact is that by the 1870s the world of clerking was evolving not so much into a distinctive, coherent profession, but into an extremely diffuse range of activities spread throughout the economy. On the one hand, productive industries, including manufacturing, which formed the beating heart of the economy and did so for another century, were drawing heavily on white-collar workers to resource their ever-expanding administrative functions. At this time, the offices associated with industrial concerns were larger than those of office-based organisations, and the office and office management, became central to the fortunes of the firm. The shift was fundamental: the latter half of the nineteenth century, industrialisation was producing some very large organisations, and larger more complex organisations require direction and management. An article in *The Office* magazine described it more graphically:

> *The office is to the business, profession or trade, what the head is to the body. The brain is the directing force of all the power exercised by the limbs of the body; so the power put forth in any department of a great business is directed from the office, where the principal or his trusted agent may be said to be the brain of the whole working body of subordinates.*[3]

On the other hand, office-based organisations – accountants, advertisers, banks, insurers, lawyers, surveyors and so on – were all expanding rapidly on the back of a rapidly industrialising economy. Clerking as a term was coming to be applied to an increasingly broad range of office jobs.

DOI: 10.1201/9781003582205-11

> *The Victorian use of the term 'clerk' was synonymous with today's 'white collar' category, subsuming everyone from quasi-managers to message boys. It is the breadth of the term that allows for confusion, for, depending on one's focus, Victorian clerks can be poor or wealthy, skilled or dullard.*[4]

The proliferation in roles during the Victorian period reflected the fact that companies and organisations were becoming larger, with more complex processes and more extensive trading activities. The rise of large national and global companies in the closing decades of the nineteenth and opening decades on the twentieth century profoundly changed the office, office work and social relations in the office.

Joint stock companies with professional managers spread, revolutionising the established owner–clerk relationship in "*the joint stock form of enterprise the relation of master and servant is immediately depersonalised and transformed into the modern relationship between capital ... and wage labour*".[5] The new companies invested in mechanical equipment and tools to increase productivity and the accuracy and reliability of output. As a result, much clerking became far more mundane, repetitive, dull and wearing. It was transformed from the specialist, generally highly regarded career which offered social mobility and financial security, that we saw in Chapter 5, into merely a 'job'. Alborn makes the point that senior management and directors had to increasingly "*maintain a delicate balance between nourishing middle-class values of independence among their employees and enforcing patterns of dependence and deference*", a balance that was altered by "*the proliferation of routinised work*".[6]

Clerks evolved from 'all-rounder' craftsmen, many of whom had close relationships with their business owners; were trusted and relatively autonomous, and who enjoyed relatively elevated social status, into lowly valued cogs in bureaucratic mazes, no longer with any overall picture of the businesses for which they toiled. Ultimately they morphed into simply office workers. The growing feminisation of the workforce also helped further transform work and social relations in offices. A male-dominated managerial class quickly evolved alongside a female workforce that dealt largely with administrative matters – secretarial, typing, data entry, filing and so on.

As office organisations underwent rapid expansion and increased complexity, so they inevitably introduced hierarchies of responsibility – particularly for male workers. At the same time office mechanisation was allowing businesses to handle huge increases in paperwork. Again, perhaps inevitably, this led to more division of labour, or more specialisation. In both cases – organisational complexity and division of labour – Klingender saw de-skilling, falling pay, the further separation of capital from labour and to the expansion of the working class. In the 1930s, against the backdrop of national and international politics, this was a very powerful message. But was it accurate?

Were clerks part of the oppressed proletariat? Or did they enjoy the privileges of middle class comfort? There is no clear, single answer to this question, because the experiences of clerks were so profoundly varied. No doubt, some were part of an exploited proletariat. But others enjoyed close and positive relationships with business owners. And right through to the end of the nineteenth century, clerking remained an aspirational vocation: one which would offer social and economic advancement. Thereafter, as 'office work' replaced 'clerking', the office workforce became synonymous with the bulging middle class in most economies.

Gentlemen and players

Gentlemen versus players was a long-running annual, 3-day first class cricket match. It was mostly played at Lord's in north London and was played most years between 1806 and 1963.

The gentlemen were amateurs, typically middle- and upper-class men, often the product of public schooling and who engaged 'for the sport'. By contrast, the players were working-class men who were paid as professional county cricketers. The amateurs received expenses, occupying as they did, a higher station in life than the wage-earning professional.

The match was a reflection of Victorian, stratified society. The term itself became a part of everyday language to distinguish between those who directed, ordered and planned and those who 'did', carried out and obeyed (or in Harry Braverman's terms, the separation of conception and execution). The question, of course, is where did firstly the clerk and then later the widely expanding office manager class sit?

As office work divided, and divided again, amoeba-like into multitudes of jobs, status and role became enormously important. Senior clerks aspired to be gentlemen, overseeing the work of the players under their command. Braverman describes clerks in mid-century as

> *semi-managerial in terms of the present distribution of functions ... The picture of the clerk as assistant manager, retainer, confidant, management trainee, and prospective son-in-law can of course be overdrawn. There were clerks – hard-driven copyists in law offices for example – whose condition and prospects in life were little better than those of dock-workers. But by and large, in terms of function, authority, pay, tenure ... prospects, not to mention status and even dress, the clerks stood much closer to the employer than to factory labor.*[7]

In addition to this semi-managerial imagery, Braverman stresses the craft skill entailed in early clerical work:

> *Clerical work in its earlier stages has been likened to a craft. Master craftsmen, such as bookkeepers or chief clerks, maintained control over the process in its totality, and apprentices or journeymen craftsmen – ordinary clerks, copying clerks, office boys – learned their crafts in office apprenticeships, and in the ordinary course of events advanced through the levels by promotion.*[8]

Similarly, Glen and Feldberg found that clerks were

> *'all-around' workers: they handled all phases of an assignment. Both organizing and executing it, and they did a wide variety of tasks ... A bookkeeper, intimately familiar with all of the financial details of a firm, might be readily consulted on financial decisions ... The organization of the work allowed the worker to develop skills and to identify as a skilled craftsman, a professional, or a part of management.*[9]

As all-rounders in expanding businesses – perhaps not so much in finance, but in the wider economy of railways and industrial manufacturing – clerks came to be seen as a part, or at least as an arm, or management – 'The Staff', a term originating in the army and referring to the senior officers with red tabs on their lapels who sat a long way from 'the action' (as we shall see, this was not the only aspect of military culture and organisation that transmuted into the workplace). In Marxian terms, the clerks were not owners of capital with the privileges that that conferred. They had no equity in their businesses and often they were beholden to the business, and their employer, for their place of living and general security. But they were somehow senior.

According to this interpretation, until the later-1800s, clerks not only saw themselves as 'all-rounders'; and proud of being such; but beyond this fact, they also drew satisfaction from being skilled and integral to the smooth running of the businesses that employed them and therefore closer to the gentlemen in their roles than to the players. If the ledgers were not kept up to date and accurate, there would be major problems for the business. Clerks routinely kept accounts, corresponded (often in foreign language) and supervised. By and large, they were also respected by their employers. They enjoyed comparative autonomy in their work and were trusted. They certainly saw themselves as distinct from, and senior to, manual workers in the wider economy. For clerks who worked in London and for banks and insurance companies there was an added level of superiority over their colleagues working in the regions and in manufacturing.

The clerk as gentleman, at least aspirant gentleman, was key to the perception of clerking. Lockwood describes how, "*clerks associate with clerks and artisans associate with artisans – but comparatively seldom with each other*", and how a clerk "*lives an entirely different life from an artisan – marries a different kind of wife – has different ideas, different possibilities, and different limitations*".[10] A contemporary account of the clerk as gentleman was given by Parsons:

> *Clerks are, as a rule, of decent address and gentlemanly habits, patient and long-suffering, not given to noisily "insisting upon their rights," and are possessed of some delicacy ... when requesting an advance of salary, not unnaturally believing that their employers ought to recognise their merits and reward their careful guardianship of trade secrets and the valuable information they frequently obtain.*[11]

Although there were some sharp income differences within the clerical class, the common values which informed their behaviour seemed at once to unite them and separate them from the wage-earning class, leading to the conception of clerking as an activity for gentlemen. Clerks were thus [in Victorian culture] 'respectable', 'honest', a 'cut above' and 'honourable'. Being of a higher status – morally, socially and financially – than 'other workers', clerks had ambition, aspiration and progression. In mimicking the dress code of their employers, clerks were conveying a sense that they were 'gentlemen in waiting', ready when asked to join the management. "*It is not by any means only a question of clothes, of the wearing or not wearing of a white shirt every day, but of differences which invade every department of life*".[12]

The reasons for the positive social standing of clerks are manifold and somewhat complex. But as Grace Coyle observed as long ago as 1929, an underlying reason was "*the ancient stigma attached to manual work, and the higher status given to any occupation supposedly mental*". Coyle also cited the fact that office occupations were usually clean; they did not require heavy manual effort and the hours were comparatively short. The income was more regular than much manual work, and there were "*certain privileges in the way of vacations, sick leave, and security of tenure usually go with them*".

> The attainment of a clerical position has come to be valued, therefore, not alone for its intrinsic advantages, economic or otherwise, but also as a symbol of other less tangible benefits.[13]

Above all, clerks were superior to most workers in so far as they, their social background and contacts permitting, had the opportunity of advancing into the employer class. In this sense they were more likely than most other workers to fulfil the Victorian dream of achieving respectability through economic independence.[14] Clerks regarded themselves as part of the middle class, rather than the working class. More so than American clerks, European clerks felt a "*sense of honourable status*", which made them "*emphatically not working class and stridently conscious of that fact*". In all other respects, British and German lower white-collar workers differed somewhat. For example, in Britain, clerks were

> *unaffected by the intrusion of a State bureaucracy, and they were not drawn by the forms of association that flourished in Germany. The British clerical worker, believing firmly in the tradition of individual mobility within a laissez faire society, did not emulate the bureaucrat as did his German counterpart but instead took the gentleman of the traditional mercantile world as his model.*[15]

Many British "black-coated" workers of the late-nineteenth century had very modest means of living, but they clung to an ideology of a previous era, where in the counting house, the relationship between clerk and employer was often a personal and particular one, when clerks 'mixed' with the sons of gentlemen, and when they were expected to behave accordingly.[16] Clerks were paid a salary, not a wage; this was a critical distinction at the time. It indicated that they could expect to work throughout an entire year and did not experience the seasonal fluctuation to which men paid daily wages were subjected.

Another aspect underlining the significance of being either a gentleman or a player was the question of what we would call today 'labour relations'. Whether specialist or all-rounder, clerks were never an 'organised' profession, neither in the sense of a profession nor in the sense of a unionised body. George Bernard Shaw served a spell as a clerk in the 1870s, and regarded the notion of joining a trade union as being a "*most ungentlemanly thing to do — almost as outrageous as coming to the office in corduroy trousers, with a belcher handkerchief round my neck*".[17]

An interesting perspective on the clerk in the context of organised labour is offered by Zakim, who argued (from an American context) that, while clerks were at the centre of a revolution

> *and was consequently branded as a subversive since he seemed to be inverting the natural order of things: His loaferism was essential to the new production regime; his deference signalled the destruction of authority; and his acquisitiveness threatened the propertied order.*

He was never seen to be part of the 'labour movement'. In fact he was omitted from most scholarly work "*devoted to charting the industrial transformations of work and the social relations that ensued*", because it was "*the clerk's very identification with capital and his determination to invent a life consonant with the practices and ethos of market exchange that have disqualified him as a subject of 'labour'*". Zakim goes on to suggest that by identifying the clerk as a central figure in the coming of the modern order suggests that

> *capital rather than labor was the agent of revolution in America, and that the social history of industrial life would be better written by business historians rather than labor historians ... Clerks certainly did not behave like the modern revolutionaries we are used to thinking about, or populating our narratives.*[18]

To a large extent, the diversity of roles, positions, and so on defied organisation. It was partly due to this lack of organised unity that clerks suffered in some areas such as remuneration and working conditions. But despite the drawbacks, clerks enjoyed a great many economic advantages not enjoyed by manual workers. Among these benefits might be numbered permanency of employment, periodical increases in salary, payment of salary during sickness and holidays, comparatively reasonable hours of work, and in certain sections superannuation. "*These advantages, chiefly matters of custom and usage, constitute a powerful common interest among clerks, and should be an equally strong reason for protective organisation.*"[19]

Managers in waiting. Clerks could not only navigate through the clerking hierarchy (say, office boy to junior to senior), but also rise from the clerking class into management, or even into ownership. There was a widespread awareness of "*the rewards to be reaped by the clerk-become-merchant and the miseries of those who were chained to their high stools for the rest of their days*".[20]

Clerking was thus seen to be superior to many, particularly manual, vocations in so far as clerks had the opportunity of advancing into the employer or even owning class. This being so, "*they were more likely than most other workers to fulfil the Victorian dream of achieving respectability through economic independence*".[21] Orchard noted that while everyone "*cannot reach the top; yet on every side are brokers, merchants, agents, who from the clerk's stool have been raised to the private room of the firm, because they deserved promotion*".[22]

Lockwood also highlights the difference between the '*clerk by profession*' and the '*clerk in statu pupillari*', the latter comprising the clerk who is placed at a desk, temporarily, to "*learn the elements of the business ... 'til he can undertake the management of the same business in its higher departments*".

To underscore their role as gentlemen, or 'bosses in waiting', many clerks adopted a particular dress code. Herbert de Fraine, who began work at the Bank of England in 1886, recalled the dress code as comprising "*silk hat, with a strip of dull cloth, three inches wide, covering the narrow silk band*", kept in pristine condition using a curved rim-brush and a velvet pad.

> *Trousers were worn without a turn-up, and everyone wore black boots, as often as not buttoned. Everyone without exception carried either an umbrella or walking-stick.*
>
> *We wore stiff-fronted shirts with stiff cuffs, and stiff white collars (not wing). We also wore tail-coats, for the short coat with the silk hat did not come in until shortly after the First World War. There was an elaborate tie with the ends of equal breadth, called a cravat, which you could build up into a formidable expanse of silk or satin, anchored in position with a gold pin.*

And to top it all off, there was the gold-chain,

> *varying in thickness with the owner's opulence, stretched across the waistcoat, with the watch at one end, and at the other a metal sovereign purse that held five sovereigns and had a spring to force up the small pile from the bottom as you slid off the top one.*[23]

In their adopted dress, clerks were making a clear sartorial statement that they were not part of the proletariat; that they were socially mobile and that they shared their cultural outlook with the owning class rather than that of the working class.

The notion of being 'managers in waiting' was also wrapped up in the Victorian concept of social mobility – a belief that hard work could lead to social and economic improvement. In the middle of the nineteenth century, the development of 'Victorian values' led to a strong sense in the value of 'social improvement' – the notion that one could (and even should) progress up the social hierarchy through the application of hard work, diligence and deference. This was nowhere more the case than among the clerking profession.

In the first instance, one could move from the lower class of manual work into the lower middle class of clerking. This was perceived by many parents as a very important step for their sons because the clerk enjoyed an elevated status in society. As the clerking profession grew in size, "*middle-class and even skilled working-class families saw in these jobs superior prospects for their sons, including a route into the mercantile class*"[24]

> *Undoubtedly the chance of achieving upward mobility in commerce was a decisive factor in attracting the sons of labour aristocrats. After all, in this period many skilled workers also began to suffer from status insecurity because, although their earnings remained good, their opportunities of rising into the ranks of the employers declined while the position they had previously occupied just below the employers was somewhat undermined by the replacement of owner-managers (entrepreneurs) by shareholders and professional managers.*[25]

Indeed, many clerks rose from clerking to ownership and the lower middle class, in particular, saw clerking as a route to upward social mobility. Anderson provides the example of Alfred Jones, "*the son of respectable, middle-class Welsh parents*" Alfred's career "*corresponded to the ideal of the self-made man*". Having started out as a junior clerk in the office of William Laird (ship builder) in 1859, by 1878 he had acquired the capital and experience necessary to set himself up as a shipping and insurance broker. "*Ultimately he became a senior partner at Elder Dempster & Company.*"[26]

There are numerous records of clerks achieving positions of seniority and then retaining their post into old age. For example, employees of the Bank of England often reached the age of 80 years or more before compulsory retirement at 65 was imposed in 1870.[27] The same was true in the insurance sector. Particularly in the early insurance houses, as they expanded rapidly in the nineteenth century, it was not unusual to have generations of family members working alongside one another.

> *The Alliance's actuary worked alongside two of his sons in the early 1860s, as did the Church of England's manager from 1857 through 1862; when one son retired that year due to poor health, a third took his place. The Clergy Mutual's founder, John Hodgson, was succeeded as secretary in 1871 by his grandson Matthew, whose brother George served as his assistant for two decades before taking over in 1893.*[28]

Many clerks trained under their fathers before succeeding them.

> *Father–son management teams spanned more than fifty years at the Scottish Provident, West of England, Standard, General, Norwich Union, and Equitable; three generations of Beaumonts managed the Provident for a century; and a trio of Newmans headed the Yorkshire for all but four of its first eighty-six years.*[29]

When firms were typically small, merchants and business people generally recruited clerks from their own social class. Often this was done by word of mouth among social groups and even families. Such social ties reinforced the paternalistic nature of many owner–clerk relationships, but it also helped maintain a sense of social standing within the clerking community. As time progressed, more formal systems for finding new recruits emerged.

Alborn observed that the appointment of salaried officers' sons was part of a more general culture of patronage, which was by far the most common method of hiring insurance clerks through the nineteenth century. He cites Atlas insurance, which established a regular rotation, whereby each director nominated clerks in turn as vacancies opened up.[30]

While social mobility was a key attraction of clerking, it was by no means a given: the fortunes of clerks varied widely, depending upon the specific circumstance of the firm in which they found themselves employed. The roles and responsibilities of clerks were highly varied, as were their pay and conditions, as was their social composition and status – perceived and real! It is therefore difficult to summarise them as a single group. To illustrate this point, the following lengthy quotes summarise two very different outcomes for what, superficially, were two similar clerks.

> *In July 1854, Peter Ewart, who had been secretary at The Standard Life Assurance Company's London branch for fifteen years, died prematurely. Assisted by a local board of directors, Ewart and three clerks were responsible for the day-to-day operations of the branch, which oversaw a rapid expansion of the Scottish company's business throughout England during his tenure there. His salary of £700 was higher than most London insurance clerks, enabling him to keep up a £5,000 insurance policy, but the proceeds from the policy proved to be insufficient to pay his debts, funeral expenses, education for his three children, and his widow's rent.*[31]

Standard Life's Edinburgh directors refused to do more than pay Ewart's salary through to November. The situation was different when William Thomson, Standard Life's head office manager, retired in 1874.

> *Trained as an accountant, Thomson had been appointed secretary in 1834 at the age of twenty-one ... and was promoted to manager three years later. Even more than Ewart, Thomson's efforts put Standard Life on the map, both as an innovator in life insurance and as a training ground for future actuaries. By 1848 Thomson earned £1,200 a year, and his salary at retirement was £3,000. He continued to receive a pension of £2,500 until he died in 1883, after which point his widow received a £500 annuity. Unlike Ewart ... Thomson turned his insurance work into an express ticket to middle-class status and prosperity.*[32]

This stark contrast in the life outcomes of two apparently similar clerks (both operating at senior level in the higher echelons of white-collar work, i.e. the insurance sector) demonstrates a wide disparity in reward. Social connections played an important role in 'getting started', and social relations with senior managers thereafter reinforced early advantage. Either way, the potential for social mobility was constrained by position in work.

Social standing. As we saw in Chapter 3, in the eighteenth and early-nineteenth century, businesses were typically small, family run affairs; owners would know everything that was happening within their business: profit and loss, their customers, their debtors, their suppliers and so on. Everything, in detail, committed to memory and very little recorded. And they would direct the business personally, issuing orders and requests to ensure the smooth, day-to-day running of the business.

> *Above all ruled the boss, often also the owner, distinguished by intuition, by instant recall, by fingertip feel. Nathan Rothschild knew his clients' affairs so minutely as to scorn records.*[33]

Nevertheless, for much of the nineteenth century, the relationship between owners and clerks was a critical one, which has provided historians and writers with a rich seam of material, illustrating as it does so much about socio-economic relations of the time. In many ways, the owner–clerk relationship was paternalistic, and sometimes even of a master–servant type, although the range of relationships was incredibly wide.

During this time, clerking was regarded as a craft, a skill to be honed and developed over years of loyal service. It was a critical task which had to keep current, accurate records of the financial and trading position of the owner's business. This in turn, demanded a level of trust not experience on the factory floor.

> *Master craftsmen, such as bookkeepers or chief clerks, maintained control over the process in its totality, and apprentices or journeymen craftsmen-ordinary clerks, copying clerks, office boys-learned their crafts in office apprenticeships, and in the ordinary course of events advanced through the levels by promotion. The work involved, in addition to ordinary bookkeeping ... such tasks as timekeeping and payroll, quality control, commercial traveling, drafting, copying duplicates by hand, preparing accounts in several copies, etc.*[34]

Writing in the 1950s, Lockwood described counting houses as creating environments which were "*generally conducive to their estrangement from the mass of working men and to their identification with the entrepreneurial and professional classes*". He noted that many of those who went into black-coated work were drawn from "*the middle class or from socially aspiring artisan families*" and that they were generally better educated than factory workers. Their work was "*clean and involved the exercise of brain not brawn*" and their dress "*distinguished them from ordinary employees and approximated that of the master class*". Lockwood suggests that clerks from this time worked in a close and personal relationship with their business owners, often being "*entrusted with confidential matters and delegated authority over other employees*". Finally, their salaries were generally secure and progressive.[35]

Alborn describes how life insurance essentially started from scratch in the early-nineteenth century, with directors learning how to run the business as 'on the job'. As part of this, they learned alongside and from the clerks "*who regularly monitored the ebb and flow of policy receipts, annual premiums, and claims, and who also generated much of the actuarial knowledge that was essential for long-term solvency*". The directors relied on their clerks "*who showed signs of being able to combine mathematical skill with sound business knowledge*", such that by mid-century, "*all British actuaries (who in most cases were de facto managers) had risen from a junior clerkship to a position of paramount authority*".[36]

The typically paternalistic relationship was born of the fact that the clerk was entrusted with all manner of knowledge about the business, even of the family, not to mention trusted to handle cash and customers. Bank clerks often lived over the bank, or shop, sometimes with their masters, or in nearby premises, courtesy of their boss. Christmas boxes, cash bonuses and non-financial support were common. Lockwood suggested that the owner–clerk relationship was, in most cases, "*a personal and particular one, strongly characterised by the exhibition of mutual trust in the form of unwritten, tacit expectations of conduct*".[37]

This early form of 'gentleman's agreement' reinforced the social hierarchy between employer and worker, and made the clerk less likely to make demands on his employer (pay, working conditions). Indeed, the 'clerking class' never formally organised in terms of labour relations, despite often widespread discontent over pay and conditions. They did not possess a common knowledge base with which to form a distinct professional group, beyond the rudimentary skills of literacy and numeracy; they lacked a clearly defined service orientation to society, and they were not bound as most professionals were by a code of ethics.[38]

The Victorian values by which most clerks lived were also reflected in their loyalty, with many clerks remaining in the same employ for decades.

> *Since life offices usually hired clerks before they reached the age of twenty, and since turnover rates were extremely low, most of these men took orders from the same two or three superiors for most, if not all, of their adult lives. In 1878, when George Cutcliffe was in his twentieth year as the Clerical Medical's actuary, half of his staff of fourteen had worked under him for the entire twenty years, and four more for at least ten. Nine of them were in their mid-thirties or older, and nine would stay at the Clerical Medical until they retired.*[39]

It was not only Victorian values in Britain that framed the employer–clerk relationship. Writing about the United States, Lewis Corey observed

> *The clerk was an honoured employee 150 years ago, and still more so in earlier times. His position was a confidential one, the employer discussed affairs with him and relied on his judgment; he might, and often did, become a partner and marry the employer's daughter. The clerk was measurably a professional and undeniably a member of the middle class.*[40]

But not all situations were infused with such trust and loyalty.

Despite the general alignment of interests between owner and clerk, not all owner–clerk relations were quite so positive. There were also many instances where paternal became feudal: in some extreme cases, the relationship was more master–servant in nature.[41] For every paternalistic relationship based on mutual trust and confidence, there was another that involved "*hard-heater employers and insubordinate or dishonest clerks*".[42]

Klingender cites a June 1846 letter, which commented upon an obituary of a clerk who had served for 50 years with the Bank of England; it stated that

> *Long periods of service on the part of clerks both in private and joint stock banks are the exception rather than the rule, owing to the system of oppression and overworking, the paltry, contemptable and mistaken economy which cuts down their reward to the lowest level of subsistence.*[43]

In his essays on the clerks of Liverpool, Orchard quoted a letter from a correspondent who claimed to represent employers.[44] The letter to the Editor of the Daily Courier made the following observations

> *Our clerks talk as if they were our partners in all but the profits, and took an equal share in the operations which create wealth; whereas they are usually little more than machines, doing routine work, and incapable of organising, calculating, and negotiating, as we do ourselves.*

With regard to pleas for more pay than artisans:

> *Extensive inquiry has satisfied me that that clerks do, on the whole, earn more than artisans … . Artisans, also, while their work generally calls for the exercise of judgement and for much bodily exertion, work more hours daily than do clerks.*

The correspondent also tore into the general competence of many clerks, stating that it is an

> *unquestionable fact that among the twelve or fourteen thousand clerks of Liverpool … a large number are miserably incompetent for the performance of any duties beyond the simplest routine – indeed, even this this appears in many cases to overtax their capabilities.*

He then questions their general honesty and sobriety:

> *Comparatively few can be trusted with money, unless strict and constant and painful supervision is exercised; and the number of unsober or unsteady young men is sadly great.*

> *The fact is, sir, that clerks, instead of grumbling in your columns, must work hard at self-improvement, make the most of themselves by cultivating their faculties, and then watch of for opportunities.*[45]

There is clearly some bias at work here, but there is an underlying message that owner–clerk relations were not as universally close as some have painted. And there is abundant corroborating evidence that this is so. For example, Orchard relayed how he was told directly by the Mayor of Liverpool that he viewed "*Clerks as little more than a lot of thievish scamps*". Orchard himself felt that clerks, as a class "*have singularly overstrained conceptions of the part they play in the great drama of commercial enterprise*" and that there is "*no cordiality between various large sections of Liverpool clerks and their employers. There is scarcely a pretence of friendliness*".[46]

Whether of the paternalistic or more oppressive type, social relations in the white-collar workplace were evolving rapidly in response to the changing form of organisations which in turn were reflecting the wider economy. As companies grew in scale, so came hierarchy, structure, systems and processes. The latter part of the nineteenth century saw the rise of management and managers, the feminisation of the workforce, and the dawn of the 'information office'.

Reward, recreation and reproach

We saw in Chapter 5 a general view that clerks were poorly paid in return for their physical and mental effort, extended hours and general loyalty. And this continued into the later decades of the century. First there is a comparison with an artisan.

> *So far as concerns clerkship, brains are less highly valued than muscles are in other employments, for the average clerk's salary is much below in amount the wage of an ordinary mechanic. Many an artisan can earn from two to four guineas per week, and is not called upon to keep up an external appearance for the purpose of impressing his neighbours that his circumstances are much better than they really are.*[47]

Then there is a plea to employers to recognise the real value of their clerks.

> *It is absolutely necessary that clerks should keep pace with the requirements of their employers, who, year by year, as trade connections extend, need assistance of a more advanced and comprehensive kind; and it is equally necessary that employers should seek that assistance with more fairness than they do at present, and offer a salary sufficient to secure a conscientious and careful clerk, in lieu of entertaining the delusion that by giving miserable wages they will obtain the able assistance which their undertakings require.*[48]

However, as with so many other aspects of clerical life, pay varied greatly according to age, experience and duties, as well as the type of firm. Kynaston points out that while higher-class firms such as the stockbrokers Heseltine Powell, whose 20 clerks and office boys had an average salary of £225 per annum by the turn of the century, or Kleinworts, where almost two-thirds of the 73 men employed were earning above the Income Tax threshold of £160 per annum by 1910, a threshold that barely a million in the whole country had crossed, in many firms "*pay was much worse and prospects uncertain*".[49]

Table 8.1 Household budget for a clerk earning £80 per year

Expense	*£*	*s*	*d*
Kept house on £1 a week	52	0	0
My lunches were 2s a week	5	4	0
Her and my clothes	12	0	0
Life insurance for £100	1	2	0
Subscriptions to Clerks Provident Association	1	4	0
Chapel expenses	2	10	0
Coal, £3; Gas, £1	4	0	0
Total	**78**	**0**	**0**

Source: Orchard (1871) p65.

Orchard provided breakdowns of household budgets from real clerks in Liverpool. Table 8.1 shows a budget for a clerk earning £80 per year. Keeping house and clothing stand out as the largest expenses, followed by lunch expenses.

In another example, he compares a clerk earning £80 per annum and another earning £150 per annum. In the case of the lower paid clerk, expenditure is clearly limited to the absolute minimum, with no discretionary spending. As the clerk himself described

> *We are both healthy, and from the outset, resolved to live on plain food ... Doing this we have escaped sickness, have been comfortable, yet never felt poor. My wife makes her own bonnets and dresses, and my shirts. She also cooks nicely.*

By contrast, the higher paid clerk enjoys more discretionary spend, with a budget comprising "*a cosy little house outside the town*", "*occasional teas up town*", "*my clothes, boots, gloves, weekly paper, etc*", "*wife's clothes*", servants' wages, summer holiday and occasional concert. He concludes

> *It is true that £150 forbids indulgence and sloth, pares off superfluities, brings a young couple to their bearings and removes them almost entirely from a thriftless circle. Nevertheless, a couple may live on £150 without hesitation.*[50]

Compared to many manual jobs of the time, including working in dirty factories, clerking in the mid- and late-nineteenth century society, carried with it many benefits, not least of which was a level of financial security. Lockwood cites a passage from the journal *The Clerk*, which stated that clerks

> *have a great many economic advantages not enjoyed by manual workers. Among them may be cited permanency of employment, periodical increases in salary, payment of salary during sickness and holidays, comparatively reasonable hours of work, and in certain sections superannuation.*[51]

This was written towards the end of the period being examined here, and no doubt working conditions for the average clerk were better at this time than mid-century but, nevertheless, it was a fair distinction with other forms of employment. It is also the case that salaries varied according to the size of the organisation and the nature of the business. Banking and insurance generally paid more than other sectors. Jeacle confirms the "*generous compensation*" afforded bank clerks, who were among "*the highest paid clerical workers and had the potential to earn significant salaries with advancing seniority*".[52]

Alborn provides an insight into pay within the insurance sector. For their daily "*seven hours' quill-driving without intermission*", insurance employees earned wages that were similar to those earned by bank clerks and higher than most other clerical workers' pay. Atlas clerks received £144 per annum on average in 1809, comparable to an average salary of £143 at the Alliance in 1827. Alborn also noted that as life offices expanded in size, some introduced tiered salary grades, enabling more promising clerks to rise more rapidly from one pay scale to another:

> *Tracking average salary by age at the Clerical Medical between 1842 and 1912, clerks earned £66 at age twenty, moved up to £111 by the age of twenty-five, £176 at thirty, £247 at thirty-five, £360 at forty, and £452 at forty-five, before levelling off, owing to salary caps on all but the highest positions.*[53]

Banking and insurance clerks not only enjoyed higher salaries but they also generally worked fewer hours and were increasingly housed in the large, purpose-built buildings where working conditions were so much the better.

Attewell pulled various sources together to provide a cross industry comparison of salaries. Table 8.2 clearly shows the higher salary bands in sectors such as shipping and mercantile. He also defines breakpoints in income to provide rough division between 'quasi-managerial' and 'other' clerks. Attewell proposes that managers, head clerks, department heads and cashiers – those whose jobs most clearly correspond to modern managerial positions – earned from at least £300 in 1870; the "*better class of clerk*" earned from £150 up to £300, and the most junior and unskilled earned under £150 per annum. The high concentrations of jobs in the lowest salary bands is striking.[54]

As noted in the introduction to Chapter 7, the later decades on the nineteenth century were afflicted by economic uncertainty, that led to business retrenchment and rising unemployment among clerks. Suddenly, the relative comfort, security and sense of superiority enjoyed by clerks over other workers evaporated. The economic uncertainties coincided with the dramatic rise in the number of female workers who, it was widely recognised, were generally paid lower

Table 8.2 Percentages of clerks at various salary levels, c1870

Sector	*£/year*	*%*	*Sector*	*£/year*	*%*
Eminent solicitors	>£300	4	Mersey Docks & Harbour	>£300	4
	£150–300	22		£150–300	12
	<£150	74		<£150	83
Corn trade	>£600	1	Life insurance	>£300	8
	£120–160	16		£100–300	30
	<£120	83		<£100	66
Shipping	>£750	1	Mercantile	>£600	1
	£250–750	12		£120–600	16
	£80–250	58		<£120	83
	<£80	29			
GW Railroad	>£150	5			
	<£150	95			

Source: Attewell (1989) p368.

salaries. It also coincided with growing mechanisation of the office and a growth in repetitive, mundane jobs which required lower skill levels (and lower pay) than traditional clerks. All in all, a highly uncertain time ensued.

While clerks had previously "*derived satisfaction from the fact that they lived and dressed more respectably than most workers, that their work was cleaner and less arduous and that it was often more secure*", it was clear that such distinctions were disappearing.

> *By the turn of the century many workers, at least socially, dressed as well as clerks. In addition their living standards and working conditions were slowly improving. Clerks, however, were in an increasingly marginal position, fitting securely into neither the working nor the middle class.*[55]

It is probably no surprise that, around the 1870s, larger employers began to engage their workers in recreational activities.

Recreation. The late Victorian period was an uncertain era for clerks. Changing economic circumstances, mechanisation, feminisation of the workforce and changing employer relations all combined to reduce the certainty and simplicity of the past. In larger, more complex organisations there was simply not enough room for everyone to progress socially or economically, let alone get on the ladder to management. It was also the case that there were fewer opportunities for personal patronage between employer and clerk and, as a result, larger organisations required more formal means of recognising performance and loyalty, and the Christmas box and cash bonus was no longer appropriate.

Company welfare schemes such as pensions, provident funds and holidays became commonplace, but employers had to find a variety of novel ways not only to encourage loyalty and team spirit, but also to compensate for reduced opportunities to progress through to management. Thus it was that the late-nineteenth century surge in popularity of team sports and other forms of recreation were exploited by larger employers who invested in buildings and facilities for a wide range of non-work activities.

Growing numbers of larger employers captured the zeitgeist with a new form of patronage, namely, organised sports, recreational activities and canteens. Such resources and amenities were often subsidised and could involve considerable investment in land, buildings and facilities. Organised sports grew rapidly in the mid- to late-Victorian era, including national leagues in football and cricket as well as more specialist pursuits such as rowing, archery and shooting. Organised sports and other leisure activities offered a real opportunity to divert "*the desire for social or economic mobility into aspiration for success on the playing field, golf course, or stage*".[56] Such activities "*helped to recalibrate the balance between deference and independence that new patterns of occupational mobility were threatening to destabilise*". In other words, they gave a sense of relief from routine and mundane work that held out little in the way of career prospects.

Insurance giant Prudential was well ahead of the game: clerks in its late-Victorian London headquarters could

- borrow a novel from the 3000 available at the company's library;
- write a short story for the in-house literary magazine;
- draw a picture for the Ibis Sketching Club;
- practice bassoon (bought on instalment from the company) at the Ibis Music Society;
- sing with the Prudential Choral Society;

- hear a missionary lecture at the Prudential Prayer Union, or
- attend a joinery class sponsored by the Ibis Technical Instruction Society.

The Ibis was the Prudential Clerks' Society. Depending upon the season, the same clerk could participate in a wide variety of sports, including cricket, rowing, running, tennis, cycling, swimming, gymnastics and chess, either intramurally or against other amateur teams from London. Funding from the Prudential Clerks Society dated back to 1871, and after 1900 the Directors added an annual grant of £1,000.

> *With a head-office staff numbering more than three hundred in the late-1870s (up to two thousand by 1918), this annual expenditure was a bargain for the company, and clerks took part in droves. In 1878, 75% of the staff belonged to the library, 55% rowed, 48% performed gymnastics, 37% played cricket, and 19% played chess.*[57]

Heller provides a few statistics to demonstrate the extent to which companies began to support sports and leisure activities. Between November 1899 and February 1904, for example, the Great Western Railway spent £2,030 on the GWR Athletic Association. This money was spent on building a sporting club and pavilion on a seven hectare (17 acre) ground in West Ealing, the equipping of the club and its upkeep. Similarly by 1905 both the London and Westminster and the London and County Banks had sports clubs in Norbury, South East London. In addition to this, the London and County had a rowing club and boat house, all paid for by the bank, and a plethora of other social and sporting organisations.

In 1911, the *County and Westminster Magazine*, the staff journal of the now merged London and Westminster and London and County Banks listed the clubs and societies of the combined operation, including 30 associations which included nine cricket teams, four football teams, four rugby teams, a rowing club, a swimming team, a tennis club, a boxing club, a rifle club, a chess team, a photographic club, a debating society and an orchestra. In addition, a Cross Country and Athletic Association was in the course of formation, and a Motor Cycle Club was being planned.[58]

The rationale for such investment was not only to distract staff, however temporarily from their typically mundane work, but also to build loyalty, a sense of shared community and nurture friendships – all of which add up to what we would call today a 'corporate culture'. But this was particularly necessary just at the time when larger, more complex and hierarchical organisations impacted greatly on the relationship between the employer and the employee. This was summed up by Heller.

> *Such support integrated the Staff more fully into the newer impersonal relations which were beginning to dominate the larger firms. It gave them a sense of a common organisational identity. In the face of such versatility and usefulness it is no wonder that such sporting and social clubs were springing up in offices in London, and why so many directors and senior management were so willing to give so generously to such activities for the 'benefit' of their staff.*[59]

While there was significant investment in 'keeping the workers on side', there was also a need, with the growing distance between employer and employee, to enforce discipline in the workplace.

Reproach. Respect for seniors was much more of a given in Victorian society than it became later in the twentieth century. Strict social codes of conduct were underpinned by the fact that there were none of today's safety nets: being fired from a job was a big deal in terms of trying to find new work, economic impact and general welfare. Clerks could invite disciplinary action from their employers for any number of transgressions, including something quite simple such as lack of punctuality. The more important rules were there to safeguard bank business. For example, to insure against bank fraud, one man's work was often checked by another, but there remained opportunities for the determined fraudster.

> *If the trust was abused the repercussions for the person concerned could be far-reaching. The clerk who violated the position of financial trust shattered his own respectability. Once dismissed, he would experience difficulty in trying to rebuild his career and reputation.*[60]

> *The ordeal of prosecution meant for the clerks involved the almost certain loss of occupation and, given their value system based on respectability, honesty and loyalty, social degradation through public exposure.*[61]

The safeguards against fraud extended to forbidding the use of erasures in ledgers. Jeacle refers to Hoare's bank where any attempt at altering bank books was viewed with suspicion and severely punished. He cites the example of one clerk "*who had altered a customer's account by one shilling in order that it balanced with a summary listing*" and who "*suffered an instant salary reduction and jeopardised his future promotion prospects*".[62]

For some bankers, discipline took the form of traditional enforcement of hierarchy. For example, around the turn of the century at the

> *Capital and Counties Bank in Threadneedle Street, Edward Merriman was for thirty years an autocratic chairman who induced many a tremble with his quill-penned letters. At the Bishopsgate office the Liverpool merchants Balfour, Williamson & CO, Robert Balfour terrified errant clerks by asking in all apparent seriousness if they had not mistaken their vocation and would be better employed as dustmen.*

Similarly at Rothschilds, "*Alfred de Rothschild was reported ... as becoming more unbearable than ever to the staff and treats men of 30 years' service like office boys*".[63] Again, Jeacle cites Hoare's where the "*level of discipline varied according to the personality of the senior partner in charge*". In the early years of his appointment, "*clerk Tilden experienced the often dictatorial rule of partner Henry Hoare*", who was

> *was a man quite of the old school, a stern disciplinarian, of whom everyone in the place, even his own family, stood in awe ... If the Family were seeking a cognomen for him he might certainly be called 'the Autocrat' for no man was ever more autocratic in his manner to those of lower grade than he.*[64]

And the net of discipline extended beyond the workplace into the private realm because many employers had "*a particular concern for two areas of personal conduct – indebtedness and intemperance*".[65]

As might be expected, the staff who resided within the bank building, on the upper floors were subject to a range of disciplinary measures. Banks would issue a list of house rules. For example, there was often a curfew with clerks expected to be home by 11.30, with any extensions requiring special permission.

Feminisation of the workforce

The later Victorian period saw one of the greatest upheavals in the history of the modern office: the arrival of female workers. Up to this point, the concept of women (universally 'ladies' in all contemporary literature) in the office workplace was entirely alien: counting houses were the exclusive preserve of men, or rather, gentlemen. So when it did happen there had to occur a massive shift not only in social relations in the office but also in the workplace environment itself. To begin with, females "*were the cause of a certain mild amusement on the part of their male colleagues, but of not real concern*".[66] This ambivalence did of course change.

The attractions to office work for female workers compared to manual work were little different to those of male clerks. Coyle suggested that while the woman office worker might not be in so advantageous a position as to salary, she usually had certain "*prerequisites that went with her status, such as vacations with pay, and payment in case of sickness or accident, all of which are more common than among manual workers. Bonus payments are often liberal and there is more security of tenure*".[67]

Early attempts to absorb women into offices threw up some situations, which seem almost unbelievable around 130 years later. The following is an extract from Alan Delgado's wonderful book, *The Enormous File.*

> *In 1889 a government department employing two ladies put them in a room on the upper floor and had a convenience (WC) fitted so that the ladies – one earning 21s a week, the other 23s – would be completely free from 'any danger of interference'. Other precautions taken in the Civil Service included locking a number of ladies in a room and serving meals to them through a hatch in the wall. The ladies left the office fifteen minute before the men.*[68]

And this was in the Civil Service – one of the more progressive areas of the economy! Delgado goes on to state that, contrary to the view expressed elsewhere that women 'had it easy', they were rather exploited: while women were employed in growing numbers because many employers considered them better adapted for the emerging work of clerks, they were paid less than clerks.

Employers and managers looked with particular suspicion upon married women. Grace Coyle observed, from an American perspective in 1929 that

> *the basis for this attitude seems to rest less upon a scientific study of the relative efficiency of married and unmarried women than on a commendable desire to defend the "American home" from subversive tendencies.*[69]

Insurance companies often segregated female workers in watertight quarters, and nearly always gave them tasks that were distinct from what men were asked to do.

> *The Refuge's new head office came with a top-floor "Harem" which had a separate entrance and an eat-in kitchen, and the same arrangements existed at the Pearl and the Prudential. Tasks for women included feeding envelopes into addressographs, taking shorthand dictation, and above all typing.*[70]

Between 1870 and 1914, many offices were transformed by the arrival of female clerks using the new technology. Figure 8.1 shows a Navy Department typing pool in the USA in 1918 and the dominance of females sat at typewriters, with largely male supervisors, is clearly visible.

Figure 8.1 Navy Department typing pool, Washington DC, 1918.

Similarly, in the UK, female clerks grew nearly tenfold between 1871 and 1911.[71] For example, at the Prudential head office,

> *Beginning with the appointment in 1871 of a female supervisor and four lady clerks, by 1891 the Prudential employed 243 women – 35% of its total staff, 42% of all women who worked for British insurance companies, and nearly three times the total number of women working for British banks.*[72]

By the turn of the century, women formed 45% of telephonic and telegraphic employees; in local government, 29% and in the Civil Service, 25%. In commercial concerns (banking, insurance and law) they were still a small minority.[73] This changed quickly, and in the interwar years the clerical industry, and the female share of it, both expanded rapidly. Again, Delgado observes

> *By the time the King died in 1910 the City of London was no longer a man's domain. It was discovered that women were better than men in operating the typewriter and working the new mechanical aids that were becoming more and more efficient.*[74]

In this way, the typewriter had an enormous impact on social relations, drawing into the office a whole new class of worker-women, whose labour had been previously confined to the factory or to domestic service. Cowan *et al.* noted that "*the number of female workers rose quickly, so that they very soon began to outnumber men in the more routine sections of the office*". Using a different data source, Lockwood notes that "*by 1951 clerical work had become one of the women's foremost occupations. One out of every five working women was employed in an office*".[75]

The sheer scale of the change is shown in Table 8.3, rising from just 2,000 (or 2% of all clerks) in 1851 to over half a million (nearly 40%) by 1921. This saw female clerks rise from less than one percent of all female jobs to 10% over the same period.[76] Within 50 years from 1851, a more or less wholly male preserve had been transformed.

Table 8.4 shows similar data for the USA. The chart shows all clerical jobs rising from just 91,000 in 1870, to 1.8m in 1910, at the same time growing from under 1% to over 5% of all jobs. Female jobs grew from just 2,000 to 677,000 over the same period – rising to nearly 36% of all clerical jobs. It is also clear that women made greater, earlier progress in government and the wider public sector than in small private sector companies, especially those in the City of London.

As Wardley notes "*relatively few female workers in Britain found employment in financial services before 1914*".[77] And when females did begin to enter the banking sector, they were generally directed into lowly paid and low skilled posts.[78] For example, many were employed to operate the new mechanised bookkeeping machines that had started to appear from the 1930s.[79]

Table 8.3 Number of male and female clerks, 1851–1931, Great Britain

Year	*No of clerks*			*Clerks as % of occupied population*		
	Male	*Female*	*Total*	*Male*	*Female*	*Total*
1851	93,000	2,000	95,000	1.6	0.7	1.2
1861	123,000	4,000	127,000	1.9	0.1	1.3
1871	194,000	7,000	201,000	2.7	0.2	1.9
1881	294,000	14,000	318,000	3.8	0.4	2.9
1891	381,000	33,000	414,000	4.3	0.8	3.2
1901	530,000	83,000	613,000	5.2	2.0	4.3
1911	677,000	166,000	843,000	5.2	3.1	4.6
1921	792,000	502,000	1,294,000	6.7	9.9	7.5
1931	882,000	583,000	1,465,000	6.7	10.4	7.8

Source: Pickard (1949).

Table 8.4 Growth of clerical workforce, 1870–1930, USA

Breakdown	*1870*	*1880*	*1890*	*1900*	*1910*	*1920*	*1930*
Total clerical workers (000)	91	186	492	770	1,885	3,311	4,274
As % of all jobs	0.7	1.1	2.1	2.6	5.1	8.0	9.0
Female clerical workers (000)	2	8	83	204	677	1,601	2,223
As % of all clerical jobs	2.4	4.3	16.9	26.5	35.9	48.4	52.0

Source: Glenn & Feldberg (1977).

Kynaston records that the first female clerk did not arrive at merchant bank Antony Gibbs until 1889, and 5 years later she was joined by two others. Inextricably linked with the coming of female clerical labour was the advent of the typewriter; but though most City offices were equipped with them by some point in the 1890s, there remained a continuing preference for handwritten correspondence undertaken by male copying clerks. A similar reticence to change marked the coming of the telephone. At Schroders, neither it nor women were sighted until the new century; while at the stockbrokers Heseltine Powell the intrusive ringing thing was banished to the basement.[80]

Women's work. We saw in Chapter 5 (and to some extent in the early part of this Chapter) how the clerk emerged and evolved as a respected, trusted and multiskilled worker. Clerking was an aspirational career, one to which literary folk gravitated as a means of supplementing their artistic income while not compromising their sense of status, as well as one to which men who might previously have left school and gone down a pit, onto the land or into the factory could find an escape route. By the start of the twentieth century this position had shifted dramatically. Lockwood cites a letter in *The Clerk* from 1925, in which a clerk who deplores the "*secondary and passive nature*" of his work, refers to his "*very unmasculine calling*" in which he and his colleagues were "*born to be men and condemned to be clerks*". He can admire the loftiest of buildings and say '*I never put one brick upon another*' just as he can watch a ship ploughing the high seas and say '*I never touched one with a tar brush*'.[81]

Lockwood also cites a letter in a regional newspaper from the time which colourfully illustrates the perceived differences between manual and office workers:

> *The office slave boasts a waist and slender figure, while the navvy rolls along, a fully developed man of bone and muscle. A young navvy at work is a sight for the gods to envy, the bowed figure of the young clerk at work must make the gods weep.*

Lockwood affirms that these "*are not isolated, but rather recurring and consistent themes*" and that the "*influx of women merely strengthened the popular stereotype of the clerk and further detracted from the prestige of the occupation*".[82]

The feminisation of the clerical workforce in both Britain and the USA was emphatic. However, the rates of change varied between different parts of the office. It was not simply a case of women replacing men. Rather, in the context of a rapidly expanding economy and a proliferation of roles using new innovations in office machinery, women made many of these new roles their own. Wright Mills wrote with feeling on the matter.

> *The bookkeeper has been grievously affected by the last century of office change: his old central position is usurped by the office manager, and even the most experienced bookkeeper with pen and ink cannot compete with a high-school girl trained in three or four months to use a machine.*[83]

Davies cites brisk growth among typists and stenographers, while there was much slower growth in bookkeeping, cash and accounting. Women also dominated switchboards and file management. Among messenger, errand and office boys and girls, females never outnumbered males.[84] Stenographers often took long hand dictation, important in an age when 'personalised' letters remained important (in contrast to a typed letter). For this skill, stenographers would generally be paid a little more than a typist.

Lowe states that many office jobs such as stenographer, typist, receptionist, office machine operator remained 'female jobs' for decades, with clerical work "*rapidly becoming the country's largest female job ghetto*".[85] And, of course, as Rosabeth Moss Kanter observed, the prototype of male domination and female subordination in the contemporary workplace is the male manager and the female secretary.[86]

Delgado reinforces this notion, stating that the reorganisation and further division of labour within the office may have abetted its feminisation: it is "*possible that if offices had simply expanded without being reorganised, women would have had a more difficult time entering clerical work*". The point being made is that reorganisation of many offices

> *often resulted in a redivision of clerical labour and in the creation of new jobs, from stenographers and typists to file clerks, billing clerks and the like. Since many of these jobs, or at least their labels, had not existed before the growth of the office, were not defined as men's jobs. Women who took such positions did not face the argument that they were taking over men's work.*[87]

But setting aside perhaps the personal secretary, it can be argued that most of the posts open to women were of the sort that required little creativity or decision-making. For example, it was not long before stenography and typing became 'women's work' in so far as they held a virtual monopoly there. William H Leffingwell, who we met in the previous chapter, suggested that a woman

> *is to be preferred to the secretarial position for she is not averse to doing minor tasks, work involving the handling of petty details, which would irk and irritate ambitious young men, who usually feel that the work they are doing is of no importance if it can be performed by some person with a lower salary.*[88]

The simple fact is that the feminisation of office work was driven by a combination of factors, the following among them. First the white-collar economy was growing at such a rate that there were not enough males to fill all the posts available; and the growing female workforce helped avert a shortage of clerical workers. Secondly, women were excluded from professional work by the expense and length of training. And they were excluded from more muscular work both physical limitations and social convention. Both meant that office work became a mainstream activity for them.

Thirdly, for their part, females were attracted to clerical work because it paid better than most traditional areas of the economy and it conferred upon them higher status, certainly compared to factory work, retail work and domestic service. Fourthly, even in this role, their chances of promotion to management were slim. As the number and range of managerial roles expanded, so the male workforce filled them, thereby gaining greater work security and prospects of further advancement. Thus it was that women became associated with the tedium of the office rather than with anything of a more senior nature.

The net result of this process is that women were discriminated against not only because they were recruited into more mundane office tasks but that, once there, they had no escape route though to management where the jobs were being taken by men who were considered more worthy of higher-level work.

The one possible exception to this process was the secretary. Personal secretaries became the apex clerical work, with their constant access to 'executives' and their inside knowledge of decisions and upcoming events. The personal secretary, or personal assistant, was often fearsome, protective of her boss and conscious of her station relative to other females in the office. They were almost universally loyal to and protective of their boss. Like the clerk of earlier times, she would be smartly dressed, mirroring the middle class deportment of her boss. She would handle confidential information, on the one hand, while collecting dry cleaning and shopping for her boss, on the other hand. But even secretarial work was subject to social layering.

> *Secretarial work was divided into two kinds: marriage-like and factory-like. The elite corps of private secretaries were directly attached to one or more bosses for whom they did a variety of tasks and from whom they derived status. Other secretarial work was done in steno and typing pools whose occupants were little more than extensions of their machines – and highly replaceable at that.*[89]

Just as a 'class distinction' between manager and worker had opened up with the bureaucratisation of business, so it did within the female workforce. The social status of the secretary was clearly identifiable from the seniority of her boss, or by her position within the office. Rosabeth Moss Kanter is clear "*Secretaries derived their formal rank and level of reward not from the skills that they utilised and the tasks they performed but from the formal rank of their bosses*".[90] And of course status was accompanied by power: the more senior the boss, the more power wielded by the secretary. For example, they 'owned' their bosses diary, controlling what went in, and sometimes what did not; managing his time and influencing his decisions.

> *They could make it easy or difficult to see a top executive. They could affect what managers read first, setting priorities for them without their knowing it. They could help or hurt someone's career by the ease with which they allowed that person access.*[91]

Rosabeth Moss Kanter is also clear that the boss–secretary relationship was often based on fealty. The secretary knew the bosses' weaknesses, the frailties behind the calm public persona; and there had to be something to bridge the often huge gap in reward and power between boss and secretary. Moss Kanter suggested this was achieved by granting secretaries symbolic rewards and making the relationship a highly personal one.

> *They were expected to value non-utilitarian, symbolic rewards – which did not include career advancement – and to take on the emotional tasks of the relationship. Secretaries were rewarded for their attitudes rather than for their skills, for their loyalty rather than their talent.*[92]

It is a key feature of the boss–secretary relationship that it ran counter to the otherwise general bureaucratisation of the office. While ever thinner layers of management developed, and while ever more specialist job functions grew, the boss–secretary relationship was more akin to the owner–clerk relationship of times passed.

Women's pay. There are many reasons advanced for the feminisation of the clerical workforce. A favourite factor is cost of labour – women were paid less than men. Other factors include greater dexterity of women (for typing and paperwork), and their ability to cope with tedium. Certainly many male clerks were convinced that lower female salaries was one of the main causes of their low income and reduced status. There was, for example, the 'pin money' image of female clerks, perceived as "*single and either financed by or living with their parents [who] used their salaries mainly as a source of personal entertainment*".[93]

Alborn cites examples which appear to support the pay differential argument. Thus according to one source, between 1910 and 1914, women insurance clerks earned on average £60 a year, compared to £139 for men. Another source stated that the Yorkshire Fire & Life paid the 12 women who were hired as typists at its head office between 1902 and 1910 six shillings a week less than male clerks when they were 20 (£43 vs. £58 a year), but within 10 years that gap grew to nearly a pound a week (£89 vs. £138 a year).

The notion of a 'wages war' between men and women is questionable on the grounds that 'women's work' and 'men's work' quickly became distinctive and so comparisons were difficult. largely misplaced,

> *For the most part female clerks did not directly threaten the position and prospects of the well placed male career clerks. The employment of female clerks was associated with the increasing use of shorthand and typewriting, which, although it had once been hoped that they might somehow revolutionise the status of clerks, were in fact little more than additional work skills.*[94]

Davies expanded on the pay issue in the USA. She began by suggesting that while, superficially, the cheapness of labour ought to explain why employers preferred women over men, women's labour in America

> *has always been come cheaper than men's, so that it is not immediately obvious why employers did not always show preference for females. There must be a further reason why employers started to favour women for certain clerical positions.*[95]

She continued to observe that the expansion of capitalist firms was such that growing numbers of men were being attracted into managerial and professional positions and that as a result, "*the supply of men available for clerical work was considerably diminished*".[96] Davies also recognised that, by 1890, the number of females receiving secondary school diplomas or better had outstripped the number of men available, and that because the typewriter was sex-neutral, without historical ties to workers of either sex, female typists did not have to counter the argument that they were operating a man's machine. Even before this, "*the employment of women as clerks in the United States Treasury Department during the Civil War established a precedent that may have eased the entrance of women into offices ten and fifteen years later*".[97]

The fact remains that, while being lowly paid compared to the men, women were confined to 'lower-level' jobs, with virtually no prospect of progress. Many female workers were young, and the expectation at the time was that they would leave the workplace at some point to raise a family and not to return to the workplace in the medium-term. Furthermore, many jobs became associated with women, and regarded as 'women's work' with their own pay scales – distinctive from 'men's work'.

Women's impact. Given the scale of growth in white-collar activities, and the scale of growth in the wider economy, it is difficult to see how this growth in female labour had anything but a most marginal impact on the opportunities for male clerks. There is also strong evidence that female salaries reflected a generally lowly status in the office compared to many male clerks.

The office economy would evolve from the turn of the century to provide the bulk of the modern middle class. But in the Victorian office, the picture was a little more nuanced. We shall see in the next chapter the arguments suggesting that clerks evolved into a new proletariat rather than a new middle class; but before that there is also a strong case to suggest a social divide between male and female. As more and more lower skilled office work became available, it was increasingly taken on by females, and as there evolved a sharp pay differential between 'female work' and 'male work', so it might be argued that women became the new office proletariat, while the men enjoyed the salary, status and benefits of the middle class. In this sense, in the context of a rapidly expanding workforce, the arrival for females in the offices actually released men from more menial work to take advantage of the rapidly expanding market for skilled workers and managers.

At one level, the feminisation of the workforce was driven by the simple demand–supply equation: workers were needed in rapidly expanding numbers and there simply were not enough men to do all the work in the expanding economy. This process has been blamed many times for 'taking men's work' and for 'dragging down wage bills' but the evidence for this is scant. For a start, many of the machines operated by women had not existed for any length of time in a male dominated workforce. The entry of women into the office coincided with the new machine rather than because of it.

At the same time, the opportunities for men expanded enormously, not just in terms of the type of posts available to them, but also in terms of the hierarchy. There can be no question that as the workforce became more gender balanced, skilled and management work was very largely a male preserve while clerical work came to be seen more and more as 'women's work'. This was a trend that would continue in the decades after the war. Gregory and Nussbaum noted that women entered the workforce in record numbers in the 1970s, at the rate of one million a year, ensuring that clerical work became increasingly 'feminised'. In 1950, 62% of all clerical workers were women; rising to 74% by 1970, and to 74% by 1980, when 80% of clerical workers were women.[98]

The death of the clerk

By the turn of the century and in the lead up to the First World War, the era of the clerk was on the wane. Saval asks us to imagine a bookkeeper pausing to look down his ledger in 1860, only to look back up in 1920.

> *He might be surprised to see that his familiar small surroundings had melted away entirely, converted into a space whose high ceilings and tall columns resembled nothing so much as a cave swollen with stalactites. His lone colleague and brother-in-arms, the clerk, was gone, replaced by dozens of unfamiliar faces surrounding him in neat serried rows of desks. The cigar-chomping partner at the nearby rolltop desk would be gone too, having multiplied into a squadron of bosses locked away in snug executive suites high up in the stratosphere.*
>
> *His work is now harried, insistent, relentless. Farewell the tranquil, languorous days of the countinghouse; greetings to the factory-like labor of the office.*[99]

Businesses grew in size and complexity; much office work became more routine and automated. The paternal relationship between employer and clerk did not so much break down, as was engulfed in a wave of social, economic and technological upheaval. By the end of the century, close owner–clerk relationships had all but disappeared. Instead of owner and clerk living over the business it was managers, often of joint stock companies. Wright Mills captured the situation brilliantly:

> *Just as the working man no longer owns the machine but is controlled by it, so the middle-class man no longer owns the enterprise but is controlled by it. The vices as well as the virtues of the old entrepreneur have been transferred to the business concern.*[100]

Larger organisations such as government, the railways, industrial concerns, banks and insurers, began to introduce their own corporate means of rewarding workers: pensions based on savings were key; widow and orphan funds, profit sharing, bonuses and life insurance all became more commonplace. They also sought to introduce a social dimension to encourage camaraderie through sports and other recreational activities.

New business structures and work processes, mechanisation and the feminisation of the workforce combined to transform the look and feel of the office; introducing a more complex set of social relations and a much wider range of opportunities. Professional managers proliferated, to occupy the middle ground between owner and worker. The immediate effect was that the social gap between owners and clerks shifted to become one between managers and workers. For those not managers, clerks saw their status diminish quickly from quasi-management to simple office worker. As the nineteenth century reached its close, clerking as a career had lost much of its lustre. Perhaps because clerking workforces were now large and regimented rather than select and superior, a different perception of clerking and social mobility seems to have emerged.

Machines began to replace people on repetitive tasks, and clerks began to resemble factory workers managing a production process. Within growing organisations, departmentalisation, specialisation and mechanisation led increasingly to an internal clerking hierarchy to describe the typical division of labour among clerks. At the base there were office boys who ran errands, and apprentices or junior clerks who did little but copy manuscripts and post or copy transactions into accounts.

From the perspective of 1929, Grace Coyle described how office work was undergoing continuous transformation through mechanisation and task specialisation of work. She describes how, in large-scale offices, mass production methods become possible, and that in many cases, including insurance companies or mail order houses, routine work is "*especially adapted to an extensive division of labor*" and that

> *there has appeared a specialization as minute as that of any industrial plant. While the tendency towards specialization has not developed so far as to reduce every clerical worker to the performance of a single minute task, there is little doubt that many jobs are increasingly narrowed in their range and that a growing number of office jobs are devoted to a single small process.*[101]

As clerical workforces grew, they became more stratified and specialised, so it became a task of management to ensure that everyone was working in concert to the firm's shared goals. Clerks were no longer loyal to an entrepreneur and his family, they were loyal to an organisation. Increasingly, 'carrots and sticks' were employed for ensuring company loyalty and discipline. The immediate and long-term impact of the management function was to force the clerking profession into one of two directions: upwards to join management; or downwards to take on the more menial and mundane tasks. Either way, the established bond between owner and trusted agent was undermined.

Certainly for those who had held positions of trust and a close relationship with business owners, 'management' diminished their role. Unlike many of their manual colleagues on the shop floor, clerks were aware that they had a closer relationship with management. They conversed far more regularly; they dressed alike and they knew more about the business. If they could not themselves join 'the management' then their career structure was fundamentally altered.

As we saw in the previous chapter, the growth in white-collar work had a transformational impact on class structure in Europe and America: a new class of worker not only emerged, but came to dominate society. The clerk was no more. He was replaced by a thousand roles, collectively, 'the office worker'. From an American perspective, Saval summarises the death of the clerk well. He notes how the growing specialisation of clerical work meant that most workers increasingly knew only one thing – be it accounting or filing or billing – and had little incentive or opportunity to learn the entire business, unlike the traditional clerk who had "*seen his intimate world dismantled and reapportioned among hundreds of desks*".[102]

Summary. By the later years of the nineteenth century, clerking was evolving into an extremely diffuse range of activities across the economy, reflecting the fact that companies and organisations were becoming larger, with more complex processes and extensive trading activities. The rise of large national and global companies at this time profoundly changed the office, office work and social relations in the office. Office-based organisations – accountants, advertisers, banks, insurers, lawyers, surveyors and so on – were all expanding rapidly on the back of a rapidly industrialising economy.

For much of the nineteenth century, the relationship between owners and clerks was a critical one. Numerous historians and writers have used it as a backdrop to descriptions of socio-economic relations of the time. In many ways, the owner–clerk relationship was paternalistic, and sometimes even of a master–servant type, although the range of relationships was incredibly wide.

During this time, clerking was regarded as a craft, a skill to be honed and developed over years of loyal service. It was a critical task which had to keep current, accurate records of the financial and trading position of the owner's business. This in turn, demanded a level of trust not experience on the factory floor.

However, as joint stock companies with professional managers spread, they revolutionised the established owner–clerk relationship. Clerks evolved from all-round craftsmen, who were close to their business owners; were trusted and relatively autonomous, and who enjoyed relatively elevated social status, into lowly valued cogs in bureaucratic mazes, no longer with any overall picture of the businesses for which they toiled. Ultimately they morphed into simply office workers.

In this period of great change, clerks were subject to a wide range of experiences. Some felt part of the oppressed proletariat while others enjoyed the privileges of middle class comfort. Nevertheless, right through to the end of the nineteenth century, clerking remained an aspirational vocation: one which would offer social and economic advancement.

The growing feminisation of the workforce also helped further transform work and social relations in offices. Female workers helped avert a shortage of clerical workers at a time of rapid growth; they were excluded from professional work by the expense and length of training and so confined to more basic office roles, and they were attracted to clerical work because it paid better than most traditional areas of the economy and conferred upon them higher status, certainly compared to factory work, retail work and domestic service.

A male-dominated managerial class quickly evolved alongside a female workforce that dealt largely with administrative matters – secretarial, typing, data entry, filing and so on. Feminisation of the workforce led to a massive shift not only in social relations in the office but also in the workplace environment itself.

Notes

1 Pickard OG (1949) Office Work and Education 1848–1948 *The Vocational Aspect of Education* Vol 1 No 3 pp221–243.

2 Anderson G (1976) *Victorian Clerks* Manchester University Press p2.

3 *The Office* October 13th, 1888 p4 Cited in Heller M (2003) *London Clerical Workers 1880–1914: the Search for Stability* PHD Thesis University College London pp141–142.

4 Attewell P (1989) The Clerk Deskilled: A Study in False Nostalgia *Journal of Historical Sociology* Vol 2 No 4 pp357–388 ISSN 0952-190.
5 Klingender FD (1935) *The Condition of Clerical Labour in Britain* Martin Lawrence, London p2.
6 Alborn T (2008) Quill-driving: British Life-Insurance Clerks and Occupational Mobility, 1890–1914 *British History Review* Vol 82 Spring 2008 pp31–58.
7 Braverman H (1974) *Labour and Monopoly Capital: the Degradation of Work in the Twentieth Century* Twenty Fifth Anniversary Edition, Monthly Review Press, New York pp294–295.
8 Braverman H (1974) *Labour and Monopoly Capital: the Degradation of Work in the Twentieth Century* Twenty Fifth Anniversary Edition, Monthly Review Press, New York pp298–299.
9 Glenn EN & Feldberg RL (1977) Degraded and Deskilled: The Proletarianization of Clerical Work *Social Problems* Vol 25 No 1 pp52–64.
10 Lockwood D (1958) *The Blackcoated Worker: A Study in Class Consciousness* George Allen & Unwin, London p28.
11 Parsons CE (1876) *Clerks; Their Position and Advancement* Provost & Co., London pp12–13.
12 Parsons CE (1876) *Clerks; Their Position and Advancement* Provost & Co., London pp12–13.
13 Coyle GL (1929) Women in the Clerical Occupations *The Annals of the American Academy of Political and Social Science* May 1929 Vol 143 pp180–187.
14 Anderson G (1976) *Victorian Clerks* Manchester University Press p129.
15 Zunz O (1987) The Collar Line: Clerical Workers in America at the Turn of the Century In: Schröder WH & Jarausch KH (Eds.) *Quantitative History of Society and Economy: some International Studies* (pp79–93) St Katharinen: Scripta Mercaturae Verl.
16 Zunz O (1987) The Collar Line: Clerical Workers in America at the Turn of the Century In: Schröder WH & Jarausch KH (Eds.) *Quantitative History of Society and Economy: some International Studies* (pp79–93) St Katharinen: Scripta Mercaturae Verl.
17 Crossick G (1977) (Ed) *The Lower Middle Class in Britain 1870–1914* Croom Helm, London p24.
18 Zakim M (2006) The Business Clerk as Social Revolutionary; or a Labor History of the Nonproducing Classes *Journal of the Early Republic* Winter 2006 Vol 26 No 4 pp563–603.
19 *The Clerk* (1916) Cited in: Lockwood D (1958) *The Blackcoated Worker: A Study in Class Consciousness* George Allen & Unwin, London p40.
20 Lockwood D (1958) *The Blackcoated Worker: A Study in Class Consciousness* George Allen & Unwin, London p24.
21 Anderson G (1976) *Victorian Clerks* Manchester University Press p129.
22 Orchard BG (1871) *The Clerks of Liverpool: Being Ten Chapters on their Numbers, Distribution, Salaries, Grievances, Marriages, etc.* p41.
23 de Fraine HG (1960) *Servant of this House: Life in the Old Bank of England* Constable, London pp25–26.
24 Attewell P (1989) The Clerk Deskilled: A Study in False Nostalgia *Journal of Historical Sociology* Vol 2 No 4 pp357–388 ISSN 0952-190.
25 Anderson G (1976) *Victorian Clerks* Manchester University Press p53.
26 Anderson G (1976) *Victorian Clerks* Manchester University Press p44.
27 WM Acres (1931) *The Bank of England from Within 1694–1900* 2 vols. Oxford University Press, London p559.
28 Alborn T (2008) Quill-driving: British Life-Insurance Clerks and Occupational Mobility, 1890–1914 *British History Review* Vol 82 Spring 2008 pp31–58.
29 Alborn T (2008) Quill-driving: British Life-Insurance Clerks and Occupational Mobility, 1890–1914 *British History Review* Vol 82 Spring 2008 pp31–58.
30 Alborn T (2008) Quill-driving: British Life-Insurance Clerks and Occupational Mobility, 1890–1914 *British History Review* Vol 82 Spring 2008 pp31–58.
31 Alborn T (2008) Quill-driving: British Life-Insurance Clerks and Occupational Mobility, 1890–1914 *British History Review* Vol 82 Spring 2008 pp31–58.
32 Alborn T (2008) Quill-driving: British Life-Insurance Clerks and Occupational Mobility, 1890–1914 *British History Review* Vol 82 Spring 2008 pp31–58.

33 Haigh G (2012) *The Office: A Hardworking History* The Miegunyah Press p95.
34 Braverman H (1974) *Labour and Monopoly Capital: the Degradation of Work in the Twentieth Century* Twenty Fifth Anniversary Edition, Monthly Review Press, New York p204.
35 Lockwood D (1958) *The Blackcoated Worker: A Study in Class Consciousness* George Allen & Unwin, London pp206–207.
36 Alborn T (2008) Quill-driving: British Life-Insurance Clerks and Occupational Mobility, 1890–1914 *British History Review* Vol 82 Spring 2008 pp31–58.
37 Lockwood D (1958) *The Blackcoated Worker: A Study in Class Consciousness* George Allen & Unwin, London p29.
38 Anderson G (1976) *Victorian Clerks* Manchester University Press p132.
39 Alborn T (2008) Quill-driving: British Life-Insurance Clerks and Occupational Mobility, 1890–1914 *British History Review* Vol 82 Spring 2008 pp31–58.
40 Corey L (1935) *The Crisis of the Middle Class* Covici Friede, New York p204.
41 Klingender FD (1935) *The Condition of Clerical Labour in Britain* Martin Lawrence, London p2.
42 Anderson G (1976) *Victorian Clerks* Manchester University Press p4.
43 Klingender FD (1935) *The Condition of Clerical Labour in Britain* Martin Lawrence, London p5.
44 The correspondent 'Benjamin Battleaxe' took complaining lower-level clerks to task in the letters column of a Liverpool newspaper in the late 1860s. After initially claiming to be an employer, he revealed himself at the end of the correspondence as 'a simple clerk.'
45 Orchard BG (1871) *The Clerks of Liverpool: Being Ten Chapters on their Numbers, Distribution, Salaries, Grievances, Marriages, etc.* pp40–41.
46 Orchard BG (1871) *The Clerks of Liverpool: Being Ten Chapters on their Numbers, Distribution, Salaries, Grievances, Marriages, etc.* p37.
47 Parsons CE (1876) *Clerks; Their Position and Advancement* Provost & Co., London p13.
48 Parsons CE (1876) *Clerks; Their Position and Advancement* Provost & Co., London p33.
49 Kynaston D (1995) *The City of London Volume Two: Golden Years 1890–1914* Chatto & Windus, London p248.
50 Orchard BG (1871) *The Clerks of Liverpool: Being Ten Chapters on their Numbers, Distribution, Salaries, Grievances, Marriages, etc.* pp63–65.
51 Lockwood D (1958) *The Blackcoated Worker: A Study in Class Consciousness* George Allen & Unwin, London p40.
52 Jeacle I (2010) The Bank Clerk in Victorian Society: the Case of Hoare and Company *Journal of Management History (Archive)* Vol 16 p6.
53 Alborn T (2008) Quill-driving: British Life-Insurance Clerks and Occupational Mobility, 1890–1914 *British History Review* Vol 82 Spring 2008 pp31–58.
54 Attewell P (1989) The Clerk Deskilled: A Study in False Nostalgia *Journal of Historical Sociology* Vol 2 No 4 pp357–388 ISSN 0952-190
55 Anderson G (1976) *Victorian Clerks* Manchester University Press p131.
56 Alborn T (2008) Quill-driving: British Life-Insurance Clerks and Occupational Mobility, 1890–1914 *British History Review* Vol 82 Spring 2008 pp31–58.
57 Alborn T (2008) Quill-driving: British Life-Insurance Clerks and Occupational Mobility, 1890–1914 *British History Review* Vol 82 Spring 2008 pp31–58.
58 Heller M (2003) *London Clerical Workers 1880– 1914: the Search for Stability* PHD Thesis University College London pp113–14.
59 Heller M (2003) *London Clerical Workers 1880– 1914: the Search for Stability* PHD Thesis University College London p117.
60 Anderson G (1976) *Victorian Clerks* Manchester University Press p35.
61 Anderson G (1976) *Victorian Clerks* Manchester University Press p37.
62 Jeacle I (2010) The Bank Clerk in Victorian Society: the Case of Hoare and Company *Journal of Management History (Archive)* Vol 16 p17.
63 Kynaston D (1995) *The City of London Volume Two: Golden Years 1890–1914* Chatto & Windus, London pp249–250.

64 Jeacle I (2010) The Bank Clerk in Victorian Society: the Case of Hoare and Company *Journal of Management History (Archive)* Vol 16 p16.
65 Anderson G (1976) *Victorian Clerks* Manchester University Press pp33–37.
66 Lockwood D (1958) *The Blackcoated Worker: A Study in Class Consciousness* George Allen & Unwin, London p122.
67 Coyle GL (1929) Women in the Clerical Occupations *The Annals of the American Academy of Political and Social Science* May 1929 Vol 143 pp180–187.
68 Delgado A (1979) *The Enormous File: A Social History of the Office* Richard Clay, Suffolk p39.
69 Coyle GL (1929) Women in the Clerical Occupations *The Annals of the American Academy of Political and Social Science* May 1929 Vol 143 pp180–187.
70 Delgado A (1979) *The Enormous File: A Social History of the Office* Richard Clay, Suffolk.
71 Lewis JE (1988) Women Clerical Workers in the Late Nineteenth and Early Twentieth Centuries In: Anderson G (Editor) (1988) *The White-Blouse Revolution – Female Office Workers Since 1870* Manchester, New York.
72 Alborn T (2008) Quill-driving: British Life-Insurance Clerks and Occupational Mobility, 1890–1914 *British History Review* Vol 82 Spring 2008 pp31–58.
73 Lockwood D (1958) *The Blackcoated Worker: A Study in Class Consciousness* George Allen & Unwin, London p122.
74 Delgado A (1979) *The Enormous File: A Social History of the Office* Richard Clay, Suffolk p38.
75 Lockwood D (1958) *The Blackcoated Worker: A Study in Class Consciousness* George Allen & Unwin, London p123.
76 Pickard OG (1949) Office Work and Education 1848–1948 *The Vocational Aspect of Education* Vol 1 No 3 pp221–243.
77 Wardley P (2006) *Women, Mechanisation and Cost-savings in Twentieth Century British Banks and Other Financial Institutions* XIV International Economic History Congress, Helsinkin Session 14 p2.
78 Batiz-Lazo B & Wardley P (2007) Banking on Change: Information Systems and Technologies in UK High Street Banking, 1919–1969 *Financial History Review*, Vol 14 pp177–205.
79 Booth A (2004) Technical Change in Branch Banking at the Midland Bank, 1945–75 *Accounting, Business & Financial History* Vol 14 No 3 pp277–300.
80 Kynaston D (1995) *The City of London Volume Two: Golden Years 1890–1914* Chatto & Windus, London p31.
81 Shall Our Sons be Clerks? *The Clerk* April 1925 p51 Cited in Lockwood D (1958) *The Blackcoated Worker: A Study in Class Consciousness* George Allen & Unwin, London p123.
82 Quoted from the *Newcastle Weekly Chronicle* by *The Clerk*, December 1930 p181 Cited in Lockwood D (1958) *The Blackcoated Worker: A Study in Class Consciousness* George Allen & Unwin, London pp124–125.
83 Wright Mills C (1951) *White Collar: the American Middle Classes* OUP, New York pp63–64.
84 Davies MW (1982) *Women's Place at the Typewriter* Temple University Press, Philadelphia p52.
85 Lowe GS (1984) The Enormous File: the Evolution of the Modern Office in Early Twentieth-Century Canada *Archivaria* 19th January.
86 Moss Kanter R (1977) *Men and Women of the Corporation* Basic Books, New York p18.
87 Delgado A (1979) *The Enormous File: A Social History of the Office* Richard Clay, Suffolk p58.
88 Leffingwell WH (1925) *Office Management: Principles and Practice* AW Shaw, New York pp621–622.
89 Moss Kanter R (1977) *Men and Women of the Corporation* Basic Books, New York p27.
90 Moss Kanter R (1977) *Men and Women of the Corporation* Basic Books, New York p74.
91 Moss Kanter R (1977) *Men and Women of the Corporation* Basic Books, New York p75.
92 Moss Kanter R (1977) *Men and Women of the Corporation* Basic Books, New York pp84–85.
93 Anderson G (1976) *Victorian Clerks* Manchester University Press pp57–59.
94 Anderson G (1976) *Victorian Clerks* Manchester University Press p60.
95 Davies MW (1982) *Women's Place at the Typewriter* Temple University Press, Philadelphia p52 p56.
96 Davies MW (1982) *Women's Place at the Typewriter* Temple University Press, Philadelphia p52 p57.
97 Davies MW (1982) *Women's Place at the Typewriter* Temple University Press, Philadelphia p52 p58.

98 Gregory J & Nussbaum K (1982) Race Against Time: Automation of the Office. An Analysis of the Trends in Office Automation and the Impact on the Office Workforce *Office: Technology and People* 1, 1982 pp197–236 Elsevier Scientific Publishing Company, Amsterdam.
99 Saval N (2015) *Cubed: a Secret History of the Workplace* First Anchor Books, New York pp33.
100 Wright Mills, *White Collar* pp108–109.
101 Coyle GL (1929) Women in the Clerical Occupations *The Annals of the American Academy of Political and Social Science* May 1929 Vol 143 pp180–187.
102 Saval N (2015) *Cubed: a Secret History of the Workplace* First Anchor Books, New York p39.

9 The machine age

Between 1870 and the First World War, the world of office work was transformed. The role of the clerk became extinct, to be replaced by the office worker and the manager class. The office environment was provided with lighting, heating, ventilation, lifts and modern sanitation; mechanised equipment arrived; the telephone and telegraph revolutionised communications, and the workforce was feminised. America saw steel frame construction and the arrival of the skyscraper. All in little more than four decades.

Typically family owned and run businesses yielded to stock companies and the manager class grew rapidly. Organisations grew larger, departmentalised, more complex and hierarchical. The role of the clerk split and diffused into hundreds of different roles, ranging from mind-numbing routine to specialist and skilled work. As we saw in the previous chapter, clerking became simply office work; an aspiring career became a job; the office workforce matured. Victorian values of loyalty, deference and service, which fitted the small and owner-run business context, began to yield to nascent corporate culture and the emergence of the manager class and company man.

Robert Thorne, London clerk. Certainly for some, in popular culture, the role of the clerk had been diminished. In his 1907 novel, *Robert Thorne: The Story of a London Clerk*, Irish writer Shan Bullock described a scene in which, one evening, the young protagonist divulges his ambitions to become a clerk to his father. Robert left his room and descended the stairs to the 'great stone-flagged kitchen', where his father (who 'kept the village school' and 'played the church organ') sat drawing on a pipe and reading 'an old Latin book'. When his father addressed him, to ask where his son had been, Robert replied that he had been studying. Rather surprised, father asked the nature of his son's enquiry, and Robert said he was studying how to be a clerk. Father recoiled in horror, asking who had been 'putting all this in your foolish head'. When Robert explained who (an acquaintance who was a clerk in the War Office), father exploded

> *You want to be like him, perched on a stool all day with your nose in a ledger. You want his pale face, and his slouch, and his simper … . Haven't I taught you that what a man owes to himself is to strive after manhood. A clerk with a clerk's narrow little soul – is that your idea of a man! … . I'd sooner see you serving cheese behind Jago's counter.*

Robert was not sure how to respond and just blurted out 'but I couldn't help myself', which just seemed to heighten father's emotions further: "*No, you can't help yourself. And why? Because yourself is not worth helping. You have the spirit of a slave, sir. A clerk! A creature with a pen behind its ear!*"[1]

This scene played out when Robert was 18 years old, somewhere around 1890. After failing his Civil Service entry exams, Robert eventually secured a place at HM Tax Office at Somerset

DOI: 10.1201/9781003582205-12

House (see Chapter 3). Bullock then proceeds to describe the office into which the young apprentice was taken.

On his due date he arrived at the imposing buildings, and was taken to the second floor, via a passage that was "*long, narrow, gloomy, on either side baize-covered doors, each numbered and bearing red slips with names in gilt letters*". He was guided to a large manager's room, with high windows and aspect over the Thames. The room had

> *a thick carpet, mahogany chairs, and tables laden with trays, papers, books. The walls were hung with pictures, photographs, almanacs, maps, and had shelves full of official literature. In the corner stood a bag of golf sticks, a Volunteer officer's sword and helmet case and a pair of Indian clubs. On the mantel was a date box, a cup and saucer, and a Japanese teapot.*

Mr Winter (the manager) "*sat working at a long polished table*". Following a formal introduction, Mr Winter called for Mr Cherry – Robert's manager – to join them. Mr Cherry – a small man, "*scrupulously neat and respectable, his benevolent face set off with tufts of whisker*" – escorted Robert to his own room where Robert would work. The room itself is described as "*small, poorly lighted by one window*", with tall oak cupboards along one wall and deal bookshelves upon another; the room itself "*needed cleaning*". The carpet was thin and shabby, the hearthrug worn through in the middle. Mr Cherry's desk was positioned next to the fire, and under the winder there were two other chairs, at which sat Robert's co-worker: "a lean sallow-faced young man" who rose to shake Robert's hand and "bent again to work".

Mr Cherry then gave Robert a copy of the office regulations, "*many pages ... expounding my personal conditions of service*". But Robert was happy to be one of Her Majesty's servants and earning an £80 salary. At lunchtime, Mr Cherry directed Robert to the luncheon room, where he was granted a 45 minute break. The room, in the basement, was a large cheerful room, where Robert found 50 or 60 clerks sitting at the tables, some in merry parties of five or six, and it cheered Robert

> *to hear the laughter, the friendly badinage, the merry sound of knife and fork; to see men enjoying good beef and mutton and bitter beer; to look up and see the sunshine upon the roofs.*[2]

After lunch, and before Mr Cherry returned, 'the sallow faced young man' who proceeded to describe to Robert what clerking for Her Majesty was *really* like. Robert noticed then "*how worn he was, how shabby too, and not very cleanly; and I also understood why he did not have even nine penny worth of food in the basement*". Oliver was very pessimistic about prospects for advancement and to make his point, showed Robert the office establishment list. Casting his finger across the list, he pointed to the names above a thick line:

> *Well, they're the top-hats of the establishment – Commissioners, Chiefs, Principals, Heads of Sections, and all the rest. Look at their salaries, anything from two thousand to four hundred.*

Robert enquired as to the reason for the thick line.

> *There it is. It's what's below them and above us, and it's what we can't pass. We're not good enough. We aren't class enough. Look at us a hundred and twenty Men clerks all in a bunch like sheep in a pen.*

Above the line, everyone was 'esquire'; below everyone was 'mister'. The afternoon wore slowly; "*the room felt hot and stuffy, smelt of coal smoke and dust*".[3]

> *One day was much like another – only worse ... I felt sometimes like a machine, a thing grinding out its daily portion, mechanically turning leaves with cold grimed fingers, writing, casting, carrying forward, the sun peering in at long intervals upon me through the grimy windows or the fog hanging drear about the candles. The work was dull, everlasting. Over it my brain toiled or some part of my brain, but it never stirred my wits, hardly my interest.*[4]

These lengthy excerpts from the book merit reproduction here because, within them, there is so much colour about ordinary office life towards the end of the nineteenth century. The extracts describe social stratification; discontent, weariness, mistrust and the impact of hierarchy. They describe the sheer poverty of the work environment; the environmental conditions and the probable pollution leading to health problems. And they show how clerking had changed: gone it appears is the gentleman craftsman, and in his place a manager supported by poorly paid office workers. Moreover, the extracts summarise perfectly the transition from the clerk stereotype of the nineteenth century to the office worker of the twentieth.

There can be little doubt that by the turn of the century the allure of clerking as an aspirational career was on the wane. Robert Thorne's father typified the view that clerking had become what we might refer to today as a 'dead-end job'. Larger organisations were spawning armies of clerical workers undertaking a wide range of tasks, with an inevitable spread between the skilled and the mundane.

Working conditions. The key practical qualities of clerks at this time included attention to detail and accuracy, together with a large helping of writing skills – both literacy and legibility – which were important in writing correspondence, contracts, sales documents and so on. Accounting skills and arithmetic were vital for accurate bookkeeping. All of these characteristics needed to be applied with conscientiousness and patience. And yet, more often than not, office work was taking place in working conditions that would not be acceptable today.

As well as the work itself, there was also the physical nature of the office workplace. Spaces were cramped; desks and chairs/stools were poorly designed and sanitation was rudimentary. Zakim provides an American perspective, describing the proliferation of furniture types that included.

> *single standing desks and double standing desks, counter desks or double counter desks that contained a lid and a drawer, single office tables and double office tables, or sitting desks with nine pigeon holes or fifteen pigeon holes or with three upper drawers and three lower drawers. There were also manoeuvrable chairs, for instance, such as Matthew W. King's patented "Tilting Chair" or Dr Oliber Halstead's "Anti-Dyspeptic Chair," the latter reaching the market in 1845 with promises to ease the affliction of those "compelled to be sedentary" by approximating the motion of equestrian exercise.*

And once sat on their chair, at their desk, the most important function of all, of course, was the scribing. For this, offices "*filled up with square inkstands and library inkstands and banker inkstands, and with a growing variety of inks, some designed for copying and others for letter writing and all guaranteed to flow freely without creating annoying globules and corrosion*".[5]

Zakim goes on to describe how steel nibs began to replace traditional quills, and were marketed by sizes and styles: "*Large no. 1 pens were designed for bold, rapid writing. The medium no. 2 and no. 3 sizes were intended for correspondence. The finer no. 4 was marketed for fine work and ladies' use*". He describes how the "*most far-reaching technical innovation combined nib, shaft, holder, and inkstand into a single tool that was now called a fountain pen*", that could be "*carried on one's person from place to place and was ever ready for use*".

The desk, chair and pen provided the bare essentials of the work itself, but the physical environment played a major part in health problems associated with clerking. Lighting was poor – often candle-lit – ventilation likewise and damp and dust were widespread. Artificial lighting enabled clerks to work longer winter hours, but gas burner technology was awkward, tiresome and polluting. The electric light of 1880 offered great promise but it was a further 20 years before the full switch took place.

It is difficult to comprehend from the perspective of the twenty-first century that nineteenth century offices were universally heated and ventilated by open coal fires. A 'typical' office, divided into multiple rooms would have a coal fire in each room. Apart from the health hazard, all of these, at least during cooler months, had to be daily emptied, cleaned and filled. The coal had to be stored and staff were employed for such duties. Of course the logistical scale of such enterprise grew alongside the increasing size of office buildings. Locker cites one extreme example, which is Salisbury House, on Finsbury Circus:

> *This building had 850 rooms with one fireplace per room. Its annual consumption of 350 tons of coal was delivered each summer and stored under the pavement. Four people in the coal stores and six more upstairs were required to distribute the coal each night. This labour intensive system continued until 1953, when central heating was finally installed.*[6]

Despite the growth in office work and the numbers of clerks and other office workers, working conditions did not always take a commensurate step forward. In wondering how much the growth and change mattered "*to the wretches actually working*" in the City, Kynaston suggested that what abided

> *day after day, week after week, year after year – was not so much the larger environment as the actual grinding, routine: the voluminous ledgers, the salient account books, the endless, pernickety correspondence. There lies the true, inner, inscrutable history of the City, as draining on human vitality as the endless pleas of endless suitors recorded on the peau de chagrin in Balzac's famous story.*[7]

The same author argued that for almost everyone

> *in the vast clerical army, daily life in the City remained manual and laborious. It is true that in 1873 the first typewriter was spotted in the square mile, but it would be many years before it became a standard feature of office furniture – not least because of an innate reluctance to employ women typists, eventual superceders (sic) of the traditional male copying clerks.*[8]

Thus, despite the growing mountains of paper, offices through to the end of the century remained quite primitive environments (as described in Chapter 5), certainly by today's standards. For example, lighting and heating in Victorian office buildings were primitive. Gas lighting had been available in factories and mills since the start of the nineteenth century, and it was adopted in office buildings as an alternative to candle light. The benefits of gas lighting over candles, however, were marginal. The fire hazard of the candle was replaced by the pollution of the gas light. But even when electric lighting arrived, it took a long time to become the norm.

Swan and Edison co-invented the carbon filament incandescent lamp around 1880 and in 1882 Parliament passed the Electric Lighting Act, which gave consent for private companies to generate and distribute electricity. But as Locker observes, the electricity was expensive, and the electric companies were not very effective: as late as 1890 there were no companies serving the City.[9] While Lloyds Bank routinely installed electric lighting after 1900 and the Midland and City Bank did likewise after 1903; it took until around 1910 before general lighting was widely installed in offices.[10]

Central heating had an equally chequered start, with progress being held back by the "*suspicion, caution, and outright abhorrence of the Victorians towards central heating*". The love affair with the open fire place persisted, but came to be criticised as being too local in effect; too wasteful of fuel and labour, dirty and unattractive, while being compared to American steam systems for venting entire office buildings.[11]

Offices were dusty, musty, smelling of gas and burning coal. They were likely draughty; despite the gas lighting, they were darkened by winter days. Women were often effectively excluded from office work due to the lack of suitable toilet facilities; the earliest flushing water closets were not available until the mid-1880s. There were no lifts.

There was much talk of proletarianisation, deskilling, mechanisation, job losses and the threat that more females in the workforce would force wages down. Of all these concerns, it was the rapid mechanisation of the office that posed that greatest and most immediate threat. Adding machines, addressographs, copying machines, mimeographs and typewriters were all exotic additions to the office landscape at this time, threatening the traditional 'craft' of clerking. Indeed, the twentieth century spawned a whole genre of writing purporting to trace the demise of clerking through deskilling and degradation. Attewell cites a raft of such authors who found in the "*vicissitudes of the clerical occupations an exemplar for the fate of work in the capitalist world: the erosion of status, income, and skill*". The story begins, he suggests in the nineteenth century

> *with the clerk as craftsman or artisan – a skilled generalist. knowledgeable, well rounded, waiting his turn to step up into a partnership. This is followed by the onslaught of Taylorism, feminization, and office machinery, three forces which reduce the clerk to a narrow machine minder in a clerical factory. The historical coup de grace is left for the present period, when computer automation removes the last vestiges of skill from a once-cherished occupation.*[12]

Different forms of this discussion rumbled on throughout the twentieth century and into the twenty-first. At the time of writing, in 2024, there is much talk about the potential impact of artificial intelligence not only on how we undertake work, but also how many people will be required. There is a widespread belief that AI will lay waste to large swathes of office work, massively reducing demand for labour and buildings.

Qwerty clatter

There was perhaps no single addition to the office that defined the end of the manual clerking era and the beginning of mechanical office work than the arrival of the typewriter. This had the twin impact of introducing more routinised work and enabling sharp increases in productivity.

> *Typewriters and shorthand stenography more than doubled the rate of writing letters from the 1880s, converting carbon paper from a curiosity into the best method for duplicating. Copying by hand and letterpress gave way to the distinctive blue or purple shades of the*

> *mimeograph; addressographs enabled clerks to address up to sixty envelopes a minute; and arithmometers enabled comparable economies in basic mathematical calculations.*[13]

Before the advent of the typewriter, offices were generally quiet. Monastery-like quiet. Clerks spoke when they needed to; but there was no gossip and chit chat, let alone laughter. Directors and partners sat in their enclosed offices dealing with matters of import; while clerks attended the meticulous detail of their ledgers and accounts. Of course, things changed when the office mechanised. The clatter, clank and general din of mechanical tools and equipment soon changed the ambience of office life forever.

Locker notes that in an age before acoustical controls, "*typewriters presented a problem in offices that could only be resolved by their isolation from other clerical tasks*", thus separating female workers from their male co-workers. "*Social conventions supported this notion; an office with separate working spaces was deemed a better place to work than one with mixed males and females. In larger organisations with clerical functions spread over several floors, the lift made it possible to establish the centralised typing room.*"[14]

One of the first typewriters to appear in London was the Fitch, made in London by the Fitch Typewriter Company from around 1890 (Figure 9.1). Up to this point, operators typed 'blind' – only seeing their work once several lines had already been typed. On the Fitch, by contrast, the type bars were located behind the paper, allowing the operator to see as they typed.

The Fitch typewriter was invented by New Yorker Eugene L Fitch (1846–1918), while he was living in Des Moines, Iowa. It was granted a patent 1886–1887. In 1888, Fitch signed a contract with Brady Manufacturing of Brooklyn to manufacture and the Fitch Typewriter Company products were available on the commercial market by 1889. Around this time, the design and patent rights for the Fitch Typewriter beyond the Americas were sold to a British company and

Figure 9.1 The Fitch typewriter, c1891.

production was taken up by the Fitch Typewriter Company in London. By May 1890, British versions of the Fitch (later referred to as the Fitch 2) were being advertised and sold in British newspapers.

Wright Mills argued that corporate growth and mergers in 1920s America "*further enlarged the unit of the business structure and entailed more extensive co-ordination*" and as a consequence, machines were needed in the office "*to keep up with the effects and management of machines in the plant*".

> *Yet we are still only in the beginning of the office machine age. Only when the machinery and the social organisation of the office are fully integrated in terms of maximum efficiency per dollar spent will that age be fully blown.*[15]

Indeed, office mechanisation played a significant role reinforcing the common notion that Taylorism led to the factory-like layout of offices. And yet the machinery was a fundamental component of business infrastructure to enable growth in scale by introducing capacity and efficiency in output. Lowe observes that machinery offers several advantages for the manager in terms of increasing the pace of production; providing a much higher degree of regulation over the labour process; and by advancing the division of labour it allows workers to be hired at lower wages. He also noted that while some technologies such as typewriters and adding machines "*precipitated little change in the actual organization of clerical routines; they merely assisted in the execution of traditional tasks*", when machines

> *were integrated into production systems, such as a typing pool or a specialized accounting division staffed by adding machine operators, the logic of efficiency inherent in machinery began to triumph over the human element.*

He cites the example of the introduction of punch card equipment during the 1920s as setting the stage for post-Second World War office automation.

> *Hollerith machines represented a quantum leap in office mechanization. The capacity for information management grew exponentially, and working conditions in Hollerith departments bore a striking resemblance to those in a factory.*[16]

Grace Coyle observed that the typewriter was simply the beginning of mechanisation, to be quickly followed by the addressograph, the autographic register, the duplicator and the mathematical machines. She went on to state that "*mechanical devices have become as universal and as essential to almost every detail of office work as it is in the factory*" and that to witness this

> *we have only to step from the bookkeeping department of any modern office with its bookkeeping machines, its billers, and statistical machines, through the stenographic bureau where automatic typewriters and dictaphones compete with the descendants of Mr. Scholes' invention, to the mailing room with its automatic folding, stamping, and sealing devices. In every corner, human labor is being replaced by mechanical devices or is being used to tend machines which seem almost human in their capacities.*[17]

Mechanisation of the office environment enabled the scaling up of businesses to handle enormously increased volumes of transactions, or trade. The arrival of the typewriter, carbon paper, copiers, printers, punch cards and the telegraph and telephone all in combination led to vast quantities of paper being produced to communicate, record, store and verify. These were the early years of information management.

Office mechanisation began in earnest in the final quarter of the nineteenth century. By the 1890s, a range of mechanical devices were established in the office, notably Morse's telegraph, Bell's telephone, Edison's dictating machine, Remington's typewriter and a host of other gadgets including adding machines, addressographs and copiers. The early decades of the twentieth century saw further mechanical innovations, including the teletypewriter, ticker tape, the electric typewriter, duplicating machines, calculators, tape recorders for dictation, offset printing presses small enough for office use and data-processing equipment operated with punched paper cards.

As we shall see in the next section, there was a prevalent political interpretation of the mechanisation – proletarianisation – but setting aside this aspect for the moment, on-the-ground responses to the mechanisation of the office varied widely. In some quarters it received a very positive welcome, being seen "*as a sign of progress and modernity*".[18] Indeed, "*The Office magazine and other similar journals almost worshipped the very idea of technology*". The opening edition of the magazine on 22 September 1888, wrote

> *Typewriting and shorthand writing are sister arts and time-savers; the copying press and other mechanical contrivances for multiplying copies of documents are of no less importance. None of these arts and contrivances have, until now, been brought to public notice and advanced and encouraged through the channel of a bitter press.*[19]

On the other hand, in some areas there was strong and long-held resistance. Kynaston cites the examples of Government brokers Mullens and bill brokers Gilletts where, before 1914 there were no typewriters; and the commodity brokers Lewis & Peat, Andrew Devitt who insisted that all copies of contracts (one to the seller, one to the buyer, one to be posted into the large contract book, each of them identical) be laboriously written out by hand.[20]

Obviously such examples of resistance were rather extreme and are not in any way representative; but they serve to demonstrate a less than universal welcome for mechanisation. Generally speaking, there was a sense that the machinery in the office would have serious negative social and economic impacts. Various sources have described the introduction of machines as leading to the demise of clerking as it was known and understood. Such sources have described the gradual process of mechanisation as transforming clerking into unskilled, under-valued, low-paid and insecure work. There was also a sense that machines would somehow 'de-humanise' work; writing in America, Lewis Corey suggested that "*a typical large office is now nothing but a white-collar factory.*"[21]

Up to the later years of the nineteenth century, companies were limited in the extent of their growth by the limits to productivity in their manual workforces. There were only so many customer accounts, invoices, ledgers and so on that could be kept up-to-date by the number of clerks employed. And the latter was capped by the profit achieved from the former. All this changed dramatically with the introduction of the typewriter, the duplicator, the vertical filing system and the calculating machine. The telegraph and telephone similarly transformed the flow of information.

Office mechanisation from the 1880s onwards allowed organisations to grow in scale and complexity. Banks were clearing more cheques, insurance companies were processing more policies and commercial offices were meeting rising orders – all were struggling to keep up with the growing demands of an expanding society that was growing accustomed to a huge increase in goods and services. Mechanisation not only allowed this growth to happen, but it also came with built-in savings in productivity: for example, an insurance business could double the number of policies it processed without doubling the number of clerks it employed. Mechanisation was labour-saving and profit lifting.

As the nineteenth century moved into its later years, the scale of operations in companies was growing significantly; they were becoming far more complex, and more bureaucratic. Mumford captured well the impact of the evolving office technology on organisational structure – internally and spatially – and the emergent office economy.

> *With the manufacture of the typewriter in the seventies, and the coincident spread of high speed stenography, more and more business could be conducted on paper. Mechanical means of communication and mechanical means of making and manifolding the permanent record, mechanical systems for audit and control – all these devices aided the rise of a vast commercial bureaucracy, capable of selling in ever remoter territories.*[22]

He not only identifies how the technologies allowed for great output but also hinted both at the concomitant rise in organisational complexity (bureaucracy) and the ability to reach ever more distant markets.

Punch cards. Herman Hollerith (1860–1929) was born in Buffalow, New York of German immigrant parents. He started at the City College of New York in 1875 and graduated in engineering from the Columbia School of Mines in 1879. After serving as a graduate assistant to his professor WP Trowbridge, he then joined the US Census Bureau as a statistician where he started to examine the problems of analysing large amounts of data resulting from the US Census.

In 1882, Hollerith joined the Massachusetts Institute of Technology where he taught mechanical engineering, while at the same time pursuing the Census problem and began his first experiments. These involved a paper tape with pins, which passed through holes in the tape to make an electrical contact, later replacing the tape with cards. Hollerith applied for his first punch card patent in 1884. He continued to refine and develop his MIT work, eventually leading to the Hollerith Electric Tabulating System. Further improvements to the card reading machines and to the pin that punched the card led to increases in the amount of information that could be stored on each card. Hollerith's system was tested on tabulating mortality statistics in Baltimore, New Jersey in 1887. Its big break came in winning a competition for use in the 1890 US Census. Figure 9.2 illustrates the Census Bureau a few years later, in the 1930s, using similar technology.

In scaling-up operations, Pratt & Whitney were commissioned to manufacture the punches and the Western Electric Company made the counting machines. Counting was completed in around 3 months compared to an anticipated 2 years if completed by hand. But as well as a much faster count, Hollerith's system also enabled the collection of more information, thereby expanding the utility of the Census.

The system was used in Austria and Norway in 1891 and in the UK Census of 1911. Further refinements included a mechanism for automatically feeding the punch cards and automatic sorting procedures.

Following widespread recognition of his work, Hollerith founded the Tabulating Machine Company in 1896. Unfortunately, Hollerith sought to exploit his virtual market monopoly by setting very high prices for the use of his system. So much so that the Census Bureau began to develop a rival system, with its own patent, under the direction of James Powers, who then established Powers Tabulating Machine Company.

In 1911, Hollerith's company merged with Computing Scale Company of America and International Time Recording Company to become the Computing Tabulating Recording

Figure 9.2 US Census Bureau employees using punch cards, 1930s.

Company (C-T-R), under the leadership of Charles Ranlett Flint. Following a decline in fortunes, the company saw Thomas Watson appointed general manager in 1914, who took complete control of the firm as President in 1924 and renamed it International Business Machines, or IBM. Hollerith served as a consulting engineer with the C-T-R until his retirement in 1921.

Adding machines. Among the earlier examples of mechanisation were adding machines. Clerks could immediately see the threat to their position which largely rested on the twin skills of literacy and numeracy. Up to this point, clerks were skilled in mental arithmetic over long columns of figures, a task referred to as casting. Machinery threatened to make the task redundant when numbers could be punched into a machine and a lever pulled to print the result.

Adding machines first emerged in the early-seventeenth century (Frenchman Blaise Pascal invented one of the first machines in 1642), but they were very cumbersome and had very basic functionality. The first mechanical calculator to enter mainstream use was not created for offices, but for military deployment. It was invented by the wonderfully named Charles Xavier Thomas de Colmar (1785–1870).

Born in Colmar, capital of the Alsace wine region, Colmar joined the French army in 1809 and rapidly rose through the ranks, to Inspector of Supply for Napoleon's army. It was in this logistics role, having to calculate the provision needs for a whole army, that he conceived the idea of mechanical calculation. Colmar created his Arithmometer in 1820, and it was very successful through to the First World War. Indeed, Colmar was made Chevalier of the Legion of Honor for his innovation.

The next major step occurred on the other side of the Atlantic. In 1877, George B Grant of Boston began producing the hand cranked Grant mechanical calculating machine, capable of addition, subtraction, multiplication and division. But then in 1886 an altogether more robust machine appeared, when William Seward Burroughs introduced the P100 Burroughs Adding Machine. The Burroughs family business, which became the Burroughs Corporation, was enormously successful; so much so that grandson William S. could be freed from the drudgery of business into a literary career that resulted in eighteen novels (including *The Naked Lunch*) and novellas, six collections of short stories and four collections of essays.

Perhaps the first office calculator was produced by Felt & Tarrant Manufacturing Company in 1887. The **Comptometer** took calculating into the push button age. Two years later the same maker introduced the Comptograph, which came complete with a printer.

In an echo of earlier innovations, adding and calculating machines "*met with very strong opposition for the first few years*", and that it was "*strongly evident that the efforts of book-keepers and counting-house clerks to prevent these machines entering their department were inspired by the fear that it would displace their services*".[23]

The first machines were built using wooden boxes, which made them light but more fragile. They were called the 'woodies'. Around 6,500 were built in between 1887 and 1903. From the Model A onwards (Figure 9.3) they had a metal casing; the Model A (produced from 1904 to 1906) also had a polished glass panel for the output display in the front.

Duplicating and copiers. The search for a clean, simple and cheap form of copying stretches back a long way. One of the earliest mechanical methods had an unlikely origin, in the famous Wedgwood pottery family. Josiah's son, Ralph (1766–1837), patented a crude duplicating tool

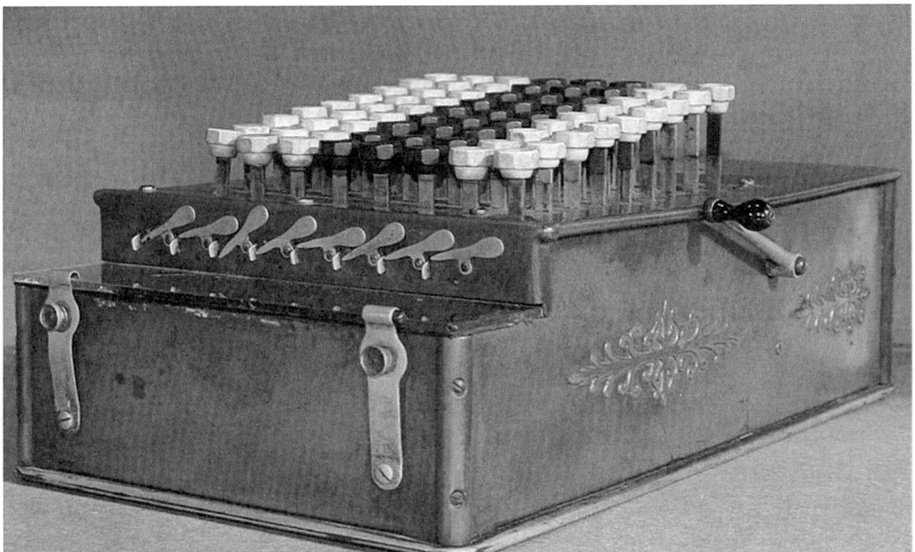

Figure 9.3 Felt & Tarrant Model A Comptometer, 1905.

as early as 1806. This became known as the Manifold Stylographic Writer consisting of two books, one of transparent sheets and what the inventor referred to as 'carbonated' paper. By using a metal stylus on a sheet of transparent paper with one of the carbonated sheets a good black copy on quality writing paper resulted which was used as the original, the copy being retained on the back of the transparent sheet readable from the other side. When a stylus was used to write on the top layer, the text was impressed on the regular paper and on the duplicate paper in reverse. Because the duplicate paper was particularly thin, the text could be read through it. Wedgwood claimed that up to six impressions could be made at once, if the technique was properly applied.[24]

As early as the 1870s, it had become commonplace to insert carbon paper or tissue between sheets of paper, while using the pressure of a pen or a pencil to produce a number of copies, up to four or five, and sometimes more. It became routine to indicate to the recipient of the 'top copy' to everyone else who had received the same document, or a carbon copy. Hence the origin of 'CC', now a ubiquitous element of business emails.

The number of people needing to be informed of 'corporate decisions' and 'corporate deliberations' grew rapidly, and new means had to be found to produce paper copies in great numbers. It was therefore not long before mechanised copying arrived. The Neostyle Duplicating machine was one such machine. It could make 1,000 to 2,000 copies from one original document, with the operator sitting at the machine, overseeing the automated supply of ink, automatic impression, and automatic copy production.

Mimeographs (or stencil duplicators) were low cost copying machines that forced ink through a stencil onto paper. Edison patented a flatbed duplicating press and an electric pen for cutting stencils in 1876, which was then licensed and improved upon by Chicago inventor Albert Blake Dick in 1887. The A.B. Dick Co. released the Model 0 Flatbed Duplicator in 1887, when it sold for the princely sum of $12.

In 1891, David Gestetner patented his Automatic Cyclostyle: one of the first rotary machines that retained the flatbed, which passed back and forth under inked rollers. This invention provided for more automated, faster reproductions since the pages were produced and moved by rollers instead of pressing one single sheet at a time.

By 1900, two primary types of mimeographs had come into use: a single-drum machine and a dual-drum machine. The single-drum machine used a single drum for ink transfer to the stencil, and the dual-drum machine used two drums and silk-screens to transfer the ink to the stencils. The single drum machine could be easily used for multicolour work by changing the drum – each of which contained ink of a different colour. Colours could not be mixed.

The mimeograph became popular because it was much cheaper than traditional print – there was neither typesetting nor skilled labour involved. One individual with a typewriter and the necessary equipment became their own printing factory, allowing for greater circulation of printed material.

Addressographs. As American corporations became larger and larger, spreading over ever greater distances, and as their customer bases broadened to include not only other businesses, but the individual customer, it was necessary to find means of communicating *en masse*. One means of achieving this was to automate the addressing of envelopes, invoices, ledgers and so on (see Figure 9.4). Orbell notes that there were two early forms, one based on stencils (offering flexibility as they could more easily be made in-house) and the other on embossed metal plates (giving the clearest print). Addressographs developed greater capacity and flexibility, and by the turn of the twentieth century treadle-operated machines were available; by

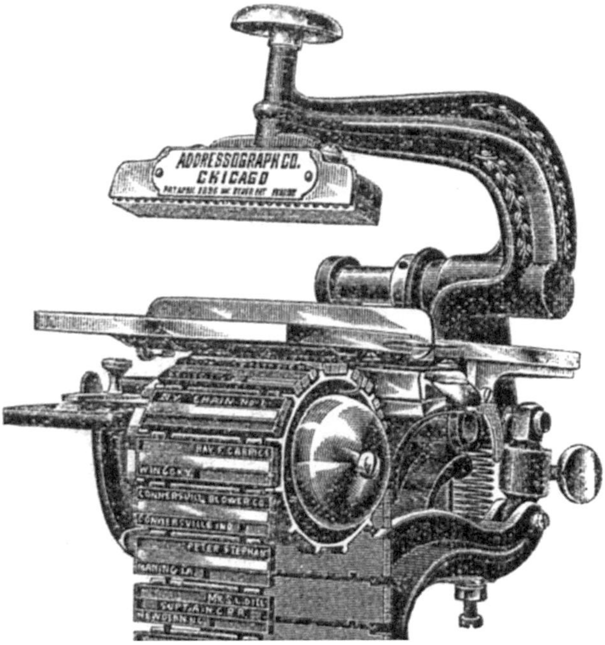

Figure 9.4 Addressograph, c1896.

the 1930s, automatic selection of address plates and selection of only certain fields for printing was possible.[25]

Joseph S. Duncan (1858–1950) was born in Pittsburgh and as an adult made his way west where he found employment as a bookkeeper in Sioux City, Iowa for the Great Northern Milling Company. Like many clerks of the time he found his work frustrating, particularly the inefficiency involved in preparing flour price lists which had to be sent to bidders on a daily basis, and the errors that could occur. He began to consider how to automate the labelling process.

In the spring of 1893, a collapse in the US economy left him suddenly out of the mill business and looking for a means of ensuring his own financial security. In the late summer of 1893, Duncan launched the Addressograph Company from his office in the old Caxton Building on Dearborn, and within 6 months, he had sold off most of his patent-pending prototypes. Over the following 2 years, Duncan started designing a more robust second model of the machine, which was manufactured in the plant of Chicago's C. H. Stoelting Company. He also found himself a business partner and salesman, John B. Hall, who took the Addressograph on a national sales tour.

The addressograph is a hybrid of a printing press and a duplicator – instead of using raised metal type or a screen, the addressograph uses embossed metal plates and an ink ribbon to create a print. The operator created the metal address card by typing on the Graphotype keyboard. The cards were then filed away in the machine so they could be accessed later for printing. To print, the card would be inked with a ribbon (as in a typewriter) and then make its impression on paper. Eventually, this technology would be used to produce military dog tags instead of prints.

Dictating machines. Thomas A. Edison invented the Phonograph for recording and reproducing the human voice in 1877, but it was not a great success. A more practical and robust version was produced by the triumvirate of Alexander Graham Bell, his cousin Chichester A. Bell, and Charles Sumner Tainter. Their Graphophone was made in 1885 and patented and illustrated in *Harper's Weekly* in 1886. In the same year, Tainter designed a dictating machine that was mounted on a sewing machine table with a foot treadle to supply power. The commercial Graphophone went into production in the fall of 1887.

Typewriters. Perhaps the most iconic addition to the modern office at this time, which somehow defined the very essence and purpose of the office, was the typewriter. Early versions of this machine included the crude and clunky Printing Machine in the USA in 1843 by Charles Thurber; the 'pterotype' in England in 1865 by John Pratt and various prototypes by Austrian Peter Mitterhofer between 1864 and 1867. The first commercially successful typewriter was patented in 1868 in America by Christopher Latham Sholes, Frank Haven Hall, Carlos Glidden and Samuel W. Soule in Milwaukee. Following changes in personnel and the appointment of sewing machine manufacturer Remington to commercialise the venture, the Sholes & Glidden typewriter was launched. Its successor, the Remington quickly established market dominance. Sales figures for the Remington No2 grew from "*146 in 1879 ... to an annual figure of 65,000 in 1890*".[26] Penman's Art Journal, in 1887, reported: "*Five years ago the typewriter was simply a mechanical curiosity. Today its monotonous click can be heard in almost every well-regulated business establishment*".[27]

By the late 1890s dozens of companies were manufacturing typewriters of diverse designs, with a variety of layouts for the keyboard and with ingenious mechanical arrangements. By 1900, more than 100,000 typewriters had been sold in America, and more than 20,000 new machines were being built each year.[28]

Nevertheless it remained a primitive piece of equipment. Wright Mills referred to the girl in the office struggling with the early typewriters, having to spend "*at least 15 minutes every morning cleaning and oiling her massive but awkwardly delicate machine*".[29] Furthermore, the operators of the early machines, as mentioned with respect to the Fitch typewriter above, typed 'blind', unable to see the emerging text. The Remington No2 set the standard for the universal key layout, the 'QWERTY system'. The Remington brand became established in Europe in the mid-1880s and 1890s. Typewriters were among the high-technology items of the era; they could be made in large numbers and at a reasonable cost only with the adoption and further development of the techniques of precision manufacturing with interchangeable parts developed by Colt and Remington for the production of pistols and rifles during the Civil War.

Of course, the introduction of the typewriter was synonymous with the rise of the female workforce. In America, female clerks rose from virtually none in 1851, to around 180,000 by 1911. In this way, the typewriter had an enormous impact on the office environment and social relations within it, drawing into the office a whole new class of worker-women, whose labour had been previously confined to the factory or to domestic service.

One of the often forgotten side effects of the typewriter was office standardisation. A standard paper size (in Europe 29.7cm by 21cm), which then helped standardise envelopes, files, folders and in/out trays, all of which in turn influence file cabinet dimensions, shelving and so on.

Telegraph. The telegraph was first used commercially on the railway, specifically between London's Euston Station and Camden Town. In 1837, William Fothergill Cooke and Charles Wheatstone demonstrated to the management of the London & Birmingham Railway company how the technology could contribute to railway safety. Cooke and Wheatstone went on to set up the Electric Telegraph Company, which would ultimately morph into BT. Telegraph cables were then laid across the English Channel in 1851 and, after a few failed attempts, across the Atlantic in 1866. Just 4 years later, in 1870, a telegraph link was completed between Europe and India. A year later, a cable reached Australia.

Significantly, a method of printing telegraph messages was invented in 1854, and developed in the USA as Hoase's printing telegraph. In 1867, the New York Stock Exchange installed the electric stock ticker. Telegraph lines multiplied in Europe and the United States in the latter part of the nineteenth century, greatly increasing the volume of work done by the office.

> *As messages came in over the telegraph they would have to be recorded and processed. And the very rapidity and ease of this new form of communication would increase the volume of messages. In order to keep up with competitors, or with the news of world events which now flowed in with increasing rapidity, an administrator must issue orders and directives more quickly.*[30]

At the same time, the telegraph was critical in making the detachment of administration and production more practical. The resulting spatial and organisational impacts reinforced the growth of the office economy as a distinct entity, particularly in the city centre.

It was not long before the telegraph had a competitor in wireless technology. Guglielmo Giovanni Maria Marconi (1874–1937) was born in Bologna into Italian nobility. Home-educated and without higher education, the young Marconi became an inventor and engineer, undertaking world-changing work on long-distance radio transmission; he is credited with inventing the radio. He found success as an entrepreneur in London, where he founded the Wireless Telegraph & Signal Company in 1897, the forerunner to the Marconi Company. In May 1897, Marconi sent his first message over the open sea – a full 6km between Flat Holm Island in the Bristol Channel and Lavernock Point in Penarth, Wales; in 1901, he despatched the first trans-Atlantic message.

Effective wireless communication whether by telegraphy or telephony was firmly established in the early years of the twentieth century, linking far-distant trading posts with head office. The two technologies co-existed for many years: "*wireless communication complemented and did not replace submarine telegraphy; when Britain's international telegraphy companies were consolidated into a single unit in 1928 it embraced both and from 1934 was known as Cable & Wireless Ltd*".[31]

The telephone was the third major office workplace innovation from this era. American-born Edison and Scottish-born Bell introduced their telephone systems in 1879 and 1880, respectively. The success was immediate: there were more than one million telephones in the US by the end of the century.[32]

Telephone. The telephone changed office activities in a similar fashion to the telegraph a few decades earlier. First, it provided people with instant contact with others located at a distance – colleagues in distant offices, customers, supply chain, factories, warehouses and so on. The main impact of distance – length of communication – had been overcome, ultimately allowing the spatial reorganisation of business. Secondly, within the office itself, the telephone helped change processes and the internal organisation of space, as well as increasing the work load.[33] Confirmation of orders, instructions, appointments and so on became one of the major components of office activity.

The first submarine telephone cable, laid in 1891, linked Paris with London. But this technology was slow to develop, and it was many years before long distance, radio-based services became available. Transatlantic telephone calls between New York and London began in 1927, quickly spreading tentacle-like across the north American continent and mainland Europe.

Figure 9.5 George McGill stapler and fasteners, 1879.

Shortly afterwards Britain and Australia were linked, but all such services were for many years prohibitively expensive and telegraphy remained the clear leader.[34]

But the overriding point was that the means had been created to link trading networks across regions and the globe, and communications were now possible on the same day, when before a written communication could take days or weeks. The telegraph and telephone symbolised a new era of global trading, and intensification of activities in trading capitals, including London. At one and the same time, the new communications overcame the friction of distance between production and administration, while encouraging the physical concentration of the growing number of specialist roles in the latter.

Other innovations. A number of the high-profile mechanical innovations have been highlighted here: they are well known and in some cases iconic. But it should not be forgotten that there were aided and abetted by countless less dramatic innovations. Rubber bands (introduced by British firm Perry & Co in 1844); brass paper fasteners with two 'legs' inserted through paper and then bent in opposite directions (US patent 1866); and envelopes which, when mass production methods were created alongside falling cost of paper, got going in the 1850s and 1860s.

In 1867, an inventor named George McGill received a patent for a machine that would press the fastener through paper. His stapler (Figure 9.5) reached the mainstream market for the first time in 1869. All this before one of humblest office accessories – the paper clip. The Gem-paperclip, the double oval shaped paperclip with which we are so familiar today, was designed and manufactured by British company Gem Manufacturing Co from the early-1870s. But it was not patented. In 1899, William Middlebrook, of Waterbury, Connecticut, patented a machine for making paper clips of the Gem design.

A new proletariat?

The telegraph, typewriter and telephone revolutionised the way in which office-based businesses communicated with one another, but one of the biggest bi-products of these innovations was paper – vast amounts of paper on which to specify, order, verify, promise, record and log countless contracts and transactions from around the world.

Adding machines, copiers, addressographs, dictating machines, typewriters, telegraphs and telephones are all examples of how the office mechanised in the later years of the nineteenth century. They brought with them mixed fortunes for office workers. In some ways, they relieved the tedium of mundane jobs; on the other hand, they created mundane jobs. What cannot be denied is that they created *more* jobs. They allowed businesses to increase the volume of their output, while at the same time, improving quality in terms of error reduction. They also accompanied a dramatically changing office demography, particularly with women entering the workforce, en masse, but also a wider range of males in terms of age and education.

All manner of writers, concerned in one way or another with social and economic relations, formed a barely contested consensus that the later years of the nineteenth century and the early years of the twentieth witnessed the transformation of clerks from respected, skilled and responsible gentlemen craftsmen who had their eyes across whole businesses on behalf of an appreciative owner, into an under-paid and over-worked proletariat, undertaking routine, mundane and repetitive tasks, having been marginalised by mechanisation, technology and female workers.

This increasingly ambiguous status of the clerk, the trajectory of clerking from skilled professional to general office worker, was first fully described by Francis Klingender (1907–1955). Born to British parents living in the Harz Mountain region of Germany, the family returned to England in 1925, where Francis attended evening classes at the London School of Economics and Political Science, studying sociology, while holding down a day job with an advertising agency to support his family. He graduated in 1930, doing field research for the New Survey of London Life and Labour, organized by Sir Hubert Llewellyn Smith (1864–1945). He was a Ratan Tata research student between 1930 and 1932, before receiving his PhD from the LSE for *The Black-coated Worker in London* (1934). A year later he published his seminal work, *The Condition of Clerical Labour in Britain*.[35]

Klingender was a Marxist and art historian whose politics in the early decades of the twentieth century made finding a stable academic post more difficult. His views were also problematic to the military, although his service there was declined for health reasons. During the war, he researched the effects of the Industrial Revolution on the fine arts in England. The result, *Art and the Industrial Revolution*, was published in 1947; followed in 1948 by his *Goya in the Democratic Tradition*. He eventually found an academic post as a lecturer in sociology at the University of Hull in 1948, where his work focused again on the labour market. In 1951, he married Winifred Margaret Kaye. Francis collapsed and died suddenly in 1955.

Klingender referred to clerical workers as the largest and most closely knit section of the lower middle class, or petty bourgeoisie, which in turn alternately sides with the capitalists and the workers: "*Clinging to the mentality and habits of the former while driven more and more down to the economic standards of the latter*".

> *The rapid expansion of capitalist production and trade during the nineteenth century brought in its train the spectacular growth of a new clerical working class. At the same time it shattered the patriarchal relations between masters and men which previously had been especially marked by this type of employment.*[36]

Klingender believed that the 'proletarianisation' of labour was as endemic in clerking as it was in the wider labour movement. But non-manual labour – and clerking was a very good example – posed questions for class theory. On the one hand, clerks identified with their capital-owning bosses and aspired, in many cases, to becoming part of that class. Klingender referred to this as a 'false consciousness' that prevented clerks from identifying with the wider, manual labour movement, while at the same time selling their labour.

Klingender highlighted the growth in the size of office buildings resulting from large-scale amalgamations, together with the introduction of machines, as key changes in the conditions of clerical work, seeing a parallel with the growth of factories, and their oppressed armies of manual workers.

> *The amalgamation process created the monster office in which vast numbers of clerks are herded together for their daily work, just as the concentration of capital in industry herded the former craftsmen or cottage workers in the factory.*[37]

He then goes on to link the clerk firmly with the industrial proletariat.

> *Just as in industry the former craftsmen were first brought together in the large scale 'factories' of the manufacturing period, before the nature of their work was revolutionised by the advent of modern machinery and they themselves transformed into proletarians in the modern sense, so it is the advent of the office machine, after the establishment of the office factory, which vitally transforms the work of the clerks and finally destroys the former craft basis of their trade.*[38]

In simple terms, Klingender's thesis was that the clerks of the later decades of the nineteenth century had experienced a prolonged period of downward mobility such that by the time of his writing they were fully working class – or part of the proletariat. For him this was an important issue because it was not simply about the continued 'oppression of workers by capitalists', but was something novel to the times and demonstrated the on-going relevance of Marx's basic construct in the evolving economy.

Klingender's analysis set the template for many studies to follow, and in the vanguard of this charge was Harry Braverman (1920–1976). Like Klingender, Braverman was a Marxist; but similarities ended there. Braverman was born in New York City to a working-class family, and he worked in a variety of metal smithing industries before becoming an editor at Grove Press, and later the Monthly Review Press, until his premature death at 55. He is most widely known for his 1974 book *Labor and Monopoly Capital: The Degradation of Work in the Twentieth Century,* 40 years after Klingender's *the Black-coated Worker.*

Braverman's criticism of the labour process of monopoly capitalism paid specific attention to the growth of giant corporations and oligopolistic industries. And into this framework fell the large white-collar corporations of the USA. Marx proposed that the emergence of machinery meant that workers' skills shifted into tending the machines. This was seen as strengthening the owners of capital since they controlled the knowledge and skills needed to operate the factory, run the machines, and employ the workforce. Braverman reinforced this thinking, suggesting that such a labour force was defined by a weakened position relative to capital. Work became transformed from being a utilisation of skills and experience into a mindless, machine-based, and powerless activity. He summarised the tenets of Taylorist scientific management as (a) dis-association of the labour process from the skills of workers; (b) separation of conception from execution, and (c) use of monopoly power over knowledge to control each step of the labour process and its mode of execution. Braverman argued that Taylor's prescriptions for the work-place were nothing less than the explicit verbalisation of the capitalist mode of production.

Braverman argued that clerical labour is procured in a similar manner to labour in productive industries, "*with wage labor purchased on a large scale ... and organized into huge 'production' machines according to the same principles that govern the organization of factory labor*". But then distinguishes the two by arguing that in contrast to productive processes, the productive processes of white-collar work (a term he disapproved of)

> *disappear into a stream of paper--a stream of paper, moreover, which is processed in a continuous flow like that of the cannery, the meatpacking line, the car assembly conveyor, by workers organized in much the same way.*[39]

Braverman correctly identified that with the rapid growth of offices in the last decades of the nineteenth century, and the change of office work from something merely incidental to management into a labour process in its own right, "*the need to systematize and control it began to be felt*".[40] And as a result

> *As this situation changed, the intimate associations, the atmosphere of mutual obligation, and the degree of loyalty which characterized the small office became transformed from a prime desideratum into a positive liability, and management began to cut those ties and substitute the impersonal discipline of a so-called modem organization.*

He goes on to lament the passing of the old arrangement.

> *The characteristic feature of this era was the ending of the reign of the bookkeeper and the rise of the office manager as the prime functionary and representative of higher management. Office managership, a product of the monopoly period of capitalism, developed as a specialized branch of management, with its own schools, professional associations, textbooks and manuals, periodicals, standards, and methods.* [41]

Braverman concludes by referring to the problem of the so-called employee or white-collar worker, which so bothered early generations of Marxists. This was that unlike earlier agrarian workers and workers in early industry, white-collar workers were not part of the oppressed and under-paid proletariat. He doubles down on his own thesis by suggesting that the debate has

> *been unambiguously clarified by the polarization of office employment and the growth at one pole of an immense mass of wage-workers. The apparent trend to a large nonproletarian "middle class" has resolved itself into the creation of a large proletariat in a new form. In its conditions of employment, this working population has lost all former superiorities over workers in industry, and in its scales of pay it has sunk almost to the very bottom.*[42]

In both Klingender and Braverman there is a sense of a 'better time' passing; of a drift towards 'wage slavery'. They saw the expansion of the 'old' industrial proletariat to encompass a growing army of white-collar workers, as skilled, senior and respected clerks were transformed by mechanisation into clerical factory operatives; members of the proletariat, undertaking mechanical, unskilled and repetitive tasks with no autonomy and with lower wages to boot.

But how accurate is this proletariat interpretation? In his classic *Life and Labour of the People of London* of 1896, Charles Booth was in no doubt:

> *The 'average, undifferentiated human labour power' upon which Karl Marx bases his gigantic fallacy does not exist anywhere on this planet, but least of all, I think, is it to be found among clerks.*[43]

This book has referred repeatedly to the heterogeneity of clerking, from the very beginning. There were never, up to this point, vast numbers of workers, all undertaking similar tasks. Put simply, there never was a unified, homogenous class of white-collar workers *undertaking mechanical, unskilled and repetitive tasks with no autonomy and with lower wages to boot.*

Nevertheless, the proletarianisation of workers was, for many observers, intrinsically part of the growing scale of business: the rise of the clerical factory with its reference to industrial assembly lines. Larger businesses were believed to create demand for mundane, unskilled work. Many observers of the changing fortunes of clerical workers have based their entire theses on the characteristics of very large offices, partly because these were seen to be 'the future' and partly because they were the most extensively mechanised environments. However such organisations were very far from typical; in fact they were in a small minority, and the very great majority of firms "*lacked the volume of routine work needed to justify the factory approach. Only in certain insurance and banking activities did one find the prerequisites for factory-like automation*".[44]

Deskilling. The proletariat discussion was fuelled in large measure by the belief that clerking was being 'deskilled' by the new machines. As organisations grew in size, employers became more cost conscious and more formal in terms of internal processes, resulting in employees being deployed more efficiently and labour being divided more rationally. A further factor was the introduction of various technologies and

> *the application of more rational approaches in the workplace. Typewriters, telephones, addressographs, adding machines, copiers, filing systems, etc., to be used most effectively needed to be used by certain groups.*[45]

Grace Coyle observed in 1929 that mechanisation in offices seemed to be "*reducing the number of skilled clerical jobs requiring initiative to a relatively few while it creates large numbers of routine or mechanical positions. Moreover, there seems to be in most offices very little provision or plan for regular promotion*".[46] The theme was picked up by Wright Mills in 1951, who argued that mechanisation was responsible for having increased the division of labour, recomposed personnel and lowered skill levels. He argued that

> *Routine operations in minutely subdivided organizations have replaced the bustling interest of work in well-known groups. Even on managerial and professional levels, the growth of rational bureaucracies has made work more like factory production.*

Wright Mills expanded on this idea, suggesting that the early stages of division of labour might lead to an increase in skill levels, but that when

> *whole operations are split and mechanized, such division develops certain faculties at the expense of others and narrows all of them. And as it comes more fully under mechanisation and centralised management, it levels men off again as automatons. Then there are a few specialists and a mass of automatons; both integrated by the authority which makes them interdependent and keeps each in his own routine.*[47]

This led to a more general concern that mechanisation, in particular, would lead to the deskilling of the clerking profession. Thus "*the image of clerical work presented by those making the case for the degradation of clerical work during this period involved an extreme division of labor, with each individual doing one narrow task on a machine*".[48] But, again, the evidence does not seem to back this up.

Kynaston emphasises the more mundane daily tasks, which involved opening and date-stamping the inward mail, clearing the partners' out-trays, attending to all sorts of callers, sending out price lists to clients, working through the day's transfer and certificate piles, copying letters without perpetrating smudges, and a hundred and one other tasks, depending on the business involved.[49] Added to this were, for the younger clerks, the daily 'walks', which included taking demand drafts which had come in with the mail to the offices of the different drawees, and later calling back for the corresponding cheques drawn on clearing banks.

In the novel *Robert Thorne*, already referred to above, the hero describes his daily tasks.

> *Being junior in our room [note use of the term 'room'] which dealt largely with correspondence, I had a junior position there, and that meant inferior duties – copying letters, indexing papers, addressing envelopes, clearing trays, carrying and fetching messages, keeping the correspondence book, besides duties of a more servile nature, such as making up the fire, attending the telephone, working the revolving stamp.*[50]

But to what extent can these be described as 'deskilling' the role of the clerk? And to what extent should they be seen as a symptom of a rapidly growing number and variety of roles in the office?

Arguably, some traditional clerking skills were deskilled: for example, the widespread uptake of adding machines. Formerly clerks were able to mentally calculate long columns of figures in a traditional clerical skill known as casting, which was made redundant by machinery. Clerks now simply had to enter the various figures (albeit accurately) and pull a lever to see the sum revealed.

But on the other hand, the number and variety of jobs grew – skilled and unskilled. Although insurance clerks' real wages continued to increase over the course of the nineteenth century, the responsibility attached to these wages declined for most clerks as office organisation grew more routinised. Typewriters and shorthand stenography more than doubled the rate of writing letters from the 1880s, converting carbon paper from a curiosity into the best method for duplicating. Copying by hand and letterpress gave way to the distinctive blue or purple shades of the mimeograph; addressographs enabled clerks to address up to 60 envelopes a minute; and arithmometers enabled comparable economies in basic mathematical calculations.[51]

It is a fact that the variety of office jobs mushroomed during this period, opening up countless new opportunities for career progression. In her wonderful account of female labour in early-twentieth century offices, Ethel Erickson gives one answer:

> *Going back only three or four decades the office function was primarily that of keeping books and handling a limited amount of rather stilted correspondence. In today's office with 50 or more clerical workers the bookkeeping and letter-writing activities frequently are overshadowed in the number employed by those engaged on the details of handling advertising, sales campaigns, market and credit analysis and collections. involved in statistical and financial reports of costs, and a variety of other recording activities.*[52]

Erickson finishes by saying that the extent of these activities and the policies and practices naturally vary and are quite unstandardised, and so it is understandable that job terminology for occupations is not comparable from office to office. But as the number of jobs grew, the fact is that there was never the exploitation (in Marxist terms) of office workers that might be attached to agricultural and early factory workers. According to Lockwood, the clerical worker "*more or less kept his position vis-a-vis the wage earner until the late thirties*".[53]

Nor was proletarianisation being nurtured by employing a growing army of lower paid, less skilled female workers. While the growth in female labour was clearly tied to the mechanisation of the office "*with the shift to typing, dictation machines, the use of forms, and bookkeeping machines*"... the "*feminisation of office workplaces did not simply lead to wage compression for males, because in most instances females were filling new roles not replacing men in old roles*".[54]

Despite the general suspicion of mechanisation and automation, there was in fact very little redundancy resulting directly from equipment replacing humans. Instead the whole process was a reinvention of the office, its function and its scope. Vast new areas of work were opened up; new skills were called for and new systems and processes required a growing army of administrative workers. Indeed, contrary to the deskilling thesis, there is mounting evidence that skills were enhanced overall. For example, clerks had to learn new skills to operate the machines effectively, and practice was key to competency. Mental arithmetic is a great skill; so is copying and ledgering. But there are natural human limits to the speed, scale and frequency with which these tasks can be undertaken. Without the new machines, there would have come a time when human limits were reached. What then? It is wrong therefore to see this kind of mechanisation as harmful to the interests of clerical workers.

Many of the new innovations meant a change in how something was done rather than the redundancy of the task. Indeed, the mechanisation of the office coincided with a massive growth in the number of clerical workers – men and women. Some innovations meant larger numbers of workers undertaking the same task, such as adding machines, addressographs, copiers, filing systems and typewriters, while other led to further division of labour and specialisation, such as accounting systems, finance, information management and telephones.

False nostalgia? Part of the deskilling argument was built upon the notion that clerks previously occupied 'craft' positions and even quasi-managerial positions. For example, Braverman paints a picture in which, by and large, "*in terms of function, authority, pay, tenure ... prospects, not to mention status and even dress, the clerks stood much closer to the employer than to factory labor*".[55] He goes on to describe the craft skill of clerks:

> *Master craftsmen, such as bookkeepers or chief clerks, maintained control over the process in its totality, and apprentices or journeymen craftsmen-ordinary clerks, copying clerks, office boys-learned their crafts in office apprenticeships, and in the ordinary course of events advanced through the levels by promotion. The work involved ... such tasks as timekeeping and payroll, quality control, commercial traveling, drafting, copying duplicates by hand, preparing accounts in several copies, etc.*[56]

It was not the case that all clerks were deskilled and marginalised as argued by Klingender and Anderson. The march of progress had not seen a total breakdown in relations between owner and trusted clerk. Instead, organisations have grown in scale, become more complex, and created a greater depth and breadth of job opportunities. All clerks had not morphed from highly skilled, trusted, well rewarded and socially mobile into a uniform underclass of poorly paid and over-worked proletariat.

Given what we know about changing organisations in the later decades of the nineteenth century, it would seem that a more intuitive position is closer to the fact. This is that clerking became more diffuse: on the one hand, there emerged clerks who were more senior, more specialist or more skilled (often from educated and higher social backgrounds) while, on the other hand, there was a growth in clerks undertaking more routine, more mundane and less skilled work.

While the former retained their advantage in income and, probably, conditions, the latter did not. A multitier labour market was evolving. And while the former developed new skills which enabled them to seek new more highly paid positions, the latter become even more dependent upon the employer.

But much of the deskilling interpretation is wrapped up in a nostalgia for a perception of a particular kind of clerk – one engaged by a family or small business who undertook a wide range of functions, was held in high regard by his employer and had a high level of autonomy. Such nostalgia ignores the harsh reality for perhaps the greater number of clerks who were being absorbed into rapidly growing businesses and for whom the daily reality was long hours, poor working conditions, mundane, repetitive tasks and unhealthy working conditions. For many of these clerks, mechanisation actually brought release. Indeed, Attewell argues that much of this historical sociology of clerical work is permeated

> *with false nostalgia regarding skill and the clerical work process. The historical comparison of the present-day clerk with her Victorian predecessor, particularly as it has been made by the more recent writers, is inaccurate, and leads to the false conclusion that the skills and knowledge of present-day clerks are far inferior to those of earlier generations, that a process of de-skilling has taken place.*[57]

It is fairly self-evident that there would have been some opposition to mechanisation: certain clerks held very specialised roles which were directly replaced: copy clerks were, in short, made redundant by copy machines. Their self-interest would have aroused fear and opposition quite naturally. But clerking duties were many and varied, junior and senior, secure and insecure, and interesting and mundane. Opposition was, in these terms, real but limited in scale.

Attewell refers to the 'false nostalgia' around this position: he argues that much clerical work was always routine, repetitious and deadening. Referring to a prevalent "*idealisation of the Victorian clerk*", he suggests that a selective portrayal emphasises the most highly skilled aspects of the work and the elite of practitioners. While simultaneously underplaying the amount of routine, deadening work and the numbers of clerical drudges in that period.[58] He also contends that

> *the de-skilling position is built in part upon a conceptual misidentification by which shifts in income, mobility, gender, and prestige are read as implying shifts in skill. Dimensions which were carefully separated by earlier writers like Lockwood (1958) – the income, prestige, promotion prospects, and skills of clerks – have become blurred. The Weberian insight that social status, income, and skill are partially independent phenomena – one dimension can alter without the others having changed – has been lost.*[59]

The final word to Attewell… ..

> *There is today a huge amount of repetitious, uncreative and uninteresting work done by clerks. This has been the case since the earliest days of clerical labor, with its interminable copying, posting, and adding work. There is little or no evidence to suggest that the proportion of clerical drudge work has increased over time. The impression of such an increase was fostered by assertion, and by the use of a historical sequence of unrepresentative ideal types: the elite clerk in Victorian commerce, the specialised office-machine operator between*

the wars, and the keypunch operator in the modem period. Each of these was based upon 10 per cent or less of clerks of the period. The occupation has always contained a heterogeneous array of functions, with some highly specialized roles and others combining multiple duties.[60]

The fact is that mechanisation was necessary to support the increasingly scale and complexity of business and while along the way, some clerical work disappeared, overall clerical work grew rapidly, created opportunities for many more people and created many roles that did not exist beforehand. Countless new specialised roles and tasks were created in the inevitable progress, and mechanisation also helped to bring about massive social change in the office.

Ultimately it laid the foundations for establishing the three hierarchies of the twentieth century office. First, there was the hierarchy in which managerial positions and roles were created and dominated by males, while females utterly dominated roles such as clerks, typists, stenographers, clerical supervisors and secretaries. Then there was the growing distinction between those who planned and managed and those who delivered output. And finally, within the managerial class, there was the ever-increasing number of layers of responsibility, allowing for social and class distinctions to be made between layers in the office hierarchy.

As a generality, and fully accepting that there were exceptions, the overall picture was one of great improvements in working conditions, rapidly expanding opportunities, higher wages and more security.

Or a new middle class?

The rise of the American corporation had a transformational impact on class structure in America: it led to little less than a reorganisation of society, as a new class of worker not only emerged, but came to dominate society. During the period 1870–1945, the clerk became extinct. In his place we had the 'office worker'; and from around the turn of the century, the office worker came to typify an emergent new middle class. Table 9.1, from Wright Mills' *White Collar*, shows the transformation in the USA: by 1940, white-collar workers rose from 15% to 56% of the new middle class.[61]

Marxist interpretations of clerking were hampered by the fact that it was an undeveloped part of the economy. In these early years, clerking was largely undertaken within family concerns and generally small-scale businesses. Clerks undertook a wide variety of functions; they were confidantes of the owners who aspired to management and even ownership; and as we have

Table 9.1 The new middle class in the USA, 1870 and 1940

Occupation	*1870*	*1940*
Old middle class	**85%**	**44%**
Farmers	62	23
Businessmen	21	19
Free professionals	2	2
New middle class	**15%**	**56%**
Managers	2	6
Salaried professionals	4	14
Salespeople	7	14
Office workers	2	22
Total middle classes	**100%**	**100%**

Source: Wright Mills (1951).

seen, they dressed and behaved as per their employer rather than as per the wider artisan class. Generally speaking – and recognising that there were some mundane and repetitive roles – clerks were in a relatively comfortable position.

Clerks did not fit comfortably into the early Marxist framework, where there were owners of capital and labourers, and nothing else. Marxist interpretations of clerking revolved around the growth of unskilled work resulting from mechanisation, but this denied the faster growth of white-collar work and the tiers of management involved.

Organisations were growing in size; office mechanisation demanded broader skills and no longer could one clerk range across the full breadth of work needed to keep the business going. Similarly more sophisticated accountancy processes and information management required more specialisation. As the all-rounder clerk yielded to a greater division of labour within clerking there emerged a hierarchy according to nature of business. For example, within the broad spectrum that constituted this clerical class, the bank clerk held a particularly privileged position. Defined in Anderson's[62] seminal work as "*the aristocracy of the clerical profession*", bank clerks occupied distinctly higher rungs on the social ladder than the rest of the clerking populace, thereby having a greater claim to being part of the middle class than their counterparts working in a variety of other contexts.

The situation in reality just reflected a more general growing separation in social relations in the office. Within organisations that were, on the whole, growing in size and organisational complexity, clerking (if ever there was such a singular activity) was diverging into a hierarchy of roles that reflected social position as much as technical ability.

The lowest ranks of clerks were involved in mainly mundane and repetitive tasks. For example, *The Bankers' Magazine* described the junior clerk's role in 1893 as including "*stereotyped*" duties – "*copying letters, entering in registers, engrossing policies, writing out receipts, keeping postages*"; which contrasted with the "*higher duties*" that came with a "proper" clerkship: "*the preparation of agency accounts, the checking and calculation of premium rates, the sending out of renewal receipts ... attending to the manifold book-keeping*".[63]

Then there were middle-ranking clerks who kept the books, and conducted routine correspondence. The highest level comprised clerks in positions of trust, such as cashiers and correspondents, and who often knew foreign languages, shorthand, and general office management.[64]

There was a further level of seniority where clerks were placed in senior positions of trust, including having responsibility for cash, working as private secretaries and undertaking rudimentary office management. Even so, it would be inaccurate to describe any of these as managers of the business – they had no input to business decisions and policy. Businesses at this time remained predominantly owner-managed, and clerks were largely task-driven rather than commercially aware; separate clerk jobs with distinct duties and skills were the norm rather than having 'generalist clerks', and within the clerking community there was a recognisable hierarchy depending upon the level of autonomy granted by the employer.

As we shall see later in this book, over the course of time, the proportion of unskilled office workers and skilled workers shifted very heavily in favour of the latter. Office work became less mechanical and mundane and more and more skilled. And in the course of this evolution, so the office hierarchy was transformed. In Harry Braverman's terms, we saw a growing chasm between conceptualisation (management) and execution (labour).

> *As has already been pointed out, the separation of conceptualization from execution-the removal of all possible work from the shop floor, the point of execution, to the office-and the*

> *further necessity of maintaining a shadow replica of the entire process of production in paper form, brings into being large technical and office staffs.*[65]

Mechanisation and new corporate structures led to increased role differentiation and social stratification in the office by distinguishing skilled from unskilled workers; 'decision-makers from order-takers' and 'managers from managed'. This division of labour within the office also further distanced the relationship between owners and workers. Klingender argues that rapid "*expansion of capitalist production*" during this period brought with it the shattering of "*patriarch relations between masters and men which previously had been especially marked in this type of employment*". It also led to the "*tendency to reduce the remuneration of clerical labour to the minimum*", which, in turn, "*achieved the assimilation of clerical wages to those of the skilled workers by the end of the period*". The key here, of course, is that up to this point, socially, "*clerks were throughout this period a section of the middle classes*".[66] So the real problem was one of social standing rather than economic standing.

The period 1870–1945 saw the transformation of the office workplace. Utterly. Completely. Economic transformation and the birth of the modern corporation (Chapter 7) led to fundamentally new work and new ways of working, and the role of the clerk – especially in the context of his relationship with mainly owner-managers – changed the nature of social relations within the office. We see in the subsequent two chapters how all of this played through in terms of the architecture of office buildings.

In this chapter we have seen how clerking quietly disappeared and the notion of office work came to dominate. More efficient processes, mechanisation and the feminisation of the workforce have all been identified as the causes of the demise of the traditional clerk. But, as we have seen there has been both some misplaced nostalgia around this account and some political myopia.

The office workforce did not become oppressed, under-paid, generally exploited or poorly regarded in society at large. Instead office work came to dominate the workforce, producing countless new roles and opportunities and creating career ladders and social mobility for millions. Of course, along the way, repetitive and mundane jobs came into being, but these were generally superseded as technology evolved.

The late-nineteenth and early-twentieth century was a period of immense social change, on both sides of the Atlantic. This was no less the case in the office. There can be no doubt that the mechanisation of the office introduced an element of unskilled labour, but this was a largely time-specific process. The far greater impact was to open up a whole new range of job and career opportunities as not only the total number of jobs expanded, but the number of *types of jobs* expanded. From this maelstrom emerged new corporate structures which sub-divided white-collar workers into ever more subtle layers of responsibility and management. In turn, status and salary came to reflect social standing, or class.

Certainly by the postwar period, office workforces were structured and stratified, and reward and status came to reflect education and aspiration, and it was a short step to managerial staff becoming more and more sharply distinguished from their subordinates, such that they became the de facto middle class of society. Manual and unskilled workers in the office came to be indistinguishable from other forms of manual worker as their subordinate status, low reward, lower education and insecurity tied them to the more general working class.

Summary. Society expanded rapidly in the later-nineteenth century; the economy became more complex with an ever-growing range of products and services to make, sell, distribute, trade and so on. Consequently the businesses undertaking these activities themselves became more complex and, at every stage of the journey adopted new tools, techniques and processes to enable them to meet the new challenges. There was never a particular slice of time when clerks were deskilled. Adding machines [and other primitive office tools] were the start of a long process of mechanisation, which was then followed by electrification and, ultimately, digitalisation. And the same discussions are being had today about the new technologies and their potential to wipe out huge swathes of jobs.

The many articles that argued that clerks were simply being deskilled by technology before and after the turn of the century can thus be challenged. More complex business led to more complex organisations led to more complex tools to complete the job – and all this in turn led to higher skilled, more specialised and better paid workers. Mechanisation of the office was a sign of advancement, innovation and modernity; it was an essential pre-requisite for growing business and a growing economy. The fact that leading edge technologies came and went over brief period in the overall process was simply the mechanism by which the overall project was achieved.

Notes

1 Bullock SF (1907) *Thorne: The Story of a London Clerk* T Werner Laurie, London pp6–7.
2 Bullock SF (1907) *Thorne: The Story of a London Clerk* T Werner Laurie, London pp35–38.
3 Bullock SF (1907) *Thorne: The Story of a London Clerk* T Werner Laurie, London pp42–43.
4 Bullock SF (1907) *Thorne: The Story of a London Clerk* T Werner Laurie, London p73.
5 Zakim M (2006) The Business Clerk as Social Revolutionary; or a Labor History of the Nonproducing Classes *Journal of the Early Republic* Winter 2006 Vol 26 No 4 Zakim pp563–603.
6 Locker FM (1984) *The Evolution of Victorian and Early Twentieth Century Office Buildings in Britain* University of Edinburgh PhD thesis p151.
7 Kynaston D (1994) *The City of London Volume One: A World of its Own 1815–1890* Chatto & Windus, London p247.
8 Kynaston D (1994) *The City of London Volume One: A World of its Own 1815–1890* Chatto & Windus, London p289.
9 Locker FM (1984) *The Evolution of Victorian and Early Twentieth Century Office Buildings in Britain* University of Edinburgh PhD thesis p311.
10 Locker FM (1984) *The Evolution of Victorian and Early Twentieth Century Office Buildings in Britain* University of Edinburgh PhD thesis p312.
11 Locker FM (1984) *The Evolution of Victorian and Early Twentieth Century Office Buildings in Britain* University of Edinburgh PhD thesis p317.
12 Attewell P (1989) The Clerk Deskilled: A Study in False Nostalgia *Journal of Historical Sociology* Vol 2 No 4 p357 ISSN 0952-190.
13 Alborn T (2008) Quill-driving: British Life-Insurance Clerks and Occupational Mobility, 1890–1914 *British History Review* Vol 82 Spring 2008 pp31–58.
14 Locker FM (1984) *The Evolution of Victorian and Early Twentieth Century Office Buildings in Britain* University of Edinburgh PhD thesis p113.
15 Wright Mills C (1951) *White Collar: the American Middle Classes* OUP, New York p195.
16 Lowe GS (1984) The Enormous File: the Evolution of the Modern Office in Early Twentieth-Century Canada *Archivaria* 19th January pp137–151.
17 Coyle GL (1929) Women in the Clerical Occupations *The Annals of the American Academy of Political and Social Science* May 1929 Vol 143 pp180–187.
18 Cited in Heller M (2003) *London Clerical Workers 1880–1914: the Search for Stability* PHD Thesis University College London p194.
19 Heller M (2003) *London Clerical Workers 1880– 1914: the Search for Stability* PHD Thesis University College London p194.

20 Kynaston D (1995) *The City of London Volume Two: Golden Years 1890–1914* Chatto & Windus, London pp251–252.
21 Corey L (1935) *The Crisis of the Middle Class* Covici Friede, New York p250.
22 Mumford L (1938) *The Culture of Cities* Secker & Warburg, London p226.
23 Turck JAV (1921) *Origin of Modern Calculating Machines* Forgotten Books, London pp144–45.
24 Delgado A (1979) *The Enormous File: A Social History of the Office* Richard Clay, Suffolk p80.
25 Orbell J (2017) Changes in Office Technology Chapter 3 pp59–60 In: Turton A (2017) *The International Business Archives Handbook: Understanding and Managing the Historical Records of Business* Routledge.
26 Daniels PW (1975) *Office Location: An Urban and Regional Study* G Bell & Sons, London p10.
27 Cited in Stephens C & Lubar S (1986) A Place for Public Business: the Material Culture of the Nineteenth Century Federal Office *Business and Economic History Series One* Vol 15 pp165–179.
28 Giuliano VE (1982) The Mechanisation of Office Work *Scientific American* Vol 247 No 3 September 1982 pp148–165.
29 Wright Mills C (1951) *White Collar: the American Middle Classes* OUP, New York p191.
30 Cowan P; Fine D; Ireland J; Jordan C; Mercer D & Sears A (1969) *The Office: A Facet of Urban Growth* Heinemann Educational Books Ltd, London p31.
31 Orbell J (2017) Changes in Office Technology In: Turton A (2017) *The International Business Archives Handbook: Understanding and Managing the Historical Records of Business* Routledge p68.
32 Daniels PW (1975) *Office Location: An Urban and Regional Study* G Bell & Sons, London p11.
33 Cowan P; Fine D; Ireland J; Jordan C; Mercer D & Sears A (1969) *The Office: A Facet of Urban Growth* Heinemann Educational Books Ltd, London.
34 Orbell, Changes in Office Technology In: Turton A (2017) *The International Business Archives Handbook: Understanding and Managing the Historical Records of Business* Routledge p70.
35 Klingender FD (1935) *The Condition of Clerical Labour in Britain* Martin Lawrence, London.
36 Klingender FD (1935) *The Condition of Clerical Labour in Britain* Martin Lawrence, London p17.
37 Klingender FD (1935) *The Condition of Clerical Labour in Britain* Martin Lawrence, London p61.
38 Klingender FD (1935) *The Condition of Clerical Labour in Britain* Martin Lawrence, London p61.
39 Braverman H (1974) *Labour and Monopoly Capital: the Degradation of Work in the Twentieth Century* Twenty Fifth Anniversary Edition, Monthly Review Press, New York p208.
40 Braverman H (1974) *Labour and Monopoly Capital: the Degradation of Work in the Twentieth Century* Twenty Fifth Anniversary Edition, Monthly Review Press, New York pp210–211.
41 Braverman H (1974) *Labour and Monopoly Capital: the Degradation of Work in the Twentieth Century* Twenty Fifth Anniversary Edition, Monthly Review Press, New York p211.
42 Braverman H (1974) *Labour and Monopoly Capital: the Degradation of Work in the Twentieth Century* Twenty Fifth Anniversary Edition, Monthly Review Press, New York p245.
43 Charles Booth (1896) Life and Labour of the People in London Vol II Quoted in Lockwood D (1958) *The Blackcoated Worker: A Study in Class Consciousness* George Allen & Unwin, London p18.
44 Attewell P (1989) The Clerk Deskilled: A Study in False Nostalgia *Journal of Historical Sociology* Vol 2 No 4 p357 ISSN 0952-190.
45 Heller M (2003) *London Clerical Workers 1880– 1914: the Search for Stability* PHD Thesis University College London p121.
46 Coyle GL (1929) Women in the Clerical Occupations *The Annals of the American Academy of Political and Social Science* May 1929 Vol 143.
47 Wright Mills C (1951) *White Collar: the American Middle Classes* OUP, New York pp226–227.
48 Attewell P (1989) The Clerk Deskilled: A Study in False Nostalgia *Journal of Historical Sociology* Vol 2 No 4 p357 ISSN 0952-190.
49 Kynaston D (1995) *The City of London Volume Two: Golden Years 1890–1914* Chatto & Windus, London p251.
50 Bullock SF (1907) *Thorne: The Story of a London Clerk* T Werner Laurie, London p139.
51 Alborn T (2008) Quill-driving: British Life-Insurance Clerks and Occupational Mobility, 1890–1914 *British History Review* Vol 82 Spring 2008.

52 Erickson E (1934) The Employment of Women in Offices U.S. *Women's Bureau Bulletin* Number 120 Washington DC US government Printing Office p5.

53 Lockwood D (1958) *The Blackcoated Worker: A Study in Class Consciousness* George Allen & Unwin, London p67.

54 Attewell P (1989) The Clerk Deskilled: A Study in False Nostalgia *Journal of Historical Sociology* Vol 2 No 4 p357 ISSN 0952-190.

55 Braverman H (1974) *Labour and Monopoly Capital: the Degradation of Work in the Twentieth Century* Twenty Fifth Anniversary Edition, Monthly Review Press, New York p204.

56 Braverman H (1974) *Labour and Monopoly Capital: the Degradation of Work in the Twentieth Century* Twenty Fifth Anniversary Edition, Monthly Review Press, New York p207.

57 Attewell P (1989) The Clerk Deskilled: A Study in False Nostalgia *Journal of Historical Sociology* Vol 2 No 4 p357 ISSN 0952-190.

58 Attewell P (1989) The Clerk Deskilled: A Study in False Nostalgia *Journal of Historical Sociology* Vol 2 No 4 p357 ISSN 0952-190.

59 Attewell P (1989) The Clerk Deskilled: A Study in False Nostalgia *Journal of Historical Sociology* Vol 2 No 4 p357 ISSN 0952-190.

60 Attewell P (1989) The Clerk Deskilled: A Study in False Nostalgia *Journal of Historical Sociology* Vol 2 No 4 p357 ISSN 0952-190.

61 Wright Mills C (1951) *White Collar: the American Middle Classes* OUP, New York p65.

62 Anderson G, *Victorian Clerks* Manchester University Press p16.

63 Cited in Alborn T (2008) Quill-driving: British Life-Insurance Clerks and Occupational Mobility, 1890–1914 *British History Review* Vol 82 Spring 2008.

64 Parsons CE (1876) *Clerks: Their Position and Advancement* Provost & Co, London p36.

65 Braverman H (1974) *Labour and Monopoly Capital: the Degradation of Work in the Twentieth Century* Twenty Fifth Anniversary Edition, Monthly Review Press, New York p165.

66 Klingender FD (1935) *The Condition of Clerical Labour in Britain* Martin Lawrence, London p17.

10 Cultural and rational filing cases

The later years of the nineteenth century through to the First World War were transformative for the office economy. Office occupiers, office work and office buildings all changed utterly. Offices quickly evolved from a minor land use in cities into the dominant downtown land use activity. Centrally located, purpose-built office buildings, as we understand the concept today, emerged and office activity become a distinct urban function:

> *The office function was clearly due to grow, and to grow quickly. And in addition it would need to separate off from other facets of urban activity, because of the growing complexity and intricacy of organisation needed to finance, administer and support the industrial revolution.*[1]

It is all the more perplexing therefore that offices and office work received virtually no academic scrutiny during the whole of this period and for some time afterwards.

Offices and office activity were simply not recognised as a distinct component of the wider economy. For example, Banister Fletcher (1866–1953) failed to cover offices as a distinct building type in his seminal work *A History of Architecture on the Comparative Method* – neither in the original 1896 edition, nor in subsequent editions through to the mid-twentieth century. Similarly, Nikolaus Pevsner (1902–1983), in his *A History of Building Types*, included separate chapters devoted to 'Exchanges and Banks' and 'Warehouses and Office Buildings', but offered no treatment of office buildings as a distinctive building form.

American sociologist Lewis Mumford (1895–1990) was one of the greatest observers of urbanism that ever lived, his long working life coinciding with the maximal period of growth in the office economy. He was also a great humanist. Leaving college early due to tuberculosis, Mumford spent enormous amounts of time studying, sketching and simply watching New York. What he saw was the dehumanising effect of modern urban life, its stultifying impact on the human soul, ultimately its denuding of human purpose.[2]

His classic work, *The Culture of Cities*, is a history of urbanism, beginning with the medieval town.[3] It has been described as "*a study of the erosion of a balanced, decentralized civilization and its replacement by one with an oppressive metropolitan centralization of power, people and culture*".[4]

So Mumford was no fan of modern cities or the planners who sought to arrange and order them. Neither was he a fan of the tall buildings springing up everywhere in America through his lifetime. He described skyscrapers as

> *files, vaults, places for live storage and dead storage, parade grounds of documents, where the records of business were alphabetically kept, with an eye to the possibility of future*

DOI: 10.1201/9781003582205-13

> *exploitation, future reference, future lawsuits, future contracts. This age found its form in a new type of office building: a sort of human filing case, whose occupants spent their days in the circumspect care of paper: numbering, ticketing, assorting, routing, recording, manifolding, filing, to the end that the commodities and services thus controlled could be sold to the profit of the absentee owners of the corporation.*[5]

The term '*human filing cases*' to describe office buildings, and the caustic '*circumspect care of paper*' highlight Mumford's severe misgivings about the concentration of white-collar work in downtown. Eighty-three years later, it also has resonance today.

One of the earliest steps towards understanding offices and office work as a facet of urban growth was taken in the 1950s by Rannells in his landmark study of central Philadelphia, *The Core of the City.*[6] This work focused on the role of economic activities in shaping the morphology of the city. He made the critical link between broad land use changes and the influence of firm-level operational decisions on the quantum and structure of final demand for space:

> *It is necessary to take in somewhat more than relates directly to land use, since changes in demand for space or location may come about as secondary results of business decisions or consumer preferences which, in themselves, have no concern with questions of land use.*[7]

His approach directly linked the decisions of countless agents (companies) with patterns of economic relationships and the physical form of the city, both spatially and in terms of building form. This became fundamental to our understanding of the evolution of the office economy, and particularly its concentration in CBDs. Rannells explained his approach

> *The underlying systems of activity, which result in varying patterns of urban land use, are organised around establishments, classified by kinds of business. These establishments, in turn, are the units for analysis of land use.*[8]

Rannells recognised the availability of suitable accommodation, services and facilities, "*as well as the associated activities of closely linked establishments or the market which their operations require*".[9] As such, his work was the first critical examination of the role of economic linkages in shaping urban structure. As he argued, the locational characteristics of activities are "*visible manifestations of the many systems of activity into which the city's life is organised*".[10] However, he took the analysis much further by examining the impact of the role of activities, or uses, in shaping the morphology of cities. He argued that activities "*carried on within an establishment determine, in general, the quantities and qualities of space used by itself for work, for display, for processing, for storage or for circulation*".[11]

The most important contribution of Rannells' work was to shift the focus of study away from the spatial delimitation of activities, to an examination of the characteristics of the activities that congregate in the urban area, and the interactions between them that explain the process of change over time. Rannells' work in Philadelphia was developed 13 years later by Peter Cowan and colleagues in London. Cowan identified the link between changing economic relations and the rise of the office (which is examined in more detail in Chapter 11), observing that the office economy was no longer simply part of other facets of urban life but a facet in its own right. Thus,

> *But if we look back to the end of the nineteenth century we can see that the office function had become well established as a separate entity in the city: new buildings were being constructed to house it; more and more of the labour force were working in offices, and women especially*

> *were beginning to go into office work in large numbers. The outlines of the office as we know it were laid down.*[12]

By the turn of the century, European and American offices were evolving in quite distinct ways. Indeed, it might be argued, following Mumford's idea, that we were seeing the emergence of cultural filing cases and rational filing cases!

While European offices were grappling with the implications of the modern office practices in terms of accommodating larger bureaucracies with mechanised work processes in aged building stock, the story on the other side of the Atlantic was rather different: the physical fabric of many North American cities was being made for the first time. The famous grid iron pattern of urban development in the USA indicated both the scale of the new buildings and the huge efficiencies that they achieved in terms of floorplate configuration. Europe was adapting its history to suit modern needs; America was creating its future with entirely novel solutions.

Chapter 7 characterised these different approaches in terms of the notion of the cultural office and rational office, respectively (Figure 7.4). The European building was cultural, reflecting the values of the family businesses that occupied the spaces; it was domestic in scale and design; it was inefficient and personal. The US building was rational, built on a large scale, with repetitive design features and standardised components; it was organised internally for efficiency and to maximise output. In this chapter we see how this distinction between the cultural and rational traditions was magnified, and how the architecture of European and US offices diverged enormously during the early years of the twentieth century. Chapter 12 then revisits the discussion in the context of the post-war period.

The cultural office

We saw in Chapter 6 that the City of London experienced its first office building boom in the 1850–1880 period. One of the most obvious features of office buildings from this period was their sheer scale. While modest by today's standards, the leap from the counting house to purpose-designed office building was dramatic and radical: "*it is hard to recapture the sense of awe and wonder that the new office buildings, blocks of flats, and hotels evoked when they first appeared, towering above their Georgian neighbours*".[13] The City was just beginning to lose some of the 'domestic' feel of its converted houses.

According to Olsen, contemporary critics were torn between admiration at the audacity and dismay at the taste of the new City architects. He suggested that the architectural profession was somewhat hostile to the traditional, self-trained speculative builder and that, in order to justify himself and his fees to sceptical clients, he had to produce something recognisably different from the non-professional builder. Hence the emphasis on the overstated, the individual, the idiosyncratic, on originality at all costs.[14]

The Architect in 1871 was more favourable than most when it contrasted the splendour and originality of their new commercial buildings with the retrograde architecture of the fashionable residential quarters:

> *There are few spaces in Europe of equal size which can compare for modern architectural interest with the mile radius round the Exchange. The edifices in which the larger city merchants are housed by day are palaces compared with the cold and monotonous dwellings which even our most aristocratic families inhabit at the West End.*[15]

The building boom also brought with it the first challenge to building height. In complete contrast to the USA, which was beginning a long period of construction and engineering invention and innovation to allow buildings to go ever higher, London began a century long denial of height. The construction of the 14-storey Queen Anne's Mansions (1876–1888) in Petty France, Westminster produced deep controversy and led to by-laws setting maximum heights on new buildings.[16]

While converted town houses continued to provide satisfactory quarters for small firms, the increasing scale of activity of the nineteenth century insurance, banking and other businesses came to demand larger and more impressive buildings, both to house their various operations and to serve as architectural expressions of their wealth, respectability and reliability. Sometimes these buildings had space to spare to let as offices to other firms; as the century progressed speculative office blocks were built wholly for subletting. The Victorian rebuilding of the City thus brought structures as arresting in their architecture and novel in their plans as they were impressive by their bulk. Typical of the era was the rebuilt Goslings Bank, at 18–19 Fleet Street, completed in 1899 (Figure 10.1).

The rebuilding was part of a longer-term transformation of the office economy in the City. During the Victorian era City floorspace increased by at least one-half, from roughly 50m sq ft (4.6m sq m) to 75m sq ft (7.0m sq m); even more tellingly, of every five buildings standing in the City in 1855, only one remained in 1901.[17]

Across the Channel, Paris was also beginning to emerge from its suspicion of banking, as the administrative institutions necessary to oversee the increasingly complex web of commercial activity matured and began to cluster, thereby becoming a physical presence as well as a commercial one. In Paris, for example, the financial district began to take shape, literally, between the Bourse and the Opera. Pelegrin-Genel cites the example of the architect Hector Horeau, who proposed in 1871 the construction of an office building entirely without living accommodation. Then in 1876, Henri Germain, the founding president of Credit Lyonnais, decided to house his Paris headquarters on the Boulevard des Italiens, "*a building that was destined to become a model for banking headquarters of the future*" and featuring a large central atrium, glazed skylight, mezzanine with reception rooms and a first floor reserved for senior management.[18] Nevertheless, Paris retained its historic street pattern and its office buildings remained modest in scale.

Despite everything, the City was being rebuilt on, essentially, a medieval street pattern, with all its inherent constraints. Even after all the redevelopment, most City offices were still of a "*highly old-fashioned, rabbit warren nature, featuring sloping desks, enormous ledgers still filled out in beautiful copper-plate, and carefully attired clerks ranging in age from 14 to 84*".[19] The level of expectation is shown in the following. "*Architecturally it has all the latest improvements*", enthused the *Bankers' Magazine* about the new building at 39 Cornhill, into which the Union Discount moved in October 1890, adding that it "*is fitted up with the electric light, has first-rate strong rooms and altogether is as airy, convenient, suitable and pleasant a building as can be wanted for the transaction of banking business*".[20]

Royal Insurance. Thus it was that the European, cultural office evolved slowly and incrementally. Even the new buildings in the late-eighteenth century, while greater in scale than their predecessors, carried with them the cultural heritage of earlier buildings in terms of internal layout and functionality. A good example of this incrementalism is the Royal Insurance Company, and its growth in Lombard Street, as told in a 1912 company publication, *Old Lombard Street*.[21] Figure 10.7 illustrates the phased growth of the company in this location

Following the formation of the company in Liverpool in 1845, it was decided to open a head office in London, and it was at 29 Lombard Street that the Royal Insurance set up shop. The

Figure 10.1 Goslings Bank, 18–19 Fleet Street.

business was successful in London, taking on more and more policies. This necessitated a move to larger accommodation, and in 1863 the company moved into a new building on the site of Nos 27 and 28 Lombard Street, built to a design by Mr J Belcher, who was assisted by his son, Mr John Belcher, the well-known architect (whose earlier work for Royal Insurance we saw on pp98). The company later also took No 3 Clement's Lane, but the business continued to grow leading to the imperative to expand its occupied space. So, in 1909 it was decided to acquire the adjoining properties, Nos 24, 25 and 26 Lombard Street, and upon the extended site to construct "*modern offices, suitable in every way to the present and future needs of the company*".[22] The new building was designed "*in the Italian Renaissance style with something of a French influence in part of the design*".[23]

The entrance hall is described as a "*handsome feature*" faced with marble, of which the principal effect is a combination of Pavanazzo and Statuary marbles, "*comprising a lofty and well-proportioned dome supported by pendentives and having balconies on each side under the dome with open balustrades supported by an entablature in statuary marble and pilasters of green Cipollino marble with bronze caps and bases*".[24]

The interior also gives an impression of grandeur. The ground floor office stands 24ft to the ceiling, and houses around 100 clerks working in the Fire, Life, Cash, Foreign and Marine Departments. The windows are "*unusually large and high, but are well divided by means of heavy bronze sashes*". The design avoided any supporting columns and other floor obstructions, with the weight of the building being carried on steel girders, largest "*being five feet deep by three feet wide and weighing fifteen tons [while] skilfully hidden through being cased with marble and made to form the base of the wall of the mezzanine floor*".[25]

The marble lining the walls, is again a combination of Pavanazzo and Statuary with a dado of Verde Antico and the general effect is not only distinctive and pleasing but materially assists in securing the wonderful natural light, which is the most striking feature of the building. The mezzanine floor contained the Board Room and rooms for the chief officers and the staff of the Secretarial Department. The first floor and part of the second floor were designed for use by tenants; with the remainder of the second floor containing the Accident and Allied Departments. Moving up the building, the third floor housed around 130 clerks comprising the Country and Branch Departments, Consequential Loss, Motor Car, Accountancy, Statistical and Fire Loss Departments. Upwards again, the fourth floor was used as dining rooms for company officials and staff, providing eating for about 150 (Figure 10.2).[26]

Another head office that speaks volumes about the different socio-economic cultures of European and American office development is that of NM Rothschild.

NM Rothschild headquarters. Nathan Rothschild registered as a merchant in London in 1804 and by 1810 he founded his bank, N M Rothschild & Sons, which was established at New Court.[27] Nathan also became a part of the social and political elite. During the Napoleonic Wars, he lent the Government funds to pay Wellington's troops on campaign in Portugal and Spain and the bank managed and financed subsidies the government sent to allies.

In 1809, Nathan acquired the lease of 2 New Court in St Swithin's Lane for £750 to function both as a home for his family and as the locus for his business interests. The property had an attached warehouse, allowing Nathan to continue his merchant business. In 1810, he founded his own bank NM Rothschild.

In 1815, the new business was expanding rapidly and Rothschild signed a new 21-year lease on the building for £175 per annum, while his family moved out of New Court to a new villa in Stamford Hill in 1816. New Court thus became the main office of N M Rothschild. It was here, in 1824, that the Alliance Insurance Company (later to become the Royal & Sun Alliance) was founded at a meeting between Nathan and his associates.

By 1850, 30–50 people worked for NM Rothschild, and salaries at New Court ranged from £50 to £500 per annum. By the 1860s, the business of government loan issues had steadily increased and the bank remained the unchallenged leader in this field.[28] In 1860, the first rebuilding of New Court began, and in 1865 the new building was completed in the style of a

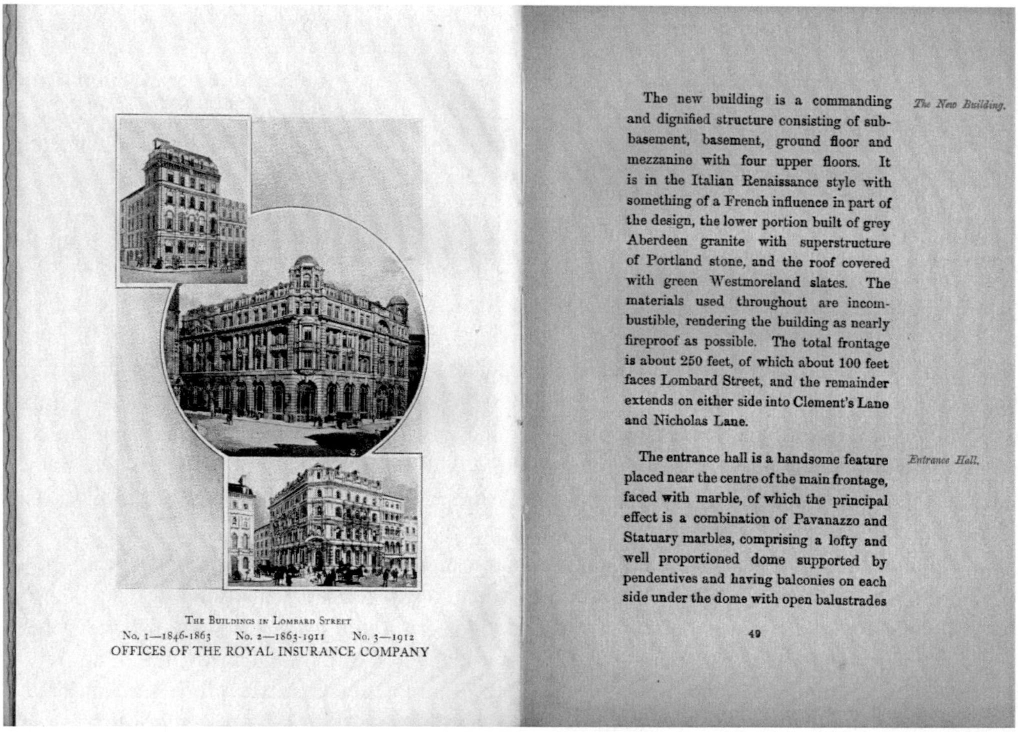
THE BUILDINGS IN LOMBARD STREET
No. 1—1846-1863 No. 2—1863-1911 No. 3—1912
OFFICES OF THE ROYAL INSURANCE COMPANY

The New Building.

The new building is a commanding and dignified structure consisting of sub-basement, basement, ground floor and mezzanine with four upper floors. It is in the Italian Renaissance style with something of a French influence in part of the design, the lower portion built of grey Aberdeen granite with superstructure of Portland stone, and the roof covered with green Westmoreland slates. The materials used throughout are incombustible, rendering the building as nearly fireproof as possible. The total frontage is about 250 feet, of which about 100 feet faces Lombard Street, and the remainder extends on either side into Clement's Lane and Nicholas Lane.

Entrance Hall.

The entrance hall is a handsome feature placed near the centre of the main frontage, faced with marble, of which the principal effect is a combination of Pavanazzo and Statuary marbles, comprising a lofty and well proportioned dome supported by pendentives and having balconies on each side under the dome with open balustrades

49

Figure 10.2 Three phases of the Royal Insurance Company in Lombard Street.

Notes: No. 1 – 1846–1863; No. 2 – 1863–1911 and No.3 – 1912.

Source: Royal Insurance Company (1912) *Old Lombard Street,* facing p49.

grand Italian 'palazzo' to the design of Thomas Marsh Nelson, of the firm Nelson & Innes.[29] Nelson had already worked on Lionel de Rothschild's London house at 148 Piccadilly. The domestic feel of the old New Court was swept away in favour of a building more imposing and business-like, and an impressive iron and enamel sign was hung over the entrance.

In 1879, Lionel de Rothschild died and the business was taken over by his three sons, Nathaniel ('Natty') (1840–1915) later to become the first Lord Rothschild, Alfred (1842–1918) and Leopold (1845–1917). In 1884, the Dividend Office was extended and Rothschild business was in such demand that in 1889, Natty had to climb into New Court via a first floor window to avoid crowds clamouring for shares in the Burma Ruby Mine issue. New Court was fitted with the most modern of conveniences, and in 1889 electric light for New Court was costing £1,000 per month. In 1899, 7 St Swithin's Lane was acquired for £8,000.

The second New Court survived the London bombing of the Second World War, despite severe damage elsewhere in St Swithin's Lane.

By the 1960s, the number of staff had increased to over 300, half of them women. The century old building was beginning to show signs of strain. Attempts were made to extend it upwards, but the building was looking dated, at a time when NM Rothschild & Sons were looking resolutely to the future.

In 1962, the bold decision was taken to rebuild New Court a second time. Senior Partner Mr Edmund de Rothschild (1916–2009), and the Partners Evelyn (later Sir Evelyn) de Rothschild (b.1931), Leopold de Rothschild (1927–2012) and Jacob (later Jacob, 4th Lord Rothschild) de Rothschild (b.1936) created a new Rothschild-owned company to undertake the development.

The architect Fitzroy Robinson was commissioned and the construction company Trollope & Colls were appointed to oversee the project. In 1962, staff said goodbye to the old New Court and left for a temporary office in City Gate House, Finsbury Square.

In 1965, the staff returned to St Swithin's Lane to a building very different to that which they had left. The main building had two floors below ground and seven above, set back from the Lane, with two wings, each of three storeys, joining the main block at right angles, arranged, in a deliberate echo of the two earlier buildings, around a central courtyard. Granite setts from the old courtyard were laid in decorative patterns in the new, and the New Court sign from the 1860 building was prominently displayed. Elsewhere in the building, historic features were incorporated including panelling from the Partners' Room and historic paintings. New features included air conditioning and a strongroom manufactured by Chubb, with Europe's then biggest strongroom door, with a lock offering over 4,000,000,000 different combinations.

The new building was the visible symbol of a trend of modernisation within the firm. The 1960s was the last decade in which the partnership operated, and in 1970 NM Rothschild & Sons became NM Rothschild & Sons Limited, with a board of directors. As business grew, extra premises were taken in Croydon to accommodate staff from administrative and accounting departments, and in 1984 an extra storey was added to New Court to create a new Board Room.[30]

In 2008, it was decided to demolish and rebuild New Court for a third time. The 1960s building was beginning to show its age and was too small to accommodate the entire staff. The internationally renowned architect Rem Koolhaas and his practice OMA were chosen to create a New Court for the twenty-first century. OMA's vision for the fourth version of New Court was inspired by the idea of 'heritage in the City'. The inspiration behind the new design for New Court came from the Palazzo Vecchio, home of Cosimo l de' Medici in Florence.

The fourth New Court was completed in 2011. The interior of the building includes many references to the company's history, and a new oak reading room for The Rothschild Archive.[31]

The history of Rothschild's London premises neatly summarises the key aspects of the cultural office. Over 150 years and more, and despite several moves, the premises remained, by US standards, small, low rise, irregular (inefficient) and strongly influenced by the social context of the family business.

Tall buildings. The rejection of height in London (and European) buildings through the nineteenth and much of the twentieth centuries is, perhaps, the most obvious difference between the cultural and the rational office. London's first tall building of any description dates back to 1873: it was the Midland Grand Hotel, part of St Pancras Station, designed by George Gilbert Scott and standing at around 76m and eight storeys high. The hotel was expensive, costing nearly £500,000 at the time, but its 300 rooms were soon fully used at prices amongst the highest in London. The building had innovative features including hydraulic lifts, concrete floors, revolving doors and fireproof floor constructions. The experience was made all the more splendid with a grand staircase, rooms with gold leaf walls and a fireplace in every room (although no en suite bathrooms at this point!).

In the same year that the Midland Grand opened, architect and building contractor Henry Alers Hankey began work on his block of flats that became known as Queen Anne's Mansions. Completed over the period through to 1889, the 47m (154ft) tall, 14-storey structure was the highest residential building in Britain. Indeed Queen Victoria objected because she could no longer see the Houses of Parliament leading to an Act of Parliament in 1890 setting the maximum height of new buildings at 90ft (27m) – later reduced to 80ft (24m) in the 1894 Act. During the War the building was occupied by the Medical Department of the Royal Navy and in 1947 the Ministry of Works retained the building on a 21-year lease and it was used as an Admiralty headquarters building. London was not the only European city with height restrictions: for example, Berlin had a limit of 72ft (22m).

The only comparable contemporary building on the continent was the Witte Huis (White House) in Rotterdam, the Netherlands (Figure 10.3). This was built to a height of 43m (141ft) across 10 storeys. This was continental Europe's first modern skyscraper.

Built in 1898 in Art Nouveau style, to a design by Willem Molenbroek, the construction overcame some major site constraints. The plot was just one metre above sea-level and there were serious doubts that Rotterdam's soft soil could, literally, support the proposed building. As a result the building is underpinned by 1,000 piles!

The building is constructed from iron, steel and concrete, and includes two thick interior walls which increase the building's strength. Unlike many other contemporary buildings, wood was not a significant construction material due to the fear of fire. The building cost 127,900 Dutch guilders, and it was a purpose-designed office building.

One reason for the lukewarm approach to tall buildings in Europe was the urban context of centuries-old street layout, where complex ownership patterns combined with small plot sizes helped defeat attempts to create larger sites for taller buildings.

One of the tallest buildings on mainland Europe was built in 1958, reflecting the North American tradition of using tall buildings to convey corporate strength, or power. The Pirelli Tower in Milan stood alone for some years, remaining the city's tallest building until the mid-1990s. This 32-storey, 417ft (127m) structure was built on the orders of Pirelli president Alberto Pirelli, with Gio Ponti as lead architect. It was built on land owned by the company where the company had its manufacturing plant. It was one of the earliest curtain wall buildings on mainland Europe. But it was very much an exceptional building rather than part of a trend.

London had to wait until 1967 for Richard Seifert's 365ft (111m) Centre Point; until 1981 for his 600ft (183m) Nat West Tower (since renamed Tower 42), and until 1991 for I M Pei's 771ft (235m) One Canada Square in Canary Wharf. Mainland Europe also shunned tall buildings until the 843ft (257m) Messeturm was completed in Frankfurt in 1990.

Even in Australia, with its relatively modern city centres, there was also a very limited foray into tall buildings, with the first tall building there being constructed in Melbourne at the end of the 1880s' land boom. By 1892 there were just 11 office buildings of eight–ten storeys, and these dominated the skyline of the city well into the 1920s, with only two surviving today.

At 12 storeys, the Australian Building office block on the corner of Elizabeth Street and Flinders Lane, completed in 1889 for the Australian Property Investment Co., was Australia's tallest. Designed in a broadly Queen Anne style, the high-pointed dormer gable and corner turret also made it the most strongly vertical of the early skyscrapers.

The introduction of a 132ft (40.2m) height limit in 1916 prevented any true skyscrapers being built in the interwar period. One of the largest was the Manchester Unity Building, constructed in 1932. The building's main roof reaches the maximum 132ft (40.2m), but its ornamental gothic tower rises a further 78ft (24m), which was permitted for non-habitable spaces in buildings. The building is clad in terracotta tiles and three high speed lifts were supplemented by two escalators, the first in Melbourne, to connect the basement, ground and first floors.

Other notable tall buildings in Melbourne included the T&G Building on the corner of Collins Street and Russell Street, built in 1929 by the Temperance & General Mutual Life Assurance Society, and rising to 220ft (67m); the APA Building on Elizabeth Street, 12 storeys and 174ft

Figure 10.3 The Witte Huis, Rotterdam, the Netherlands.

(53m) at the tip of its corner spire when it was completed in 1890, and the art deco Russell Street Police Headquarters, built between 1940 and 1943, and rising over 16 storeys to 223ft (68m).

Corporate buildings. During the interwar years, London saw the development of a number of large buildings, by its own standards, but none of these was high rise in a US context. Most were idiosyncratic in some manner, many were built to the specific demands of their sole occupiers and few exuded the slightest air of efficiency in terms of layout. Some of these were public sector buildings that came to be synonymous with their famous and long-term occupants.

- Australia House for the Australia High Commission on Aldwych (1918).
- Broadcasting House, the BBC's iconic building at Langham Place (1931).
- India House for the High Commissioner on Aldwych (1930).
- County Hall for the London County Council (later GLC) on the South Bank (1922).
- Broadway House for London Passenger Transport's (forerunner of London Underground) on Broadway, Victoria (1929).
- Senate House for the University of London on Malet Street (1932).

Many others were designed and built for private sector occupiers.

- Daily Express on Fleet Street (1932), the UK's first curtain wall office building.
- Daily Telegraph on Fleet Street (1928).
- Midland Bank on Poultry (1939), the biggest building in England at the time.
- National Westminster Bank on Threadneedle Street (1931).
- Reuters & Press Association on Fleet Street (1938).
- Shell-Mex House on the Strand (1931).
- Unilever House on the Thames at Blackfriars (1931).
- Wellcome Building on Euston Road (1931).

None of these buildings copied the efficient floor layouts universally applied in American towers. Instead, their floors were typically much smaller and irregular (reflecting the ancient street pattern) than the average skyscraper; they were highly cellularised, with long, narrow corridors. They were steeped in the history and cultures of their occupants. They were larger versions of the buildings they replaced; they were not a new building form.

It was not until the 1990s that London was to directly experience tall buildings built on the rational American model. Conceived and delivered by Americans and Canadians, the first building, One Canada Square, was an alien addition to the London skyline. As Canary Wharf took shape with more buildings the whole scheme came to be known as 'Manhattan on Thames'. The buildings offered regular, contiguous, deep floorplans and were primarily occupied by large investment banks (for example, Credit Suisse and Morgan Stanley).

The potted history of the Rothschilds' buildings; the muted response to tall buildings in the UK, Europe and Australia, and the growth of the tailor-made corporate buildings are all deeply rooted in economic and cultural factors, not to mention the physical constraints imposed by the complex and aged nature of Europe's city centres. Free from such constraints, the US embraced the new office economy, driving many of the technological advances that were to enable the modern office.

While the arrival of the skyscraper was necessitated by the reorganisation of business, it was enabled by new technologies, not the least of which included lifts, telephony, steel frame construction and new sanitation techniques. In a process not dissimilar to that which is changing the office environment today, the birth of the modern office in the late-nineteenth century was enabled and driven by a number of technological innovations.

The rational office

There can be few better places to begin a discussion of the rational office than with the account of the early days of Chicago's rebirth provided by Earle Shultz and Walter Simmons in their wonderful book, *Offices in the Sky*.[32] In a situation reminiscent of London in the late-1660s, during the mid-1870s Chicago was coming to terms with the impact of a devastating conflagration. The great fire of 1871 spread across more than three square miles, destroyed over 18,000 buildings, including almost the entire business district, and killed 300 people. In a further parallel with London, the spread of the fire was assisted by the fact that most buildings were made of wood. In fact, over two-thirds of the structures in Chicago at the time of the fire were made entirely of wood, with most of the houses and buildings being topped with highly combustible tar or shingle roofs.

Even after Chicago's Great Fire, when rebuilding was swift, the new buildings remained quite primitive: they were poorly serviced, there was no electricity, no lifts/elevators, and they were constrained in height to the limits of daily human exertion.

> *Plodding upward, wheezing and snorting, this luckless fellow carried fuel for his stove or fireplace, for there was no central heat. He usually took care of his own cleaning. Often the water supply consisted of a common sink on each floor. Each office was supposed to have its own washbowl and pitcher. Most buildings had only one or two toilets.*[33]

Owen Aldis was a lawyer and just 24 when he arrived in Chicago in 1875. In 1879, he was appointed to act as agent and attorney in the management of the Portland Block, occupied by the John Marshall Law School. The six-storey, two elevator, masonry building of 28,460 sq ft (2,650 sq m) was built in 1872, to a design by Major William Le Baron Jenney. Aldis was to have a greater influence over the modern shape of the city of Chicago than perhaps any other individual.

When the building's owner, Peter Brooks, set out on his next major development project, the ten-storey Montauk Block on Monroe Street, he had no hesitation in choosing to work with Aldis. In a move that some might have regarded as risky, Aldis turned to an untried young firm of architects, Burnham and Root.

Daniel Burnham and John Root formed their partnership in 1873 and they were firmly part of what became known as *The Chicago School*. When appointed by Aldis, the firm had not yet built a large commercial building. The first issue to overcome was how to deliver a ten-storey, wall-bearing masonry building on Chicago's infamously unstable clay subsoil. Working closely together, Aldis and Root came up with an innovative solution, by devising a floating raft: a single concrete slab two to four feet thick, reinforced with steel beams and covering the entire lot, thus distributing the weight of the building across the whole plot.[34]

Completing in 1882, the building had a 90ft (27.4m) frontage, and a depth of 180ft (54.8m). Its ten storeys stood 130ft (40m) high (Figure 10.4). It offered 150 individual offices, and boasted a form of central heating instead of open fireplaces, as well as hot and cold running water to supply radiators, wash basins and toilets. The building was described as "*strikingly original, with little precedent behind it*" and praised for its "*new technical and aesthetic characteristics*". Nevertheless it was demolished twenty years later: the First National Bank needed the plot for

Figure 10.4 The 'rational office': Montauk Building, Chicago, 1882.

a large new building and paid Brooks "*$500,000 in cash for the land and building that had cost him $325,000*".[35]

Aldis was just 30 when the Montauk Building completed, and he was a partner in law firm Padock, Aldis and Wright. He had become "*a giant figure in the office building field and was striding from one success to another*". In the two decades from the Montauk in 1882, he helped create and manage one million square feet of office space.[36]

Aldis made one further contribution that helps underpin the notion of the American 'rational office'. In 1895, with architect Holabird and Roche, Aldis completed the Marquette Building, "*the most modern structure of its kind in Chicago and probably the world*", an outstanding example of the Chicago School, with features contained in no other building "*including a floor plan well in advance of its time*". The building resulted from what we would call today a Design Brief, in which he "*laid down the fundamentals of office building design*" most of which "*are valid today*". The Brief was as follows.

> *First: the office building that gives up the most light and air is the best investment.*
>
> *Second: second-class space costs as much to build and operate as first-class space.*
>
> *Third: the parts every person entering sees must make the lasting impression.*
>
> *Fourth: generally office space should be about 24 feet deep from good light.*
>
> *Fifth: operating expenses must be constantly borne in mind.*
>
> *Sixth: carefully consider and provide for changes in location of corridor doors, partitions, light, plumbing and telephones*
>
> *Seventh: arrange typical layout for intensive use (several small occupiers more desirable than one large tenant)*
>
> *Eighth: upkeep of an office building is most important. Janitor service must be of a high quality, elevator operators of good personality, management progressive.*[37]

Written around 1894, the Brief was an extraordinary template for office buildings. It was a rational approach to office provision, based on the requirements of occupiers, on economics, and on investment prospects. It exemplified the North American rational office that contrasted so fundamentally with the European culture office.

The Montauk exemplified the new, rational approach to building design. Without the trappings of centuries of tradition both in building design and the means of procuring them, America's economy made great strides in buildings that were built for the age.

As the nineteenth century drew to a close, and the twentieth got underway, the huge American economy was thundering along at great speed, and laying down a very different urban structure. The process involved the American economy switching structural models from an entrepreneurial, owner-managed form characterised by many layers of production and distribution, to a model dominated by large, vertically integrated companies that gained significant economies of scale and stripped out the add-on costs of intermediaries, brokers and distributors. The new economic model, and the rise of 'corporatism' gave rise to a fundamentally new building form: the skyscraper, with its distinctive modern form of regular internal space and repetitive facades. The buildings soon dominated the skylines of Boston, Chicago, New York, Philadelphia and the other great American cities (Table 10.1).

Multinational corporations deployed armies of clerical workers – and then office workers more generally – to administer and manage their ever more complex organisations. By the First World War, 'downtown' dominated most large American cities and in Europe, office activities were firmly established as a distinctive form of economic activity rather than as an adjunct to

Table 10.1 The rise of skyscrapers in America, 1870–1931

Building	*City*	*Completion*	*Metres*
Equitable Building	New York	1870	43
Western Union Building	New York	1875	70
Tribune Building	New York	1875	79
Boreel Building	New York	1879	30
Montauk Block	Chicago	1882	39
Mills Building	New York	1882	47
Temple Court	New York	1883	45
Counselman Building	Chicago	1884	44
Pullman Building	Chicago	1884	49
Home Insurance Building	Chicago	1885	44
Tacoma Building	Chicago	1889	50
World Building	NYC	1890	94
Monadnock Building	NYC	1893	60
Reliance Building	Chicago	1894	61
Manhattan Life Insurance Building	NYC	1894	106
American Surety Building	NYC	1896	103
Park Row Building	NYC	1899	119
New York Times Tower	NYC	1905	220
Singer Building	NYC	1908	187
Metro Life Insurance Building	NYC	1909	213
Bankers Trust Building	NYC	1911	164
Woolworth Building	NYC	1913	241
Manhattan Municipal Building	NYC	1914	177
Equitable Building	NYC	1915	169
Tribune Tower	Chicago	1925	141
Bank of Manhattan Trust	NYC	1930	283
Chrysler Building	NYC	1931	319
Empire State Building	NYC	1931	443

manufacturing. During this huge convulsion of change, the relationships between the economy (businesses), buildings (space) and work (people) were transformed.

After the devastation of the 1871 fire, the demand for new buildings in Chicago was intense, causing land prices to soar and the pressure to build taller to intensify. The solution was found by the wonderfully named Major William LeBaron Jenney (1832–1907), designer of the Portland Block referred to above. Born in Fairhaven, Massachusetts, Jenney went to study engineering and architecture at the École Centrale des Arts et Manufactures in Paris in 1853, where he discovered iron construction techniques (Gustave Eiffel, of the eponymous tower, graduated from the same school in 1855). He graduated in 1856 and returned to the USA in 1861. Following a spell in the army, working as an engineer, Jenney moved in 1867 to Chicago where he opened an architectural practice to work on commercial buildings.

The First Leiter building marked Jenney's foray into commercial buildings, located at Washington and Wells Streets. Completed in 1879, it was renovated and extended in 1888 (and demolished in 1972). Jenney designed this building, as a department store for Levi Z Leiter. Originally five storeys high (later extended to seven), the building marked a significant milestone in architectural engineering by bringing together an iron skeletal frame; terra cotta fireproofing materials on all of its structural members, and vertical transportation in the form of elevators. It also utilised a new type of glass in its windows.

Jenney designed both the Ludington Building and the Manhattan Building in Chicago (both delivered in 1891), but his name and reputation are firmly attached to the Home Insurance Building there, built between 1884 and 1885. This 144ft (44m), ten-storey building was the first high-rise building in which an iron skeleton of vertical columns and horizontal beams carried the whole load of the building, and in which the external walls were reduced to panel infillings between the members of the frame. Saving on expensive foundations owing to the reduced weight of the building, and "*the commercial possibility of buildings with more than twelve storeys, owing to the saving of space and increase of light on the lower floors which went with the elimination of load-bearing masonry, provided the economic basis for the skyscraper.*"[38]

As skyscrapers grew in popularity, many were commissioned by business owners seeking to demonstrate their corporate strength and to use their structures as giant billboards, nowhere more so than in New York. Building patrons were themselves often exuberant, larger-than-life characters. Frank W Woolworth teamed up with architect Cass Gilbert in 1910 to create the Woolworth Tower. Its opening ceremony, on 24 April 1913, created a little history. President Woodrow Wilson pressed a button in the White House to activate 80,000 light bulbs in front of 900 assembled guests. One of them, the prominent clergyman Samuel Parkes Cadman, was said to have cried out "*the cathedral of commerce!*".

Fenske and Holdsworth describe the business background to the skyscraper. Woolworth opened his first successful five-and-ten-cent store in Lancaster, Pennsylvania, in 1879. Before he incorporated his company in 1905, Woolworth had purchased several five-and-ten-cent stores as well as chains from a number of competitors. In 1909, the year before he began acquiring parcels for his building project, Woolworth opened his first store overseas in Liverpool, England. Within 2 years, he opened 12 stores in England.

Woolworth began construction on the building's foundations in 1910, but this did not deter him from continuing to enlarge the site, which, by early 1911, included the entire Broadway block front, a feat in the eyes of New York real estate dealers. Gilbert, with a series of proposals, increased the scale of the design accordingly. The enlargement of the project occurred just before the F. W. Woolworth Company underwent its 1911 merger, which involved the acquisition of four sizable five-and-ten-cent chains, causing it to nearly double in size.[39]

A few years later, Walter P Chrysler set his architect a very simple brief for his own New York tower: make it taller than the Eiffel Tower. William Van Alen's design for the Chrysler building typified many buildings of the period with its ornate design, intricate patterns, mixing of materials and complex structural style. As time passed, the buildings became much simpler, more economic and utilitarian in their design, but their iconic status simply grew.[40]

The search for immortality as expressed through named buildings was such that building patrons often occupied only a fraction of the space they built, and filling the remainder of their buildings with numerous smaller enterprises in their own supply chains – lawyers, accountants, surveyors, architects, engineers and so on – with whom they shared the central business district. For example, the New York Tribune occupied only the top two floors of its 1875, nine-storey building; Singer occupied the top four floors of its 1899, ten-storey building, and FW Woolworth occupied a little less than two floors of its iconic 50-storey tower.[41] Fenske and Holdsworth observe that the towers were rarely of any true functional use to their corporate builders:

> *this was shown time and time again, whether by the American Surety Company, which never made full use of, and eventually vacated, its tower; by the Singer Company, which hardly occupied its tower; by the Woolworth Company, which had difficulty leasing its tower; or by the Banker's Trust Company, which refused to take space in its tower before opening ancillary offices in midtown.*

Fenske and Holdsworth argue that the "*only justification for the towers was an enterprise's augmented visibility on the urban scene*", with the reasons why some enterprises sought such visibility including "*advertising and public relations, establishment of a physical presence for an unmaterial business such as insurance, and the assertion of an individual ego*".

> *In general, the visibility afforded by the towers expressed, and ultimately, legitimized a newly powerful commercial order. Unabashed appropriation of a form typically reserved in earlier civilizations for ecclesiastical or civic purposes vividly conveyed the dominance of commercial affairs in American life.*[42]

Thus it may be argued that the relentless search for height was not simply a question of land economics, but also one of corporate ego. In a fascinating paper, Domosh described some of the 'softer' symbolism of tall buildings in late-nineteenth century America. She suggested that "*skyscrapers expressed the dominance of commercial values in New York*" and that "*as forms of ostentatious display and advertisement, skyscrapers perfectly fit the needs of New York's nouveau riche*", albeit viewed with dismay by a rump of New York's 'old' money class, who simply had to "*submit to the dictates of economic forces*". To illustrate her case, Domosh cites the World Building in New York.

She begins by quoting a speech made by Senator Chauncey Depew at the laying of the building's foundation stone, quoted in the *New York World*, 11 October 1889).

> *Around and in sight of us are buildings which express more eloquently both the intellectual and material progress of the country than its stately capitals or splendid palaces, its furnaces or factories, its mills or its railroads.*
>
> *... these splendid structures built by the modern newspaper, are the results of a combination of brains and business, of mental vigour and culture, of ability in the conduct of affairs, of statesmanship and common sense, which makes possible American literature and perpetuates American liberty.*

The World Building was delivered on Park Row in 1890 to house the offices of Joseph Pulitzer's newspaper, *The World*. Park Row, the traditional home of the news industry in New York. *The World* had outgrown its previous quarters on Park Row, and in contemplating the location and design of a new structure, Pulitzer "*wanted his newspaper to be housed in a building 'appropriate' to the image of the paper, and one that served his promotional intents*". To this end, Pulitzer turned to renowned New York architect, George Post, instructing him to "*design a building that would be considered a cultural adornment to the city*". He further instructed Post that "*the building must be taller than the Tribune and, preferably, the tallest in the city*". Domosh described the resulting 16-storey building as designed in the French Renaissance Revival Style, ornamented with four torch-bearers representing art, literature, science and invention.

Domosh concludes that tall buildings resulted for reasons more fundamental than land economics and in response to developments in construction techniques. Instead, she suggests that that such buildings

> *served as material expressions of social legitimacy and economic power for New York's nouveau riche class, and thereby were functional responses to a far more varied set of needs than is usually assumed.*

In the case of the World Building, "*Joseph Pulitzer, who lived within a particular socio-economic context, made a decision, and selected an architectural form*". Domosh explains the decision in the context, first of "*the larger social and economic context*" which included his "*need for legitimacy*" which was "*a response to the unstable class structure Pulitzer was a part of, and to his search for material expressions of his status*" and secondly his "*desire to create a symbol that would serve as advertisement and a sign of economic power*" in the context of the highly competitive newspaper industry in New York.[43]

As new technologies including fire-resistant materials, lifts, lighting, telephones and sanitation made great strides, so buildings became become larger and taller still. For example, the first passenger elevator was installed by Elisha Graves Otis in New York in 1857.

The incandescent electric lamp was invented around 1870. Sanitation too was critical to the development of the modern office buildings. Alexander Cummings' 1875 patent for a toilet with a 'stink trap' that prevented sewer gases from backing up through drains was an important development. And as we have already seen, steel frame construction was born, enabling tall buildings, free from the limitations of load-bearing concrete structures. The Equitable Life Assurance Society Building was the first modern skyscraper, built in New York in 1870.

The speed with which steel frame construction and tall buildings fundamentally altered the skyline in New York and Chicago was dramatic. Small, narrow buildings accommodating small-scale economic activities were swept away as the huge towers replaced a dozen or more front doors with one at the base of a vertical village.

By the turn of the century, tall buildings were commonplace in New York and Chicago. Indeed, in New York, the "*dark shadows of great buildings lay across one another. Men in offices switched on electric lights to ease the gloom*". Such was the problem that building owners purchased adjacent properties "*at great expense to prevent competing skyscrapers from snuggling even closer*".

The impact of the innovations in office design and servicing was so profound that it might be argued that the subsequent century of evolution in building design involved mere refinements on this early model. But more broadly, skyscrapers came to symbolise the new corporatism that was sweeping through the American economy, and many became synonymous with the corporations that occupied them. In New York, for example, there were Bank of Manhattan Trust, Bankers Trust Building, Chrysler Building (vehicles), Equitable Building (insurance), Manhattan Life Insurance, the Metropolitan Life Building (insurance), the Singer Building (sewing machines) and the Woolworth Building (retail).

The phase in the development of American offices described here saw the establishment of a new building form that was unique and entirely new. In stark contrast to the European approach of muddling through and incremental design change, in America a whole new building type was created – the rational office.

Steel frame construction. The fundamental innovation that allowed buildings to go taller was steel frame construction. Concrete structures were limited in height due to the weight bearing on the lower levels demanding much thicker walls. The Home Insurance Building is generally considered the world's first skyscraper; designed by architect William Le Baron Jenney and built in 1885 in Chicago. The ten-storey building stood 138ft (42m) in height and served as the headquarters of a home insurance company.

It was supported by a revolutionary steel frame consisting of vertical columns and horizontal beams, which allowed for much greater stability.

Though it was expensive, the use of steel posed a number of advantages aside from stability. Thinner walls made for more spacious interiors and more windows, and fire posed less of a hazard than it did to older masonry structures

The light masonry walls were put on the steel framework as curtains (there was no longer a need to increase the wall thickness to make a high-rise building more stable). Thanks to its steel frame, the ten-storey building did not pose a risk of toppling under its own weight. The rooms were also brighter in buildings with such framework, as the steel made it possible to significantly increase the number of windows.

Building the first of a new generation of steel-framed skyscrapers set the standard for various other innovations in construction, including elevators, roof structure and plumbing. The Home Insurance Building stood until 1931. Later, another skyscraper was constructed in its place: the 45-storey LaSalle National Bank Building.

Built in 1902, the Flatiron Building is an architectural landmark of the borough of Manhattan in New York City. For more than 100 years, it has stood as a symbol of the Big Apple. This unusual office building consisting of 22 storeys was built in an extremely short time of just one year. The skyscraper's steel frame was made by American Bridge Company, which specialised in bridge construction. The steel structures were partly pre-assembled and delivered to the construction site. This allowed the construction of the Flatiron Building to proceed at an amazing speed for its time, at a pace of no more than a week per floor.

Initially, the building's unusual triangular shape evoked mixed responses from the public. There were concerns that such a strange building, which was situated in a rather windy area, would quickly fall over. Local bookies were even taking bets on how far the building's debris could be found after its collapse. However, the architects calculated that the steel framework enabled the building to withstand four times the amount of wind force than it could ever be expected to endure.

Aside from its triangular shape, this office building had other specifics. For example, it lacked an air conditioning system, had a single staircase and the oddly shaped rooms formed a draughty labyrinth. Nevertheless, the value of the Flatiron Building was estimated at US$190 million in the 2010s.

The Home Insurance Building is accepted as the first true steel skeleton building (finished in 1885). The Tacoma Building claimed the title until various committees determined that it was not as fully of steel construction as Home Insurance, though it did prove to be the first with a riveted steel frame. Once the system for building higher was established, there began a sort of architectural arms race; although tellingly, the first steel frame building in Europe was not delivered until 1904, and that was not for an office but rather a hotel, the Ritz Hotel in London.

Lifts (elevators). Of course tall buildings would not have been possible without safe passenger lifts, or elevators. According to Turvey, lifts were operating in London hotels from the 1850s, but the first office lifts were later:

> *in 1873 a hydraulic lift was installed in the eighteen-sixty-seven office block of Palmerston Buildings between Old Broad Street and Bishopsgate at a cost of £750. It enabled the letting of offices on upper floors that had never before been tenanted.*

Turvey cites further early lift installations in the Foreign Office in 1875; at Maw & Thompsons's Aldersgate Street offices in 1877 and at Leadenhall House in Leadenhall Street in 1879.[44] However, it was not in London, but in America that the lift was, literally, to reach new heights.

The early elevators using hoist ropes were severely limited in terms of the height they could reach. The next step was the introduction of the hydraulic elevator in 1870, which went through several iterations before overcoming design flaws. In 1877, German Friedrich Koepe developed an alternative for winding cables into large drums. He introduced a friction-driven hoist: a large wheel with a dove-tailed groove in which the hoisting cable was laid. This friction-driven hoist substantially reduced the risk of cable breakage and allowed for greater heights.

Elisha Graves Otis demonstrated his safety brake for freight hoists in New York in 1853. The early steam-and-drum machines were primitive by any standards and were superseded by the rope-geared hydraulic elevator, capable of more floors and greater speed. The development of geared hydraulic elevators necessitated a new kind of safety device, which, in the event of excessive downward speed, would grip the vertical guides and bring the elevator to a gradual stop.

Steel skeleton construction quickly reached the operational limits of hydraulic lifts just two decades after the completion of the Home Insurance Building. They were limited by the physical laws of the pressure of the water column: an elevator could be lifted by hydraulic force to a height of 18 or 20 storeys. To be freed from this constraint, major innovation was required. This came in 1878 when German industrialist Verner von Siemens designed, built and presented the first electric elevator. This opened the door for high speed elevators and more height.

From then on, the height to which office workers was no longer limited by natural laws people could be lifted mechanically was no longer restricted by physical laws, and the sky became the limit.

> *The invention of the direct drive gearless electric elevator and the caisson-type foundation suggested a spectacular remedy for this headache, and almost overnight it was possible to double and triple the heigh of office buildings. The tower was conceived.*[45]

In 1904, gearless traction elevators were introduced, in the Beaver Building in New York City. The invention of the direct drive gearless electric elevator and the caisson-type foundation made it possible to double and triple the height of existing tall buildings.

Buildings were not only getting taller, their plot sizes were also increasing, which meant that their floor size grew, bring ever larger number of people together onto single floors. Two key enabling technologies resolved the issue: lighting and air handling.

Lighting. Many early American buildings were gas lit, which, as in London, carried with it safety risks. But from around 1877, electric lights gained popularity and, eventually, the widespread availability of the incandescent light bulb changed the way offices were lit. As early as 1860, in England, Sir Joseph Wilson Swan (1828–1914) developed a primitive electric light which used a filament of carbonized paper in an evacuated glass bulb. But the lack of a vacuum helped reduce its reliability and lifetime. Twenty years later, Edison patented an incandescent bulb with a filament that gave off light when heated to incandescence by an electric current.

Initially Edison used a carbonised thread for the filament, before experimenting with card paper and then carbonised bamboo. Extruded cellulose filaments were introduced by Swan in 1883. The next significant innovation came with the invention of the tungsten filament in Europe in 1904. The new filaments lasted longer and were brighter. But it was the introduction of the fluorescent lamp in the 1930s that revolutionised office lighting: early fluorescents provided more than twice as much illumination per watt as incandescent lamps, without the heat output.

The first credible fluorescent lamp is usually credited to Edmund Germer (1901–1987), born in Berlin, Germany, and educated at the University of Berlin, where he wrote a doctorate in lighting technology. Together with Friedrich Meyer and Hans Spanner, Edmund Germer patented an experimental fluorescent lamp in 1927. However, Germer's invention was not successfully commercialised. This was not to occur until scientists at General Electric in the USA, led by George Inman, created a practical fluorescent lamp, first sold in 1938.

Air handling. The recognised inventor of office air conditioning is New Yorker Willis Haviland Carrier (1876–1950), who graduated from Cornell University in 1901 with a master's degree in engineering. After graduation, Carrier accepted a job at the Buffalo Forge Company, drafting plans for heating and drying lumber and coffee.

In the spring of 1902, consulting engineer Walter Timmis visited Buffalo Forge's sales office in Manhattan to meet with head of office J Irvine Lyle. Timmis was acting on behalf of his client, Sackett & Wilhelms, whose Brooklyn plant suffered from a level of humidity that caused problems in its printing process, with low quality production, waste and downtime. Timmis approached Lyle in the hope that he could help find a solution, and it was at this point that Lyle consulted with colleague Willis Carrier.

Carrier made haste to the plant and began work immediately, with data gathering, testing and questioning. Following an aborted test recommended by Timmis, Carrier set about his own experiment, using cold water flowing through heating coils, balancing the temperature of the coil surface with the rate of air flow to pull the air temperature down to the desired dew point temperature. It was late summer 1902. Making use of *fans, ducts, heaters, perforated steam pipes for humidification, and temperature controls, Carrier drew cooling water from an artesian well to maintain a constant humidity of 55% year-round and have the equivalent cooling effect of melting over 100,000 pounds of ice per day.*

During autumn 1903, Lyle reported Carrier's cooling coils had given excellent results. In fact, Carrier had demonstrated an exceptional engineering intellect and vision, creating an entirely novel approach to air handling. Carrier had done nothing less than invent air conditioning. Eight years later, at the Annual meeting of the American Society of Mechanical Engineers, he presented his '*Rational Psychrometric Formulae*'. Carrier's paper explained how relative humidity, absolute humidity and the dew point temperature could be harnessed to build air handling systems that could provide constant, managed temperatures in buildings – most significantly here, in office buildings. Carrier's patent solved the problem of humidity removal by condensing the water vapour on droplets of cold water sprayed into an airstream.

Carrier founded the Carrier Engineering Corporation with a group of colleagues in 1915, and continued to develop his industrial applications. But it was clear that other applications were being called for, and in 1922 Carrier created a smaller and more powerful centrifugal refrigeration compressor, which became the precursor to modern air conditioning systems. Carrier installed the first air conditioning system in an office building at the 21-storey Milam Building in San Antonio, Texas, designed by George Willis, in 1928. It had a central refrigeration plant in the basement that supplied cold water to small air-handling units on every other floor; these supplied conditioned air to each office space through ducts in the ceiling; the air was returned through grills in doors to the corridors and then back to the air-handling units.

The first air-conditioned office building in London was the sprawling Broadcasting House, for the BBC, designed by G Val Myers. Completed in 1931, the building set particular challenges for studios which needed acoustic treatment and air conditioning, and administrative offices which needed to maximise daylight. The air conditioning system was designed and installed by Carrier Engineering.

The physical organisation of work and our ability to control the working environment are intrinsically linked. People need lighting, air handling and furniture all of which need to be laid out efficiently and effectively. No matter where a desk is situated – in the centre of a floor, on the façade or somewhere else – it needs to be heated, cooled and lit in the same way as every other desk in the building. This requires co-ordination between the organisation of work and the services needed to support it.

Thus it was that office building technologies such as air conditioning and lighting combined with office equipment and improvements in office management and processes to enable the growth of very large, corporate organisations. However, the two World Wars, the Spanish Flu, the Great Depression following 1929 and a worldwide economic malaise combined to mean that the office economy changed relatively little during the first half of the twentieth century. Certainly compared to the tumultuous years of the later-nineteenth century, it was a quiet period for the office economy.

While office buildings in America were breaking records in terms of scale and setting new benchmarks for innovation in their internal servicing and layout, progress in Europe was rather different. Even at the back end of the nineteenth century, the US economy was much greater than that of Britain, and the vertical integration of industry, together with the rapid spatial growth, led to the creation of corporate behemoths that simply did not exist in Britain or across Europe. The challenge in London and other European capitals was about how to blend new businesses, workplaces and workers with an updated version of the culturally influenced nineteenth century office.

When rationality drops its guard

The tall buildings and skyscrapers discussed above exemplify the rational approach to building and workplace design taken in North America. We shall see in the next section this has been widely explained in terms of the introduction of scientific management by Frederick Taylor which, in the view of this author, is an oversimplification. Two buildings, built 35 years apart, underscore this oversimplification: the Larkin Building and the Johnson Wax Building.

The Larkin Building. The Larkin Soap Company was a successful business with a good record for corporate responsibility towards its staff. Founded in 1875 by John D Larkin (1845–1926), by 1900 the business, located in Buffalo, New York, was growing and diversifying, and established a pioneering national mail order business. Growth was such that the firm identified a need for an administrative building, separate to its production plant, but on the same industrial estate. Larkin's co-directors, Darwin Martin and William Heath, both suggested Frank Lloyd Wright (he had designed houses for both of them). Thus it was that Wright won his first commercial commission. Completed in 1904, Wright designed the building to focus on a five-storey court, or atrium in modern terms.

The administration floors could accommodate around 1,800 secretaries, clerks and executives. The building had plenty of natural light and air conditioning. Metal office furniture was deployed for the first time in America. Figure 10.5 is of the fourth floor (east side) of the building and shows the rigid layout, with supervisors in the bottom right of the image and four rows of largely female workers. Such imagery reinforced the 'factory' reputation of such buildings and underscores the rational tradition in American offices.

Wright worked closely with the Larkin executives to reinforce the corporate culture through the design of the new space. In one example, the executives occupied the main floor with Martin and Heath and their subordinates occupying an entire floor beneath the skylight and also under the balcony, leaving them with no privacy. Such arrangements were highly unusual for the time, but consistent with the image Larkin wanted to project of open, familial relations. Similarly, the uppermost level included a bakery, kitchen, dining rooms, rest rooms, library and a conservatory – all major benefits for the staff.

Johnson Wax Building. Thirty years after completing the Larkin Building, Frank Lloyd Wright was once again commissioned by a manufacturer to build its administrative headquarters. This time it was SC Johnson & Son in Racine, Wisconsin. Again there was strong involvement from a determined client, this time the company's president, Herbert F. "Hib" Johnson.

Figure 10.5 The Larkin Building, Buffalo.

The low rise building shown in Figure 10.6 was constructed between 1936 and 1939 (the Research Tower shown in many photos was added in 1950). The new building was in an industrial zone, and so Wright decided to create a sealed environment lit from above, as he had done with the Larkin Building.

Throughout the Great Workroom, which has no internal walls, the thin, white dendriform columns rise to circular 'lily pad' tops that form the ceiling, with the spaces between the circles consisting of skylights made of Pyrex glass tubing. The Great Workroom is the largest expanse of space in the building, and was intended for the secretaries of the Johnson Wax company, with a mezzanine for administrators. Wright designed not only the building but its furniture, which was manufactured by Steelcase.

Both the Larkin Building and the Johnson Wax Building were commissioned by private (cash-rich) companies, and by CEOs who were interested in legacy. Neither played by the same rules as commercial developers. The main floor of the Larkin building is often shown as an example of Taylorism, and yet it was completed 7 years before Taylor published his magnum opus. Furthermore, the facilities elsewhere in the building were anything but Taylorist – they were designed to create a sense of community. Similarly, Johnson Wax defied the strictures of Taylorism with its low density, relaxed layout.

Figure 10.6 The Johnson Wax Building, Racine.

Rational management

The emergence and growth of the modern corporation in the later years of the nineteenth century and the early years of the twentieth, both in Europe and America, led to remarkable changes in the office – both as a place and as an activity. However, as we have seen, the routes taken on either side of the Atlantic were very different, influencing office design in each through to the twenty-first century. In short, the American experience was revolutionary: rapid, transformative and entirely new; whereas the rise of corporatism in Europe was evolutionary: taking place incrementally over a long period of time and carrying cultural baggage.

The American corporation evolved incredibly quickly, in a very short space of time and, crucially, without the influence of centuries of evolution expressed in the social complexity of master and servant, of social class or of rank. Unencumbered by the socio-cultural overlay of European companies, late-nineteenth century US corporations took a far more rational turn. Their managers quickly began to rationalise production processes in order to minimise manual input and to maximise product output.

The US offices of the later-nineteenth century that were attached to railroads and industrial concerns were a rational response to rapid and fundamental change. The American office was from the early stages, conceived to accommodate a particular form of organisation, which itself was the child of the economic revolution. The unquestioned rationality of production inevitably found its way into the factory office which Harry Braverman described as a "*paper replica of production*", whereby management exercises control by requiring "*that every activity in production have its several parallel activities in the management centre: each must be devised, precalculated, tested, laid out, assigned and ordered, checked and inspected, and recorded throughout its duration and upon completion*".[46]

Braverman argued that the novelty of the divide between the management and the process (or between conception and execution) lay not in the separation itself, "*but the rigor with which they are divided from one another, and then increasingly subdivided, so that conception is concentrated, insofar as possible, in ever more limited groups within management or closely associated with it*". Furthermore,

> *in the setting of antagonistic social relations, of alienated labor, hand and brain become not just separated, but divided and hostile". He argued that the "separation of hand and brain is the most decisive single step in the division of labor taken by the capitalist mode of production.*[47]

Thus it was that factories, railroads and other industrial concerns set the template for the organisation and management of offices before offices existed in their own right, rather than as adjuncts to production processes. It was not long before this established search for rational organisation found expression in a wholly new branch of management theory and practice: *scientific management.*

Frederick Winslow Taylor (1856–1915) was born in Germantown, Philadelphia, into a wealthy, highly educated and liberal-minded Quaker family. His father was a lawyer and graduate of Princeton, and his mother an abolitionist and feminist. Following a 3-year 'Grand Tour' of Europe, Frederick enrolled into the highly respected Phillips Exeter Academy in New Hampshire. However, his health was poor and despite passing the Harvard entrance examination, he left Exeter prematurely.

After several months of recuperation, and in his 18th year, he chose a vocation in engineering, joining Ferrell and Jones (later the Enterprise Hydraulic Works). Despite his social background and obvious intellectual strengths, Taylor followed the pure and principled nature of his upbringing and started his new vocation as a humble pattern maker's apprentice. On completing his apprenticeship in 1878, Frederick moved to the Midvale Steel Company to take a post as a machinist.

His inquisitive and innovative mind soon marked him out for greater things, and he scaled the corporate ladder rapidly, rising to research director and then chief engineer. His dedication was to be demonstrated by his ability to earn an engineering degree from the Stevens Institute of Technology in New Jersey, at the age of 25, while working full-time. Taylor was an innovator and a rationalist, and he was imbued with a belief in the role of science and technology in the advancement of mankind.

While at Midvale, Taylor studied industrial productivity and methods for measuring output. He introduced piecework, developing detailed systems aimed at maximising the efficiency of men and machines in the factory. His work involved time and motion studies to determine the best methods for performing tasks in the least amount of time. His basic thesis was that incentive-based wages were ineffective unless they were used alongside carefully planned and managed tasks, backed up by supportive and co-operative management. He argued for the careful selection and development of people, identifying clear goals and tasks, organising

Figure 10.7 Frederick Winslow Taylor.

feedback, providing training and encouraging group support. His methods were successful and output rose significantly. Scientific management was born.

Taylor became a consulting engineer at the age of 37, often making enemies in the companies where his methods resulted in job losses aimed at increasing efficiency. His most important client became the Bethlehem Iron Company, but following various disputes with the management, he was fired in 1901. This was to be his last job. His most famous published work, *The Principles of Scientific Management*,[48] was based largely on transcripts of talks given long after leaving employment. He contracted influenza in 1915 while on a speaking tour and died the day after his 59th birthday.

'Taylorism' is a term used today in a faintly pejorative manner, reflecting the popular perception that Taylor introduced 'time-and-motion' principles into the office; establishing the principle of office workers arranged like workers on a production line. He is caricatured as having a singular focus on efficiency in production, and even of having influenced industrial production thinking in Nazi Germany and the Soviet Union. However, these are either gross oversimplifications of the truth or historically inaccurate, for at least four reasons.

First, in his evidence before a Special House Committee in 1912, Taylor demonstrated a concern for workplace harmony (between workers and bosses), which could be nurtured, he argued, through more equal responsibilities and opportunities for workers to suggest improvements to conditions and processes. He stated categorically that scientific management was not about holding a stop watch to a worker and called for a complete mind shift by management.

Secondly, most of Taylor's work was on the factory floor, not in the office; his influence on office processes was modest to say the least. Despite countless references to the long shadow of 'Taylorist planning' in office workplaces, and while recognising Taylor's role as the founder of scientific management, it was in fact his followers who brought the discipline to corporate offices.

Even then, and thirdly, it is arguable that scientific management failed to be adopted in Europe. Nazi Germany and the Soviet Union post-dated Taylor's death by around two decades, and despite some records of French factories adopting scientific management there is precious little evidence elsewhere. This is addressed more fully below, but there is meagre evidence of London office-based organisations adopting the rigours of scientific management, preferring instead to keep with their cultural tradition of muddling through.

Finally, the 'rational' template for the US office – regular floorplates, factory-style layout and bureaucratic management – was laid down before Taylorism was popularised. Table 10.1 listed a sample of 28 skyscrapers built between 1870 and 1931. Taylor was not yet 20 years old when the New York Tribune and Western Union buildings were completed in 1875, and 21 of the buildings were completed *before* Taylor published his seminal work in 1911. Duffy was thus incorrect to state that

> *Taylorism was the dominant management philosophy when the office as a building type was created, so the particular values that Taylor emphasised – order, hierarchy, supervision, depersonalisation – became an integral part of the architecture of those initial, pioneering, turn-of-the-century North American buildings.*[49]

Lowe makes the point that the new 'science' of management was founded on the principles of efficiency and rationality, which became entrenched in systems of administrative control, and that most efficiency programmes, from scientific management to cost accounting,

> *were initially devised for the factory and shop. But after the First World War these schemes were applied with equal vigour to the office. Offices were thus subjected to increasing rationalization, bureaucratization, and mechanization. As clerical work became more standardized, regulated, and fragmented, working conditions became like those in a factory.*[50]

It is therefore somewhat problematic to suggest that Taylor had such a major and long-lasting influence on architecture and design in both America and Europe. The rational approach to management co-evolved with America's industrial complex before Taylor was active, and the basic template of its office buildings was determined more by land economics and construction innovation than by management theory. And indirectly, the office tradition in the US carried none of the two centuries' worth of socio-cultural baggage that was hard-wired into European offices.

None of this is to deny that scientific management was influential; but rather that its influence has been misunderstood, particularly the notion that it was so influential in office design and the production line style layouts of offices. To better appreciate the impact of scientific management it is important to review the work of some of Taylor's followers.

Taylor's early disciples included Henry L Gantt (1861–1919), he of the eponymous charts, who published *Work, Ways and Profits* (1910) and *Organising for Work* (1919); Harrington

Emerson (1853–1921) who published *The Twelve Principles of Efficiency* (1912) and C Bertrand Thompson (1882–1969) who published *The Theory and Practice of Scientific Management* (1917). Thompson had a particularly lofty view of his and others' endeavours:

> *The substitution of a basis of scientific law and principles for guesswork or tradition reminds one strongly of August Compte's theory of progress from the theological, through the metaphysical to the positive or scientific state of thought.*[51]

All of this work related to factories. The first tentative steps to transfer scientific management from factory to office were taken by Morris Llewellyn Cooke (1872–1960), who first met Taylor in 1903. He set up his own consultancy business in 1905, applying scientific management to printing and publishing and office firms. In 1911, he was appointed director of the Department of Works, where he implemented his version of scientific management.

Cooke laboured at the Department for 4 years, where he found widespread inefficiency and plentiful opportunities to bring about improvements. At the Department of Works he established new approaches to financial planning, complaint handling, inventory recording, contracts and public relations.

Another important Taylor disciple was Carl Copeland Parsons (1874–1947), about whom very little has been published. What we do know is that Parsons lectured at the University of Michigan where he was appointed to run the Commerce and Business Management course. Before this, he had been an accountant and a branch manager for an office equipment business. More significantly here, Parsons published an extensive guide to his principles, *Office Organisation and Management*.[52] He believed that the office was a critical organisational component for every organisation, having the ability to influence efficiency everywhere. He also saw that the office and its records, reports and data offered managers the vital tools with which to make intelligent decisions.

Parsons saw how scientific management in the office could be applied to the selection and training of workers, the utilisation of office machinery and, critically here, how it could influence the proper layout of office clerical work, with work flow based on a straight line principle. To achieve this, he preached the importance of appropriate heating, lighting and ventilation, while minimising background noise.

Cooke and Parsons were pioneers in thinking about efficiency in the office, and how it should be organised and managed. But the biggest impact of scientific management in the office was made by William H Leffingwell (1876–1934), author of numerous articles and two very long books: *Scientific Office Management* in 1917[53] and the mammoth 800-page *Office Management; Principles and Practice* in 1925.[54]

Leffingwell was born in Oxford County, Ontario to Wendell Phillips and Mary Catherine Leffingwell, both Americans. He was trained as a stenographer and throughout the 1920s, he was a key figure in the Taylor Society. In his 1917 book, Leffingwell was clear in distinguishing his work from that of Taylor.

> *many businessmen, after analysing the remarkable results secured by applying Fredrick W. Taylor's system of scientific management in factories, have asked whether or not similar betterments could not be obtained in offices with the system. Their questions can now be answered, for the main principles of the Taylor system have actually been adapted and applied in office work.*[55]

In his preface to *Office Management* of 1925 he demonstrated clearly that he saw himself as being on a mission, to advance Taylor's legacy, and to bring 'the word' to a broad audience of professionalising business managers:

> *a pressing need exists for a thorough understanding on the part of business men in general, and office managers in particular, of the fundamental principles underlying the work of that pioneer of scientific management, the late Frederick Winslow Taylor … . I have attempted in this work to explain the scientific basis of office procedure, and at the same time accord to the profession of office management the dignity and position it deserves. I am not aware that any writer has previously attempted this task.*[56]

Leffingwell was near evangelical in his belief that office management was the way forward; he just needed to convince managers that it would improve their bottom line.

> *There is … a brilliant future for those office managers who have arrived at the recognition of the office as a major function of business and the equal importance of the work of office management with every other business activity. When these men set themselves to discover the range of possibilities of scientific office management, and make it their profession, they will substantially advance the interests of their company, effect large savings, and thereby greatly add to the profits of the business, while at the same time forcing the more or less reluctant admission of the vital importance of their work from other executives.*[57]

With an air almost of disbelief that no-one should have noticed before, Leffingwell observed that there are "*millions of unnecessary motions, and when one begins to investigate an office with an eye for these alone, one comes to believe that most of them are in the office*" and as an example he offers the following.

> *Watch a clerk rushing through his work, throwing the papers in a disorderly heap as he goes and then when he has finished, watch him spend a few minutes straightening things up. It never occurs to him to pile them up in an orderly manner in the beginning.*[58]

The levels of detail to which Leffingwell and other scientific management gurus delved seemed to know no bounds. Consider the following two examples.

> *In many offices little attention is given to the selection of pencils. In some kinds of work a soft lead pencil is required., in others a medium, and in still others a hard lead. Sometimes an eraser is necessary.*[59]
>
> *The average person should drink water at least five or six times a day. If each of one hundred clerks in an office were compelled to walk fifty feet to, and fifty feet from, the fountain, five times a day, each one would walk five hundred feet a day. Multiplied by one hundred clerks the distance travelled would be fifty thousand feet, or nearly ten miles! Multiplied by three hundred working days, the clerks would be walking three thousand miles for water in a year.*[60]

One can immediately see the fact that the techniques were directly transferred from the factory floor. How else could the grade of graphite be considered of import? And from this point, it is easy to understand why, when scientific managers turned their detailed scrutiny to the office, they focused on productivity. Their approach to the organisation of the office were guided by two objectives: to make the paperwork flow as efficiently as possible, and to encourage workers to concentrate on their tasks. Even when they discussed such physical improvements as better ventilation or lighting, they were more concerned with increasing output than with the intrinsic value of a healthy working environment.[61]

Davies argued that the preoccupation with inefficiency was caused by two connected problems. First was the dramatic increase in the volume of office work to the point where the older methods of organising production could not keep up with the increased workload, and secondly the inability of many employers to discipline clerical workers effectively.

> *If the amount of paperwork had not grown so large, a lack of discipline on the part of clerical workers might not have seemed so serious … . The response of most businessmen to these problems was either to hire more clerical workers or to replace those who were "lazy" or "insubordinate." But there were also some employers, at first few in number, who responded by restructuring the office itself. By the second decade of the twentieth century, their efforts came to be referred to as the scientific management of the office.*[62]

The year after Leffingwell's *Scientific Management* another key text was produced. Lee Galloway (1871–1962) was an American educator, publisher and most importantly here organisational theorist. He was a Professor in the School of Finance and Commerce at New York University. Galloway's *Office Management, its Principles and Practice* sets out a comprehensive approach to scientific management. In the introduction he makes clear his belief that the purpose of management is control over operations. He argues that as businesses grow in size and complexity, so

> *the more difficult and important become the problems of management. Orders must be given to employees by the managers, and reports of work performed must be recorded. Inspectors, superintendents, foremen, senior clerks, and office managers increase in number---their function being to keep the employees and machines working harmoniously.*

He then describes the growing complexity of communications:

> *At first one of these supervisors can give instructions verbally and keep the details in his memory, but as the subdivisions of work increase the necessity grows for continual communication between the various ranks of authority. Letters and memos, production orders and work tickets, speaking tubes and telautographs, cost statistics and controlling accounts, time clocks and messenger boys, multiply to keep pace with the growing complexity of business and to save the time of executives and workmen alike.*[63]

A few pages later, Galloway underlines the importance that he attached to control:

> *Execution implies control – control of the factory organization – control of the financial organization – control of the marketing organization. It is the work of the office organization, under the supervision of the office manager, to devise records, methods, and systems for carrying out the function of control and for co-ordinating the activities of one department with those of another.*[64]

Scientific management was rationality in extremis. Its obsession with the division of labour meant that for many workers their work was reduced to the repeated performance of very limited tasks: opening envelopes, typing letters, filing records and so. It standardised and routinised. Departmental organisation rationalised internal processes and systems; layers of management interlocked and information flowed.

The extent to which scientific management was implemented beyond a relatively small number of large firms is to be debated; but the number might be relatively small. And of course the very process itself was not without its critics, Harry Braverman among them.

> *When Leffingwell says, for example, that 'the output of one clerk was doubled merely by the re-arrangement of the work on the desk', we may understand this was an effect of close and frightening supervision rather than a miracle of efficiency; this was understood by the managers as well, although concealed beneath a "scientific" mystique.*[65]

He goes on to state that from the early days, office managers held that all forms of clerical work, not just routine or repetitive ones, could be standardised and 'rationalised', and that for this purpose

> *they undertook elaborate studies of even those occupations which involved little routine, scores of different operations each day, and the exercise of judgment. The essential feature of this effort was to make the clerical worker, of whatever sort, account for the entire working day. Its effect was to make the work of every office employee, no matter how experienced, the subject of management interference. In this way, management began to assert in the office its hitherto unused or sporadically exercised right of control over the labour process.*[66]

For Braverman, scientific management was about regimentation, uniformity, process and control. It was not about individuality, trust and autonomy. His perspective was underscored more recently by Lowe, who argues that control was the decisive force in terms of coping with a rapidly growing labour force. The essential function of the office, he argues, was the provision of information for managerial decision making but that as it expanded so "*it fell prey to a range of inefficiencies*", and that larger and more complex organisations

> *made it exceedingly difficult to achieve Max Weber's ideal of a well-oiled, finely tuned bureaucracy. Clerical costs threatened to reduce profits or, in the case of government, unduly burden the treasury … . By 1914, office managers were realizing the advantages of applying some of the new principles of "progressive" administration to white-collar work. They became convinced that by rationalizing the office, administration would become a more effective tool in their hands.*[67]

Whatever the American experience with scientific management, it seems that barely none of the thinking or practice landed on European shores. The rise of the corporation and the explosive growth in white-collar work in the US in the latter decades of the nineteenth century contrasted strongly with the European experience. Scale of operation was undoubtedly a factor in the broader uptake of scientific management in the USA – enterprises in Europe remained typically smaller and less capital intensive that in America – but there were also socio-cultural factors at play. Spiegelberg observed that, in the UK clearing banks, "*the benefits of management sciences has been a slow and agonising process … Clearing bankers still regard themselves as bankers rather than managers*". Spiegelberg goes on to suggest that by comparison with other industries, "*the Big Four's application of management techniques has been little more than a flirtation*".[68] By contrast, US firms were not so embedded in the mores of their history as were so many European enterprises.

The contrasting features of the cultural and rational traditions in office work and workplace design continued after the Second World War. In America, the situation was largely one of 'business-as-usual' but, of course, in Europe and elsewhere large-scale devastation had to be dealt with while national the economies remained in a parlous state, coping with debt accrued during the war and societies enduring rationing and general privation.

It is curious coda to the rational and cultural discussion that Peter Drucker (1909–2005), a European working in America, set out a perspective on business that was, in some ways, the antithesis of the scientific management in his host country. Born in Vienna, Drucker graduated in 1927 and worked as a journalist in Germany, leaving for London in 1933, where he worked as a chief economist in a bank. In 1937, he moved to the USA to become an academic, including a spell at New York University as professor of management, from 1950 to 1971. In 1954, he published his masterpiece *The Practice of Management*.[69]

Scientific management saw companies as profit-maximising entities organised, essentially, around production lines – both factories and offices – where the efficient organisation of company assets and resources was the key driver, and where management oversaw work processes broken down into clearly defined tasks based on centralised, hierarchical decision-making. By contrast Drucker regarded the firm as a function of society, not merely focused on profit maximisation, but also as a social agent, providing the means for people to develop careers in a secure environment.

Drucker believed that the firm existed at three levels. The first is as an economic institution that creates value for society and shareholders. The second is as a human and social organisation that employs people, develops them, pays them, organises them to increase productivity, and as such requires effective governance and management. The third is as a "*social institution that is deeply embedded in society and values and as such is affected by public interest discussion, debate, and values*."[70] In these ways, Drucker had more in common with the cultural office than the rational office.

Summary. While it took until the mid-twentieth century for geographers, economists and planners to recognise the office economy as a distinct facet of cities, office buildings as we understand them today were beginning to emerge at the end of the nineteenth century. During this time, the architecture of commercial office buildings and their internal design, servicing and layout all changed dramatically. We saw in the final section of Chapter 6 the arrival of purpose-built office buildings and speculative development in the mid-nineteenth century. Initially, the buildings themselves were mostly small in scale; they were up to three, sometimes four storeys high, they were solid structures, heavily sub-divided internally and they lacked any services such as electricity, air handling or lighting. But as well-capitalised joint stock companies grew, as the economy evolved and as competition increased, so office buildings began to take on their more modern look.

During the early years of the twentieth century, the increasing scale of activity of banking and insurance firms came to demand larger and more impressive buildings, both to house their various operations and to serve as architectural expressions of their wealth, respectability, and reliability. The Victorian rebuilding of the City brought structures as arresting in their architecture and novel in their plans as they were impressive by their bulk.

Nevertheless, the European, cultural office evolved slowly, incrementally, with companies more often than not adapting to existing buildings rather than, like America, creating new buildings for a new age. The rejection of height in European buildings through the nineteenth and much of the twentieth centuries is, perhaps, the most obvious point of separation with

American offices. Interwar years London saw the development of a number of large buildings, by its own standards, but none of these was high rise in a US sense. Most were idiosyncratic in some manner, many were built to the specific demands of their sole occupiers and few exuded the slightest air of efficiency in terms of layout.

By contrast the later-nineteenth century American economy was expanding at a much faster rate than European economies, and at an altogether different scale. In so doing, it set down a very different urban structure. The new economic model, and the rise of 'corporatism' gave rise to a fundamentally new building form: the skyscraper. A new architecture emerged to meet the needs of a rapidly and radically evolving economy in which corporations gathered together armies of clerical workers to administer and manage their ever more complex organisations. Innovation in office design and servicing established a pattern of development that would survive to the current times.

Scientific management in America underlined the widening gulf between rational American and cultural European offices. The role played by Taylorism in office design has been exaggerated as indeed, in a European context, has that of his successors. Scientific management had a minimal impact on European offices, as bankers and insurers regarded themselves simply as bankers and insurers rather than as managers.

The contrasting features of the cultural and rational traditions in office work and workplace design continued after the Second World War. In America, the situation was largely one of 'business-as-usual' but, of course, in Europe and elsewhere large-scale devastation had to be dealt with while nationally the economies remained in a parlous state, coping with debt accrued during the war and societies enduring rationing and general privation.

Notes

1 Cowan P; Fine D; Ireland J; Jordan C; Mercer D & Sears A (1969) *The Office: A Facet of Urban Growth* Heinemann Educational Books Ltd, London p28.
2 Harris R (2021) *From Clerical Factory to Digital Hub: London's Global Office Economy* Routledge, London p51.
3 Mumford L (1938) *The Culture of Cities* Secker & Warburg, London.
4 Miller DL (1989) *Lewis Mumford: A Life* Weidenfield & Nicolson, New York p355.
5 Mumford L (1938) *The Culture of Cities* Secker & Warburg pp227–228.
6 Rannells J (1956) *The Core of the City* Institute of Urban Land Use and Housing Studies, Columbia University Press.
7 Rannells J (1956) *The Core of the City* Institute of Urban Land Use and Housing Studies, Columbia University Press p5.
8 Rannells J (1956) *The Core of the City* Institute of Urban Land Use and Housing Studies, Columbia University Press p17.
9 Rannells J (1956) *The Core of the City* Institute of Urban Land Use and Housing Studies, Columbia University Press p54.
10 Rannells J (1956) *The Core of the City* Institute of Urban Land Use and Housing Studies, Columbia University Press p35.
11 Rannells J (1956) *The Core of the City* Institute of Urban Land Use and Housing Studies, Columbia University Press p36.
12 Cowan P; Fine D; Ireland J; Jordan C; Mercer D & Sears A (1969) *The Office: A Facet of Urban Growth* Heinemann Educational Books Ltd, London p34.
13 *The Builder* in 1857. Cited in Olsen DJ (1976) *The Growth of Victorian London* Penguin Books p64.
14 Olsen DJ (1976) *The Growth of Victorian London* Penguin Books p72.
15 *Architectural Magazine* Vol V 1871 p167 Cited in Olsen DJ (1976) *The Growth of Victorian London* Penguin Books pp121–122.

16 Olsen DJ (1976) *The Growth of Victorian London* Penguin Books p65.
17 Holden CH & Halford WG (1951) *The City of London: A Record of Destruction and Survival* Architectural Press, London p173.
18 Pelegrin-Genel E (1996) *The Office* Flammarion, Paris p26.
19 Kynaston D (1995) *The City of London Volume Two: Golden Years 1890–1914* Chatto & Windus, London p31.
20 Kynaston D (1995) *The City of London Volume Two: Golden Years 1890–1914* Chatto & Windus, London p27.
21 Royal Insurance Company Limited (1912) *Old Lombard Street* Sir Joseph Causton & Sons, London.
22 Royal Insurance Company Limited (1912) *Old Lombard Street* Sir Joseph Causton & Sons, London pp41–42.
23 Royal Insurance Company Limited (1912) *Old Lombard Street* Sir Joseph Causton & Sons, London p49.
24 Royal Insurance Company Limited (1912) *Old Lombard Street* Sir Joseph Causton & Sons, London pp49–50.
25 Royal Insurance Company Limited (1912) *Old Lombard Street* Sir Joseph Causton & Sons, London p50.
26 Royal Insurance Company Limited (1912) *Old Lombard Street* Sir Joseph Causton & Sons, London pp51–52.
27 The Rothschild Archive, business premises https://guide-to-the-archive.rothschildarchive.org/the-london-banking-house/depts/manchester-business
28 The Rothschild Archive, business premises www.rothschildarchive.org/business/business_premises/first_new_court
29 The Rothschild Archive, business premises www.rothschildarchive.org/business/business_premises/second_new_court
30 The Rothschild Archive, business premises www.rothschildarchive.org/business/business_premises/third_new_court
31 The Rothschild Archive, business premises www.rothschildarchive.org/business/business_premises/fourth_new_court
32 Shultz E & Simmons W (1959) *Offices in the Sky* New York Bobbs Merrill.
33 Shultz E & Simmons W (1959) *Offices in the Sky* New York Bobbs Merrill p19.
34 Shultz E & Simmons W (1959) *Offices in the Sky* New York Bobbs Merrill p27.
35 Shultz E & Simmons W (1959) *Offices in the Sky* New York Bobbs Merrill p31.
36 Shultz E & Simmons W (1959) *Offices in the Sky* New York Bobbs Merrill p33.
37 Shultz E & Simmons W (1959) *Offices in the Sky* New York Bobbs Merrill p33–34.
38 Bowley M (1966) *The British Building Industry: Four Studies in Response and Resistance to Change* CUP p10.
39 Fenske G & Holdsworth D (1992) Corporate Identity and the New York Office Building: 1895–1915 In Ward D & Zunz O (1992) (eds) *The Landscape of Modernity: Essays on New York City, 1900–1940* Russell Sage Foundation, New York pp129–159.
40 Harris R (2021) *From Clerical Factory to Digital Hub: London's Global Office Economy* Routledge, London p43.
41 Fenske G & Holdsworth D (1992) Corporate Identity and the New York Office Building: 1895–1915 In Ward D & Zunz O (1992) (eds) *The Landscape of Modernity: Essays on New York City, 1900–1940* Russell Sage Foundation, New York.
42 Fenske G & Holdsworth D (1992) Corporate Identity and the New York Office Building: 1895–1915 In Ward D & Zunz O (1992) (eds) *The Landscape of Modernity: Essays on New York City, 1900–1940* Russell Sage Foundation, New York.
43 Domosh M (1989) A Method for Interpreting Landscape: a Case Study of the New York World Building *Area* (1989) Vol 21 No 4 pp347–355.
44 Turvey R (1993) London Lifts and Hydraulic Power *Transactions of the Newcomen Society* Vol 65 No 1 pp147–164.
45 Gottmann J (1966) Why the Skyscraper? *Geographical Review* Vol 56 No 2 pp190–212.

46 Braverman H (1974) *Labour and Monopoly Capital: the Degradation of Work in the Twentieth Century* Twenty Fifth Anniversary Edition, Monthly Review Press, New York p86.
47 Braverman H (1974) *Labour and Monopoly Capital: the Degradation of Work in the Twentieth Century* Twenty Fifth Anniversary Edition, Monthly Review Press, New York p86 p87.
48 Taylor FW (1911) *The Principles of Scientific Management* Dover Publications, New York.
49 Duffy F (1997) *The New Office* Conran Octopus Limited, London p17.
50 Lowe GS (1984) The Enormous File: the Evolution of the Modern Office in Early Twentieth-Century Canada *Archivaria* 19 (January) pp137–151.
51 Thompson CB (1917) *The Theory and Practice of Scientific Management* New York, Houghton Mifflin p12.
52 Parsons CC (1918) *Office Organisation and Management* LaSalle Extension University, Chicago.
53 Leffingwell WH (1917) *Office Management; Principles and Practice* AW Shaw & Co, Chicago.
54 Leffingwell WH (1917) *Office Management; Principles and Practice* AW Shaw & Co, Chicago.
55 Leffingwell WH (1917) *Office Management; Principles and Practice* AW Shaw & Co, Chicago p5.
56 Leffingwell WH (1917) *Office Management; Principles and Practice* AW Shaw & Co, Chicago pIX.
57 Leffingwell WH (1917) *Office Management; Principles and Practice* AW Shaw & Co, Chicago p115.
58 Leffingwell WH (1917) *Office Management; Principles and Practice* AW Shaw & Co, Chicago p214.
59 Leffingwell WH (1917) *Office Management; Principles and Practice* AW Shaw & Co, Chicago p19.
60 Leffingwell WH (1917) *Office Management; Principles and Practice* AW Shaw & Co, Chicago p11.
61 Davies MW, *Women's Place at the Typewriter* Temple University Press, Philadelphia pp98–99.
62 Davies MW, *Women's Place at the Typewriter* Temple University Press, Philadelphia pp100–101.
63 Galloway L (1918) *Office Management, its Principles and Practice: Covering Organization, Arrangement and Operation with Special Consideration of the Employment, Training and Payment of Office Workers* Ronald Press, New York pvii.
64 Galloway L (1918) *Office Management, its Principles and Practice: Covering Organization, Arrangement and Operation with Special Consideration of the Employment, Training and Payment of Office Workers* Ronald Press, New York pvii.
65 Braverman H (1974) *Labour and Monopoly Capital: the Degradation of Work in the Twentieth Century* Twenty Fifth Anniversary Edition, Monthly Review Press, New York p213.
66 Braverman H (1974) *Labour and Monopoly Capital: the Degradation of Work in the Twentieth Century* Twenty Fifth Anniversary Edition, Monthly Review Press, New York p13–14.
67 Lowe GS (1984) The Enormous File: the Evolution of the Modern Office in Early Twentieth-Century Canada *Archivaria* 19 (January).
68 Spiegelberg R (1973) *The City: Power Without Accountability* Blond & Briggs, London pp131–132.
69 Drucker PF (1954) *The Practice of Management* Harper & Row, New York.
70 Zahra SA (2003) The Practice of Management: Reflections on Peter F. Drucker's Landmark Book *The Academy of Management Executive* Vol 17 No 3 August 2003 pp16–23.

Part IV

1945–1990

The corporate office

IBM Selectric II Typewriter (1970s)

Office Era	Time period	Work & Place Themes	Office Technology
Market	**Part I: 1670-1770**	**Trade**. Goldsmiths as bankers. Central banking. Coffee Houses. Joint Stock Companies. Exchanges. Insurance.	Coal fires. Candles. Copying. Ledgers. Paper. Bills of Exchange. Quill pen. Cash book. Shop counter.
Market / Factory	**Part II: 1770-1870**	**Organisation.** Stock Exchange. Clearing House. Private and merchant bankers. Joint Stock Banks. Office buildings. Speculative development.	Gas lighting. Copying. Ink Pen. Ledgers. Paper. Stenography. Polygraph. Correspondence book.
Factory / Corporate	**Part III: 1870-1945**	**Production**. Rise of corporation. Division of labour. Work: fixed, manual, repetitive and mechanical. Female workforce. Rise of professions.	Mechanisation. Adding machine. Electricity. Telegraph. Telephone. Typewriter. Copiers. Elevator. Air conditioning. Fluorescent light.
Corporate	**Part IV: 1945-2000**	**Process**. Globalisation. Scale. Bureaucracy. Command & Control. 'Corporate man'. Work landscapes. Birth of digital office.	Photocopier. Mainframe. System Furniture. Golf Ball Typewriter. PC. MS Office. Mobile phone. Internet. Email.
Digital	**Part V: 2000 onward**	**Knowledge**. Agile work. Data. Power.	Smart phone. The Cloud. Social Media. Augmented Reality. Automation. Artificial Intelligence. Robotics. Virtual Reality.

11 Blitz to Big Bang

Balzac wrote *Les Employés* (1836) as one of his vast collection of interlinked novels known as *La Comédie Humaine* (48 volumes in total!). The novel is about life and advancement in the government bureaucracy of 1820s France, under Charles X. *Les Employés* are the bureaucrats, and Balzac was quick to see the ramifications of a sprawling government: "*And thus bureaucracy, the giant power wielded by pygmies came into the world.*"

Honoré de Balzac (1799-1850)

Almost a century after Balzac's big government came big business, the rise of corporatism, to turbo charge the proliferation of white-collar work: magnifying the power invested there and fuelling a huge growth in pygmies!

The rise of the bureaucratic behemoths in the post-war era was a central feature of the economic reorganisation that saw the coming of age of the office economy. The number of large bureaucratic organisations mushroomed as increasing specialisation in work led to the rapid expansion of gossamer-like layers of workers to manage, supervise, co-ordinate and administer the new routines of colleagues. The workplace rapidly ossified into a rigid (social) hierarchy, graded according to the authority granted to each level to plan and execute their own work while planning and supervising the work of others, down to those who did not plan work or make decisions.

DOI: 10.1201/9781003582205-15

Indeed, office workforces resembled military regimentation, with officers, NCOs and men each respecting of the others' rank and duties. The men were assembled in large platoons, undertaking tiny, abstract, aspects of the overall operation, without sight of the greater plan. They were corralled and directed by NCOs who received daily operational orders from the officer class, some distance away on another, usually higher, floor of the building.

> *On the shop floor, a corps of supervisors acted as an information conduit between management and problems arising in the production process… . Information was starkly centralised in a bureaucratised hierarchy, workers' informational input to the production process having been brutally reduced by de-skilling and a minute division of labour.*[1]

The post-war years, from the Blitz to Big Bang, saw the rise of the corporate organisation, managerialism and of 'company man'. The period witnessed a complete transformation in the economy, which, in little more than four decades, transitioned from predominantly manufacturing to predominantly service. And within this service element, office activity was the central growth pole. Banking and finance, legal and accounting, advertising and media, real estate, information and communications, along with others, all grew rapidly. And the growth in these sectors was over and above that in sectors including education, energy, government, health, manufacturing, retailing and transport.

The American invasion. The 1960s and 1970s also witnessed the internationalisation of the rational office, with the spread of large American corporates to Europe. While there was an established US presence in London, up to this point, the City of London remained dominated by dynastic families, Oxbridge graduates, gentleman's clubs, bowler hats and the '*my word is my bond*' culture. This was a world away from the corporate structures, company man, anonymity and rational business culture of Wall Street. UK clearing banks avoided change in the post-war period; as Spiegelberg observed: "*without much more than a twitch or a sigh [they] survived more than a decade of radical change in virtually every other form of British institutional life*", largely achieved because they exuded

> *caution, conservatism, stone facades, marble halls, safety, solidity, stability and anything else one might associate with crystallised immutability.*[2]

In this one quote we see the interaction of organisation and space: the banks reflected their organisational conservatism through the solidity of stone and marble; and we see the essence of the cultural office in 'crystallised immutability', as the clearing banks sought to defy the inevitable (global competition) by doubling down on their cultural tradition.

But despite the conservatism of London banks, change was coming and a key feature of the City's transformation was the Eurodollar market. The 1947 Exchange Control Act was intended to safeguard Britain's currency reserves and prevent any deleterious movement of capital away from sterling. However, while the regulated, domestic market was protected, the side-effect was to encourage a parallel, unregulated market trade in dollars between non-residents. The Bank of England was unable to interfere with the development of this hugely important market, which effectively changed the face of the City of London.

The Eurodollar market led to a rapid expansion of international banks in the City, particularly American banks in the 1960s. At the start of the decade in 1960 there were 77 foreign banks in the City, of which 14 were American. By 1970, these figures had grown to 159 and 37, respectively.

> *It was an emblematic moment when, in 1968, Bankers Trust moved the head office of its international department from New York to London. That year no fewer than eight US banks ... opened branches or representative offices in the City, followed in 1969 by another eight.*[3]

Kynaston noted that American banks were 'utterly dominant' in the Eurodollar market by 1965 when, in London, they held overseas deposits to the equivalent of £1,010m – almost double that held by London's own merchant banks. The Americans were

> *Borrowing heavily in Eurodollars in order to repatriate them to the US, where their principal customers were the large corporations and multinationals anxious not to be stunted by the latest American credit restraint measures.*[4]

By 1969, Citicorp employed 700 people in London, followed by Morgan Guaranty with 325. The American invasion led to Moorgate being dubbed 'Avenue of the Americas', and Richard Spiegelberg drew the contrast in stark terms.

> *Enter the central banking hall of a large American bank on Moorgate, and you will find a large expanse of open-plan office, girl receptionists to direct you and heavy advertising of the bank's services. In a British clearing bank a few doors away, the walls are lined with marble, there is a great mahogany and glass divide between the customer and the bank officials, and the use of display is ascetic.*[5]

The particular relevance of this development to our story here is that the large-scale inflow of American banks exposed, finally and in some ways, cruelly the clash between the rational and the cultural office. London found itself hosting business that was conducted in altogether different ways and, inevitably, there was some tension. In 1969, an article in The Banker from a general manager of a leading clearing bank began by listing the widespread undercurrent of objections to the American invasion: "*pushing up City rents, high enough even before; 'touting of business'; poaching staff; and syphoning away Eurodollars with the effect of artificially increasing Eurodollar rates*". However, the anonymous banker then admitted to the coming cultural shift that lay at the heart of the issue.

> *Over many years the market dominance of the Big Five clearing banks led to a complacency which has now been rudely shattered ... The days of waiting for customers or potential customers to come to the bank for help are over: increasingly the need is seen to approach the customer, telling him what services are now available.*

The banker finished his piece with "*it ill becomes the City, with its reputation for financial flexibility, to complain about competition*".[6] The cultural and rational traditions were exposed. The same differences in business tradition also permeated work and the workplace. The cultural office with its reflection of history, tradition, personal relationships and class structure expressed through layers of deference, clashed mightily with the rational model of bureaucracy, standardisation, and individual submission to 'the corporate way'.

European cities such as Paris and Frankfurt were not affected by the American invasion in the same way that London was. Indeed, while offices in London came to resemble a 'cultural/rational hybrid', European office markets largely retained their distinctive cultural features. The North American model brought with it two fundamental characteristics that would help define the post-war office economy: managerialism and company man.

The triumph of managerialism

The latter part of the nineteenth century and the early years of the twentieth witnessed a growth in scale in all aspects of society and the economy. Churches, schools and universities, social clubs, shops, organised sport and entertainment were all "*affected by the drive towards large-scale*".[7] In the UK, industry and commerce also expanded rapidly and at the same time consolidated through amalgamations to create organisations of greater scale and complexity, including regionally or nationally organised headquarters, branches and institutions. The trend was clearly seen in economic reorganisation.

> *In 1880, when joint stock organisation was still in its infancy, the hundred largest firms in Britain probably accounted for less than 10 per cent of national production. In 1909, after the first great wave of industrial mergers [they] probably accounted for about 15 per cent of manufacturing output. By 1930 at the end of the second wave the largest one hundred firms accounted for 26 per cent ... by 1970 ... they produced no less than 45 per cent of total output.*[8]

In the services sector, consolidation was particularly sharp. Clearing banks were reduced to just four by 1968; and insurance, accounting and legal all followed similar paths with ever more activity becoming concentrated in a rapidly diminishing number of firms. Perkin argues that the driving force behind consolidation generally was not technological necessity, nor competitive economic advantage, but was instead driven by "*the very human collective self-interest of the managers in the process of professionalisation*".[9] In other words, the desire of the emerging manager class for ever-more influence and power – driving an ever greater wedge between capital and control (a process reinforced by the growing influence of institutional shareholders, notably pension funds and insurance companies). As Berle remarked in 1954: "*The capital is there and so is capitalism. The waning factor is the capitalist*".[10]

A similar pattern was emerging in the USA. In Chapter 7 we saw how the US economy vertically integrated to create large organisations that controlled supply chains from production to sales, and how the national economy evolved from a "*society of island communities into a homogenous national community*".[11] The implications for the office economy were profound – both in the US and worldwide as other national economies mimicked the American corporate model. After 1945 these trends continued, only with greater speed and at greater scale.

Large, inflexible, departmentalised, hierarchical and slowly moving businesses dominated by clerical, process-based work and with clear demarcations between grades and functions of work grew and dominated the office economy in the twentieth century. The headquarters of large manufacturers, banks and insurance companies, advertising and media businesses, lawyers and accountants and so on all concentrated into large buildings. Layer upon layer of management evolved, and companies divided like amoeba into complex departmental structures. This was the age of corporatism, and 'corporate man'.[12]

Of course, the flip side of the rise of corporatism was a centralisation of power in industry in a process parallel with the trend in the UK. During the course of the final three decades of the century, large-scale integrated businesses came to dominate the US economy. By 1904, after a period of consolidations, 318 firms controlled nearly 40% of America manufacturing assets.[13] Theodore Roosevelt (1858–1919) had an early insight into what was happening. On 31 August 1910, in a highly charged speech in Osawatomie, Kansas, the former president heaped praise on the growing power of the American economy, but at the same time took aim at a less visible feature of this power base. He warned that America's economy was controlled by a handful of corporate giants that were generating unparalleled wealth for a small number of people while exercising growing control over American politics. Roosevelt cautioned that a country founded on the principle of equality of opportunity was in danger of becoming a land of corporate privilege, and pledged to do whatever he could to bring the new giants under control.

Roosevelt's speech sounds as fresh today as on the day he delivered it. Eerily so. A small number of giant companies are once again on the march, tightening their grip on global markets, merging with each other to get even bigger, and enjoying vast profits. As a proportion of GDP, American corporate profits are higher than they have been at any time since 1929. Amazon, Apple, Google and their peers dominate today's economy just as surely as US Steel, Standard Oil and Sears, Roebuck and Company dominated the US economy of Roosevelt's day. Today's companies are very different to those of a century ago in terms of their internal organisation and their workplaces, but their significance to the economy is no less so.

Sampson cites journalist Vance Packard who described, in *The Status Seekers* (1959), how big corporations had created a new national class structure by constructing layers of management and status which were almost as rigid as ranks in the army or the civil service: they had devised intricate varieties of status through the size of offices and desks.[14] Sampson suggested that by the early-1960s,

> *a new species of company man was evolving in the West with migratory habit which had been unpredictable twenty years earlier. He could fly non-stop across the Atlantic in seven hours; he could stay in a foreign country for only one night, at a hotel which looked indistinguishable from its counterpart at home; he could talk to his home office or his wife without delay by a direct-dial telephone.*[15]

Also in Chapter 7, reference was made to Peter Drucker's effusive description of the importance of management. He went on to state that management would remain "*a basic and dominant institution perhaps as long as Western Civilisation itself survives*".[16] Quite a claim! Of course, as we have seen in the history of offices presented here, we had management and managers before corporatism reached its phase of maximal expansion in the late-twentieth century. What Drucker was referring to was 'managerialism', a much broader concept than management.

When managers and management first appeared in the mid-nineteenth century, they signalled not simply the separation of capital and labour, but the creation of an intermediary form of worker, the salaried class for whom, in Marxist terms, supervision is their established and exclusive function. They answered to the capital owners and shareholders and were charged with the tactical operation of their businesses. They represented the start of the shift towards complex, bureaucratic and departmentalised organisations, and their evolving role led them to rationalise office practices and procedures. The scientific management discussed in the previous chapter was one symptom of this process in America. Managerialism however is more than the implementation of management practices and procedures; it is about more than the organisation of corporations.

> *It is, rather, the widely-held belief that all organisations can only work properly if decision-making is centralised in some manner in the hands of professionally trained and 'objective' managers. Furthermore, this view contends that not only is this organizational form desirable for efficiency but it is also part of an historical progression toward more efficient and effective social structures.*[17]

It is thus a wider belief set, or as Drucker called it 'an institution'. It is not questioned, it is simply accepted. Managerialism is regarded as the next step of the progress or evolution of business; it is, if you will, a natural development. In many ways, managerialism became a defining feature of office organisations and office work during the post-war decades.

James Burnham (1905–1987) was an American philosopher and political theorist; a radical activist and a prominent leader of the American Trotskyist movement in the 1930s. However, in May 1940 he parted company with Marxism, rather abruptly in a forthright letter to the National Committee of the Workers Party, in which he rejected the notion that 'socialism is inevitable'. He argued instead that, on the basis of emerging evidence, a new form of exploitive society was not only possible but a more probable outcome than socialism. He turned to conservatism, writing regularly for the *National Review* on a variety of topics.

Burnham's central claim was that instead of the binary struggle between capitalism and socialism, there was a new economic system emerging, 'managerialism', which he described in his 1941 *The Managerial Revolution: What is Happening in the World.*[18] He argued that capitalism was waning and that the world was moving into the hands of the 'managers' – the administrators in business and in government. The 'new world', according to Burnham, was a kind of planned, centralised society in which the capitalists had lost their controlling influence. The central thesis is said to have influenced George Orwell in his writing of 1984.

The means of production in Burnham's emerging society would be under the control of an undemocratic matrix comprising business executives, technicians, bureaucrats and soldiers, lumped together by Burnham as 'managers'. These managers, Burnham argued, would organise society such that all power and economic privilege remained in their 'class', over a mass of workers. While he made some questionable claims around the abolition or private property rights and a free press, his central thesis of managerial influence was a prescient insight that would see some realisation half a century later.

Burnham was writing at a point in history when government and private corporations were becoming ever larger, more complex, more opaque, and less accountable. Anthony Sampson picked up the thread in his *Company Man.*

> *The managerial image of the corporation was reflected at annual general meetings which were increasingly publicised and prearranged with elaborate public relations. The chairman would proclaim unanimous decisions, and announce new directors chosen by existing officers.*[19]

Dissident individual shareholders would question the corporate plan, but could not carry the institutional investors who preferred the *status quo*. The managers "*fortified their kingdom to perpetuate themselves and reinforce their promotion and security*". By 1955 half of the large American corporates held management development programmes "*ostensibly to ensure the continuity of the company, but primarily to perpetuate the managerial class*".[20]

The professionalisation of management, or the separation of capital and control, was analysed by Adolf Berle and Gardiner Means in *The Modern Corporation and Private Property* in 1932, a book that became one of the most cited works in management studies.[21] Their central thesis of corporate governance was to challenge the generally accepted notion that the principal purpose of a corporation was to create value for its shareholders. They warned that the concentration of economic power brought on by the rise of the corporation led to the emergence of a powerful class of professional managers, insulated from the pressure of owners and stockholders.

The book argued that capital in the US economy had become enormously concentrated which in turn afforded a small number of corporations great power; and as this proceeded, the influence of their owners waned in relation to the dilution of their shareholdings. And as the number of shareholders grew, so the power of the corporate managers grew. Over time, so the argument went, the managers came to work for their own interests, with diminishing accountability to shareholders, or to wider society. Berle and Means were concerned about the influence of managers who could act without regard either to shareholders or to their employees and, by extension, the communities in which they operated. Further, the corporations that they managed were concentrating economic power which could compete on equal terms with the state.

> *Where its own interests are concerned, it even attempts to dominate the state. The future may see the economic organism, now typified by the corporation, not only on an equal plane with the state, but possibly even superseding it as the dominant form of social organization.*[22]

Berle and Means underlined the broader, societal significance of managerialism when they argued for a greater public service ethos in business culture. They called for managers to place great priority on furthering the interests of a wider range of stakeholders, including workers, consumers and the state. Berle and Means promoted the philosophy of 'managerialism', the dominant early- to mid-twentieth century doctrine of corporate governance prevalent in America.

Smith *et al.* argued that Berle and Means' stakeholder model of corporate governance was influenced by their experiences during the First World War, seeking to introduce into the business world the ethos of professionalism and disinterested public service that characterised the US Army's officer class.

> *In most modern professional armies, the dominant ethos is one of altruism and devotion to the national interest: individuals are expected to serve the nation selflessly and at modest rates of pay.*[23]

Smith *et al.* suggested Berle and Means were promoting a paternalist version of stakeholder theory "*in which senior managers were to expect unconditional obedience from subordinates in the same way that early twentieth century U.S. military commanders expected soldiers to respect the chain of command*". Thus, Berle and Means were calling on top managers to selflessly disregard their pecuniary interests and those of shareholders so that they could serve a wider range of stakeholders, including workers, consumers, and crucially, the nation-state. Berle and Means outlined a vision of corporate governance in which business executives serve the national interest.

The public service ethos that Berle and Means promoted in their book was dominant in US business culture until about 1980, when a new generation of business leaders and intellectuals resurrected the idea that the sole purpose of a corporation was to maximise shareholder value. By the 1990s, shareholder value ideology was hegemonic in US business culture, and it has had a significant impact on business decision-making.[24]

Perkin extended the managerial debate to discuss the emergence in the UK of the new middle class or, as he calls it, *the salariat.* While employers and proprietors declined from 888,000 in 1951 to 812,000 in 1971, higher and lower professionals and administrators and managers grew by 70% from 1,964,000 to 3,355,000.[25]

Company chairmen, directors and top managers of corporations had median salaries of £66.50 and £69.60 per week in 1971; while medical doctors and lawyers claimed the next highest status and incomes, and top civil servants and academics were not far behind, with median salaries of £58 a week or more. Architects and planners, accountants, engineers, scientists and technologists, college lecturers and a range of middle managers earned weekly salaries from

£41 to £53. School teachers, technicians and draughtsmen, welfare workers, office supervisors, works foremen, senior clerks and sales representatives were all clearly a grade below with weekly salaries in the £30s. All were clearly above the manual and routine office workers, with weekly earnings in the £20s.[26]

These income differentials "*were reflected in the size of house, the sort of neighbourhood, the make of car, the kinds of holiday, the type of children's schooling and so on which each occupational group could afford*". Perkin argued that the chief difference between the salariat and the working class was the

> *psychological one that flowed from greater job security and rising scales of pay. They found it easier to plan for the future, to obtain mortgages and loans for buying cars and consumer goods, to invest in their own and their children's education, to commit savings to holidays and leisure pursuits.*[27]

The rise of company man

As suggested by Ward in the previous quote, the salaried strata of managers also reflected wider societal changes. A particularly astute observation of this process was made by David Riesman in his classic text, *The Lonely Crowd: A Study of the Changing American Character*.[28] In his fascinating book, Riesman talked briefly about an older 'tradition directed' character before focusing on what he described as a transition from an 'inner directed' to an 'other directed' American character. Riesman's controversial thesis argued that the inner directed character was, in essence, enduring, self-sufficient, enterprising. The inner directed character was

> *characterised by increased mobility of people, by rapid accumulation of capital teamed with marked technological changes, by almost constant expansion in the production of goods and services, and by extensive exploration, colonisation and imperialism. The greater demands on initiative and the wider range of choice in this society lead to a dominance of character types who can manage to live socially without strict and self-evident tradition-direction.*[29]

By contrast, the other directed character, the one that Riesman saw as being increasingly dominant in American society in the late-1940s, was a response to a new age,

> *where fewer and fewer people work the land, mine metals or engage in manufacturing, and where working hours are shorter, people have both more material abundance and more leisure. The problems of life are different, for the solution of old problems has but given rise to new ones. People find themselves in a highly centralised and bureaucratised social order ...*[30]

Riesman's thesis also suggested that the outcome is that the dominant social character – the other directed character – leads to a society where more people require the approval of and direction by other people. Of course, in the context of the corporate office bureaucracy this means fealty to the corporate body as goals and direction are provided by 'others', rather than being driven by the inner self. Riesman was describing the essential characteristics of 'company man'.

Micklethwait and Wooldridge refer to the "*triumph of managerial corporatism*" in which the "*multidivisional firm was an important innovation ... because it professionalised the big company and set its dominant structure*". While referring to largely industrial conglomerates, including King Gillette, William Wrigley, HJ Heinz and John D Rockefeller, the authors were targeting their office activities. They "*hired hordes of black-coated managers to bring order to their chaotic empires*", and American cities were "*redesigned to provide these managers with a home – the new vertical filing cabinets knows as skyscrapers*".[31] The authors suggested that company man's character had been formed as early as 1920, to reflect two characteristics: professional standards and corporate loyalty.

> *He was part of a professional caste and adopted Frederick Taylor's motto that there was 'the one best way' for organising work and sneered at rough-hewn entrepreneurs for not knowing it.*[32]

The following pages examine the rise of corporate man through the eyes of three great thinkers on twentieth century economic and organisational change.

C Wright Mills. One of the greatest expositions of the origin, nature and outlook of corporate man was written by C Wright Mills (1916–1962), an American sociologist and a professor of sociology at Columbia University from 1946 until his death in 1962. His book *White Collar*[33] provides a deeply insightful account of the growth of the middle classes in America – particularly as represented by the burgeoning office workforce.

Wright Mills suggested that in the managerial demiurge, "*the capitalist spirit itself has been bureaucratized and the enterprise men are cogs*".[34] He further suggested that, while modern managers may be neither greedy nor aggressive as persons, they nevertheless man the machines that operate in a greedy and aggressive manner. "*The men are cogs in a business machinery that has routinised greed and made aggression an impersonal principle of organization*".[35]

Wright Mills recognised that the employees composing the new middle class did not make up a single stratum, having "*been shuffled out simultaneously on the several levels of modern society*"; he conceptualised them as "*a new pyramid within the old pyramid of society at large, rather than a horizontal layer*".[36]

Wright Mills saw the rise of thousands of bureaucracies and the elaborate specialisation of the system as a whole creating "*the need for many men and women to plan, co-ordinate, and administer new routines for others*", with "*larger and more elaborate units of economic activity*" leading to increased proportions of employees being "*drawn into co-ordinating and managing*".

> *Managerial and professional employees and office workers of varied sorts – floorwalkers, foremen, office managers – are needed; people to whom subordinates report, and who in turn report to superiors, are links in chains of power and obedience, co-ordinating and supervising other occupational experiences, functions, and skills.*[37]

As one of the first authors to directly describe the growing masses of office workers as 'middle class', Wright Mills argued that we could circumscribe the new middle class in terms of skill and function, but also in terms of income (normally carrying a quota of prestige on and off the job) and in degrees of power over other people.

> *Occupations are thus tied to class, status and power as well as to skill and function; to understand the occupations composing the new middle class, we must consider them in terms of each of these dimensions.*[38]

He also posited that corporate bureaucracies overwhelmed workers by turning them into a compliant, doleful and self-satisfied stratum of society. He then argued that they are kept in their unchallenging position through a combination of coercion, authority and manipulation. He begins by describing the growing armies of lowly office operatives in "*the enormous file of the office*" who

> *grind along, loading and emptying the filing system; there are private secretaries and typists, entry clerks, billing clerks, corresponding clerks – a thousand kinds of clerks; the operators of light machinery, comptometers, dictaphones, addressographs; and receptionists to let you in or keep you out.*[39]

And from this sobering observation he then transposed his analysis of the organisation into an altogether more all-encompassing reflection of societal trends in America, which contained more than a hint of an acknowledgement of Riesman's theses. He argued that by examining white-collar life,

> *it is possible to learn something about what is becoming more typically 'American' than the frontier character probably ever was. What must be grasped is the picture of society as a great salesroom, an enormous file, an incorporated brain, a new universe of management and manipulation.*[40]

Wright Mills was describing, as he saw it, a change in national culture, or character: from the expansionist, innovative, entrepreneurial, creative frontier culture of the eighteenth century to a culture dominated by middle managers, with no power, who he disparagingly referred to as coupon clippers and parasites.[41] As bureaucracies grew, not exclusively in private sector corporate organisations, but in every sphere of modern society, so the managerial men became more important in the total social structure. These men, according to Wright Mills were the outcome of a century long shift in society, and they operate within bureaucracies which "*select them for their positions and then shape their characters. Their role within these bureaucracies, and the role of the bureaucracies within the social structure, set the scope and pace of the managerial demiurge*".[42]

The very term 'demiurge' comes loaded with meaning, brilliantly capturing the suffocating oppression and even coercion of management.

> *Seen from below, management is not a Who but a series of Theys and even Its. Management is something one reports to in some office ... it is a printed instruction and a sign on the bulletin board; it is the voice coming through the loudspeakers; it is the name in the newspaper; it is the signature you can never make out, except that it is printed underneath.*[43]

Even without the loudspeakers and bulletin boards of yesteryear, management remained an 'it' rather than a 'who' in many corporate offices. Referring back to his 'new pyramid' concept Wright Mills saw the managerial cadre as a hierarchy, graded according to their authority to initiate tasks, to plan and execute their own work and freely to plan and order the work of others. Each level of the cadre's hierarchy is beholden to the levels above. Manager talks with manager and each manager talks with his assistant managers and to the employees, that is, those who do not plan work or make decisions, but perform assigned work.

> *There is no doubt that managers of big business have replaced the captains of industry as the ostensibly central figures in modern capitalism. They are the economic elite of the new society; they are the men who have the most of whatever there is to have; the men in charge of things and of other men, who make the large-sale plans. They are the high bosses, the big money, the great say-so.*[44]

In moving from smaller and simpler organisations to larger and complex organisations, growing numbers of workers were drawn into activities associated with planning, co-ordinating and managing. Wright Mills described managers as "*people to whom subordinates report, and who in turn report to superiors*", "*links in chains of power and obedience, co-ordinating and supervising other occupational experiences, functions and skills*".[45]

There is a hint here of the role of office work in helping to reorganise social class, as discussed in Chapter 9. Wright Mills suggested that class, in its simplest objective sense has to do with the amount and source of income:

> *Today, occupation rather than property is the source of income for most of those who receive any direct income: the possibilities of selling their services in the labor market, rather than of profitably buying and selling their property and its yields, now determine the life chances of most of the middle class.*[46]

He goes on

> *The occupational groups forming the white-collar pyramids, different as they may be from one another, have certain common characteristics, which are central to the character of the new middle class as a general pyramid overlapping the entrepreneurs and wage workers.*[47]

Wright Mills refers to white-collar pyramids and the notion that occupations help circumscribe the new middle class. As specific activities, occupations entail various types and levels of skill, and their exercise fulfils certain functions within an industrial division of labour. As sources of income, occupations are connected with class positions; and since they normally carry an expected quota of prestige, on and off the job, they are relevant to status and position. They also involve certain degrees of power over other social areas.[48] People in mid-twentieth century white-collar occupations claimed higher prestige than wage-workers, and while as a general rule they could not be defined as "*one compact horizontal stratum*", in general "*their functions are similar to those of the old middle class*".[49]

William H Whyte. Wright Mills' work was soon followed by another masterpiece, written by American journalist, urbanist and organisational analyst William Hollingsworth Whyte (1917–1999). Born in West Chester, Pennsylvania, Whyte graduated from Princeton University in 1939 and served in the Marine Corps between 1941 and 1945, including time as an intelligence officer in the Guadalcanal Campaign. In 1946, he joined *Fortune Magazine,* where he spent a good deal of time talking to corporate CEOs. Whyte was responsible in 1952 for coining the term 'Groupthink', to refer to a form of rational thinking leading to conformity with a set of group values that work in a positive manner for greater good. All of this was a prelude to his best-known work, *The Organization Man.*[50]

Published in 1956, the book was a mammoth exposition on conformity, particularly (echoing Wright Mills) that type of conformity by a generation of risk averse middle managers to their corporate employers. Whyte was not really concerned with the superficial conformity of grey hats and trilby's, but a much deeper form of conformity: the ceding of the Protestant ethic to a social ethic, which subjugated the individual to the group.[51] In this shift, average American workers had voted with their feet for a collectivist ethic rather than the frontier individualism of their grandparents. The new generation had sacrificed individualism on an altar of corporatism: organisation knows best. In the opening chapter, Whyte defines what he means by the social ethic.

> *By social ethic I mean that contemporary body of thought which makes morally legitimate the pressures of society against the individual. Its major propositions are three: a belief in the group as the source of creativity; a belief in "belongingness" as the ultimate need of the individual; and a belief in the application of science to achieve that belongingness.*[52]

For Whyte this was all in sharp contrast to America's 'frontier culture' and the 'can-do' spirit of individualism and self-reliance. He argued that it was in America that the contrast between the old ethic and the new, current reality was most apparent "*and most poignant*" because "*of all peoples it is we who have led the public worship of individualism*".[53] He even cited Alexis de Tocqueville:

> *though our special genius — and failing — lay in cooperative action, we talked more than others of personal independence and freedom. We kept on, and as late as the twenties, when big organization was long since a fact, affirmed the old faith as if nothing had really changed at all.*[54]

It is, of course, no coincidence that Alexis de Tocqueville (1805–1859) was a young French aristocrat and civil servant – in fact one of Balzac's pygmies! He made a year-long trip through eastern America in 1831, and published *Democracy in America* in 1835. He admired libertarian American individualism and egalitarianism and warned against forces that sought to make society more uniform.

The contrast between the cultural context and the bureaucratic reality of corporate America was startling. Whyte's organisation men are not "*the white-collar people in the usual, clerk sense of the word. These people only work for the Organisation. The ones I am talking about belong to it as well*". He goes on to say that they

> *are the ones of our middle class who have left home, spiritually as well as physically, to take the vows of economic life, and it is they who are the mind and soul of our great self-perpetuating institutions.*[55]

But even more significantly, and picking up de Tocqueville's key point, the road to conformity was neutering the tradition of individualism. The new workers might be the 'mind and soul' of their organisations, but the latter were self-perpetuating! Individualism is surrendered in return for a comfortable life. The workers are a *passive* mind and soul, not an *active* mind and soul. Corporate recruitment and training were far less concerned with natural ability, technical skills or experience, and far more instead on 'cultural fit'. Corporate man settled instead into a comfort zone with salary, prospects, suburban home and a good quality of life

Whyte observed that, contrary to the Protestant Ethic, success was due to neither luck nor the environment "*but only to one's natural qualities*", in the organisation, those who survived and succeeded "*were not necessarily the fittest but, in more cases than not, those who by birth and personal connections had the breaks*". One of Whyte's main themes was individualism versus collectivism, and how organisation man surrenders his individuality (through for example, decision-making) to the organisation in return for stability and security, or 'an easy life'.

This disturbed Whyte, greatly. America, after all, had grown great on the back of individual endeavour. He believed the Protestant ethic of individual endeavour trumped the collectivist processes of the social ethic. He was suspicious that the latter led to risk averse decision-making. When added to job security and collective responsibility, the result was the individual subsumed within the organisation. Cue tier upon tier of management, all organised to avoid risk taking and individual responsibility. Atrophy. The collectivist, social ethic dominated corporate life (and wider society) for four post-war decades, through to the mid-1990s.

Anthony Sampson. Wright Mills and Whyte were both reflecting on twentieth century corporate America. They were both writing in the shadow of the Second World War and, in simple terms, sought to demonstrate how flawed individuality was being sacrificed on the altar of the apparently more perfect decisions of the huge corporations rising up across the USA. The third writer on corporate man considered here is Anthony Sampson (1926–2004), who published *Company Man: the Rise and Fall of Corporate Life* in the mid-1990s.[56]

Several decades on from Wright Mills and Whyte, Sampson concurred with much of their thinking but also had the benefit of a lapse of time, and the experience of the early-1990s recession, to be able to speculate on the eclipsing of corporate man by the harsh wins of the first white-collar recession. Before drawing upon Sampson, it is helpful to describe the economic context to his 1995 text.

Sampson's expansive text ranges from the East India Company to Microsoft, tracking the evolution of business from family concerns to modern corporations. Sampson describes company man in the mid-twentieth century as having become a stable fixture in the economy, with the big corporations seemingly as permanent as nations, with many confidently preparing for their centenaries.

But he then describes how this was an illusion of stability. In fact, by the mid-1970s, the "*words capitalism and capitalist made a triumphant comeback*"; and the "*capitalist was reinterpreted as a heroic life-force*". He records how the "*central hero of this new drama was the raider*" who sought to demand "*higher share values and dividends, attacking the strongholds of big business*", whereupon "*company men became suddenly vulnerable*".[57]

Asian companies were expanding rapidly in the free markets of America and Europe and computer technology was beginning to make itself felt in terms of starting to replace company man functions and roles. Then, suddenly, in 1990 the economy slipped into recession – deep, severe and, unlike in the early-1980s when manufacturing was decimated, spreading to white-collar work. The pressure was on to reduce costs and as the chill winds of change gathered pace, thousands of jobs disappeared. Sampson observed how companies everywhere found themselves scrabbling to make efficiency savings: job security disappeared virtually overnight.

Sampson started his final chapter

> *The traditional company man with his confidence in annual increments and a growing pension is as extinct as an eighteenth century clergyman. His decline has taken nearly everyone by surprise ... yet it is causing one of the biggest social upheavals of the twentieth century.*

Most corporations no longer felt paternal and began to replace loyal employees with short-term specialists and temporary consultants.

> *The promise of corporate loyalty began to look like an absurd indulgence which preserved incompetents and built up extravagant bureaucracies. What were they all doing through the fifties and sixties – those co-ordinators, liaison officers, in-house specialists, advisors and lawyers – if they could eventually be dispensed with so easily?*[58]

The wholesale shift in corporate culture was underpinned by a plethora of management books, all in various ways proclaiming to have found an angle on the emerging economic landscape. New wave management theory and the names of Drucker, Hammer, Handy, Peters, Porter and Senge, became very familiar to anyone with even a passing interest in what was happening to the seemingly permanent corporations. Their consistent messages were about downsizing, outsourcing, delayering, re-engineering, mobility and virtual work: in short, methods for becoming leaner, more efficient, more flexible and more competitive.

Outsourcing spread rapidly as companies sought to shrink to a newly defined core of activities, relieving themselves of the overhead burdens associated with support activities such as catering, cleaning, building maintenance, security, IT and even accounting and personnel. Demergers multiplied as the fashion for breaking down the lumbering leviathans took hold: BAT, BTR, Courtaulds, Forte, Hanson, ICI, Racal and Thorn EMI all got in on the act. Others, such as AT&T, Eastman Kodak and General Electric, delayered; while yet others, including Asea Brown Boveri, Rank Xerox and Shell, underwent full-scale business process re-engineering.

While workforces were rightsized, outsourced and re-engineered, the iron grip of management somehow seemed to strengthen. The drive for efficiency, alongside new workplace technologies, contributed to the drive to create large flexible floorplans, in hermetically sealed buildings with all environmental services (lighting, temperature and ventilation) controlled centrally. Densities were increased to maximise people per unit area in the rush to make real estate cost savings. Dumb terminals linked to central mainframes, and then the growth of laptops, led to the introduction of 'free addressing' and hot desking (imported from Naval submarine practices), which had a further demeaning impact on workers. Any sense of corporate loyalty to employees was swept away, and workers came more to resemble units of production to be managed with mechanistic efficiency.

Corporate CEOs and CFOs pursued efficiency measures with an almost evangelical fervour, interweaving the cost cutting agenda with theories about the nature of companies in the emerging office economy. We were told, for example, that as the efficiency measures took hold they would lead to horizontal networks replacing vertical hierarchies, and highly skilled, group-based knowledge workers replacing clerical, departmentalised information processors. Peter Drucker went so far as to suggest that knowledge had become the primary resource for individuals and the economy, and that the traditional factors of production favoured by economists – land, labour and capital – would not disappear, but would become secondary.[59] Robert Heller picked up the theme, arguing that since horizontal business processes are all multifunctional and multidisciplinary, the manager is being led in similar directions. Business Week argued that horizontal corporations should organise around processes rather than tasks; flatten hierarchies; use teams to manage everything; let customers drive performance, and reward team performance.[60]

In tune with Wright Mills and Whyte, Sampson suggested that the corporate organisations that emerged late in the nineteenth century in the USA, with their massive structures, rigid hierarchies and life-time employment, created the professional manager, or company man, thereby allowing the middle class to grow immensely. By the 1920s and 1930s, he argued, British corporations were emerging along similar lines with their own armies of loyal company men occupying managerial roles in corporate organisations typified by the likes of BP, Shell and Unilever: corporate leviathans that bestrode the economic landscape for much of the post-war period.

These new company men were beginning to appear as 'world managers'. As Richard J Barnet and Ronald G Miller described them in 1974 in their book *Global Reach*, they were "*the first in history with the organisation, technology, money, and ideology to make a credible try at managing the world as an integrated unit*".[61]

But the chill winds of change were gathering pace, and the stable, almost paternal corporate environment began to change from the late-1980s when the security of Company Man began to be undermined, and a changing economic landscape sounded the death knell for large, slumbering, multilayered and complacent organisations. Sampson's particular contribution, distinct from Wright Mills and Whyte, was that he was writing from the perspective of the 1990s, and he didn't so much trace the evolution of company man – although his work in this respect was excellent – but rather he was one of the earliest to foresee Company Man's demise. Sampson began his final chapter with the following:

> *The traditional company man with his confidence in annual increments and a growing pension is as extinct as an eighteenth century clergyman [Sampson might have said nineteenth century clerk!]. His decline has taken nearly everyone by surprise, particularly politicians who preferred not to face up to it; yet it is causing one of the biggest social upheavals of the twentieth century, and its repercussions are still spreading.*[62]

He observed that the large corporations of the 1970s seemed as permanent as nations but that some (e.g. Pan Am and Gulf) had disappeared altogether. Critically, he noted the wholesale shift in corporate culture, whereby most no longer felt paternal towards their employees. Lifetime employees were replaced with consultants and temps who did not share the lifer's ethos. Corporate loyalty "*began to look like an absurd indulgence which preserved incompetents and built up extravagant bureaucracies*".

> *What were they all doing through the fifties and sixties – those co-ordinators, liaison officers, in-house specialists, advisors and lawyers – if they could be dispensed with so easily?*[63]

As job insecurity escalated across economies the company men "*who had been clambering up the corporate ladders were now back on the pavement, with the doors shut against them*".[64] The scale of the change is shown by data from the US economy. During the 1980s, the Fortune 500 companies cut their workforces by 3.5m, while small companies created more than 20m new jobs; and by the early-1990s, corporate giants employed only about 10% of the workforce compared with 20% in the late-1960s and early-1970s.[65] The 'corporatism' of the post-war years was over. The era in which the key requirements of corporations over their employees were conformity, uniformity, docility and fealty were about to change forever. Sampson argued that a combination of economic and technological change would usher in an era when corporations would become more dependent upon a narrower band of most senior managers. His key point was that the leaner organisations that had cut back "*their layers of fat*" and came to look "*less like republics, and more like monarchies*".[66]

The economic restructuring of the late-1980s and early-1990s marked the rise of globalisation and the deregulation of markets in the US and Europe, and with geo-political changes such as the dismantling of the Berlin Wall, this all led to radical changes in trading patterns. But it was the computer in particular and communications technology generally that was to have the most profound influence on change in the office economy.

The new technologies changed both the nature of the work itself and the nature of the physical surroundings in which office work was undertaken. They were to have a profound impact on the design, management and occupation of buildings in coming years. The final section of this chapter looks at the early stage of this transition which began with Big Bang in the City of London in 1986.

Something happened!

Charles Babbage (1791–1871) was a mathematician, philosopher, inventor and mechanical engineer, and famously he introduced the concept of a digital programmable computer. In his *Economy of Machinery and Manufactures* (1832) he set out what is now called the 'Babbage principle'. His thesis was that there was a relationship between innovation and the technical division of labour. More specifically, he argued that a worker who performed high-skill tasks received a commensurate reward, despite the fact that he might also perform lower-skill level tasks. Thus by dividing the labour process according to skill (and pay) level, efficiencies could be achieved in output and costs. It was but a small step to Taylor's scientific management.

Babbage's more famous prediction started to become reality in the 1970s and 1980s, with the introduction of microprocessors and then in 1981, IBM's first personal computer (PC). This was the time when speculation mushroomed about the potential impact of technology on office work and the office workforce. In short, the new technologies were widely expected to allow the reinvention of white-collar organisations and lead to widescale job losses as the potential power of computers began to be realised by managements eager to introduce efficiencies. Given that the early-1980s extend to no more than one full working career from the present day, some of the observations and claims from the time look incredibly naïve today. But, at the same time, they show how far we have travelled in a relatively short period of time.

One article that gives a very clear insight into thinking from this time was published by Gregory and Nussbaum in 1982, and reflected emerging technologies in the USA.[67] In an example of leading edge technology from the time, the authors cite a White House initiative involving the testing of Xerox's Advanced Multi-Function Workstation, code named 'Alpha',

> *which consists of a copier with a keyboard and video display screen. Besides enabling the user to do electronic editing of documents before they are committed to paper, the system can also create charts and graphs from keyboard instructions, print letters in a variety of typefaces, electronically store thousands of pages of information and print paper copies of graphs created on display screens.*

Noting that the speed and viability of electronic mail had increased and that in 1979 approximately 200,000 facsimile transmission units were used, sending messages at an average speed of 20 seconds per page.

> *What will the "office of the future" be like? "Electronic", "paperless" and as "people-less" as possible, according to many automation advocates. Units with desktop terminals will replace the desk and typing stand. File cabinets will be nowhere in sight. Interoffice memos will be composed, edited and filed by computer terminals which can "talk" to each other. Letters and documents can be sent by electronic mail across the country.*

In echoes of earlier times, the authors predicted that clerical workers will be disproportionately affected by technology (through automation), suggesting that the occupations which were especially vulnerable included file clerks, bookkeepers, secretaries and typists, bank tellers and insurance workers, at least 90% of which were female. They suggested that the new technologies would give managers a way to enforce discipline and standardisation in the office, and eliminate unnecessary managerial layers of supervisors whose primary activities are monitoring others. Furthermore, automated office systems would create new office hierarchies and maintain the sex-stratification in pay, power and prestige long familiar to women office workers. Despite these forecasts, the size of the office economy has continued to grow rapidly and take a growing proportion of total jobs, while offering a wide variety of opportunities irrespective of gender or skill level.

One of Gregory and Nussbaum's more prescient conclusions related to the potential for remote work. The authors noted that companies were experimenting with having computer work done by employees at home, what they described as a return to the cottage industries of the 1880s. Portable terminals, they argued, would allow people to work at home, on the road or even in a mountain cabin. The basic grasp of the implications of digital technology for office work described here, in 1982, underline how quickly the 'digital revolution' has happened.

In the early-1980s, London was, in many respects, a successful city. It was a major centre of finance and insurance and an economic powerhouse; its political influence across the globe reflected its imperial past, and its cultural heritage attracted great numbers of visitors. London was already a World City.[68] Arguably it was the first, having wrapped its political, cultural and commercial tentacles around the globe in the latter part of the nineteenth century. It grew throughout the twentieth century, pulsing with the economic cycle, but forever broadening and deepening its international relationships.

However, London was at this time a very different place to the city we know today. The City of London was a much smaller, more insular place. Its most exciting building was the newly completed Nat West Tower (today, Tower 42), and the skyline was dominated by the 1970s rent slabs along London Wall. It was a world of fixed commissions, of jobbers and brokers, and of

'Bank Walks'. Bowler hats remained in vogue, and there were very few shops or hotels. The 1984 *City of London Draft Local Plan* designated much of the City as a Conservation Area.

In the West End, today's prestigious shopping areas of Bond Street, Knightsbridge and Oxford Street were all successful, but around them the picture was rather different. Midtown was a ghost town; much of the South Bank was badly neglected (remember 'cardboard city'?), and Victoria was populated by mandarins and the Petro-chem industry.

Outside of the central area, the effects of the 1970s' economic decline were obvious. In the east, London's docks were closed, having succumbed to the technology of containerisation; and large parts of Outer London had taken a most severe economic battering. Following 150 years of continuous growth, London's population had been in decline for 30 years, mirrored by a fall in the number of jobs. Not only had London shared in the collapse of manufacturing, but the government had been actively encouraging companies to *leave* London.

In the mid-1980s, the fax machine and golf ball typewriter were leading edge technologies. Personal computers were not widely available, and we were yet to experience the internet, email, mobile phones, social media, and digital media technology. In London, the time pre-dated Canary Wharf, the Channel Tunnel, the Jubilee Line, the M25 and Terminal Five; and Big Bang was a cosmological theory, not a seminal economic event. London cuisine was, at best, mediocre; and coffee culture was a quaint French custom.

Then, something happened!

In 1984, Graham Lowe began his brief history of the twentieth century office in Canada by stating that an

> *electronic revolution of staggering magnitude is sweeping the office. The integrated electronic office is becoming a reality made possible through the merging of telecommunications and computer technologies. Filing cabinets, office memos, and written reports are being replaced by a network of computer terminals with a huge capacity for generating, storing, and processing data.*[69]

In 1986, office buildings, office workplaces and office work were all revolutionised by a single event. It is rare for such a thing to happen; as shown throughout this book, normally a whole series of events coalesce and bring about change over a period of time. In 1986 it was different; and the singular event was known, appropriately, as 'Big Bang'. This was the deregulation of financial services coming into effect on 27 October, bringing about an unprecedented wholesale reorganisation of business activities in the City.

The pressure to deregulate financial services had been building for some time. Internationalisation saw the number of foreign banks represented in the City rise nearly fourfold since 1970, to almost 500. These included large US and Japanese integrated merchant banks and securities companies (for example, Goldman Sachs, Merrill Lynch and Nomura Securities). The New York stock market had already deregulated in 1975. Figure 11.1 shows the New York exchange in the 1960s.

In 1979, the new Conservative government abolished exchange controls and, in its wider agenda of removing restrictive practices across the economy, it placed the London Stock Exchange on the Office of Fair Trading's Restrictive Practices Register. Consequently, in the early-1980s, the competition authority was threatening to take the Stock Exchange to the Restrictive Practices Court. The restrictive practices included the Exchange's fixed minimum commissions; the single capacity rule that separated the activities of brokers and jobbers; the requirement that neither could be part of a broadly based financial group, and the Exchange's exclusion of overseas firms from membership.

Figure 11.1 New York Stock Exchange, 1960s.

The then chairman of the Stock Exchange, Nicholas Goodison, took the view in 1983 that rather than engage in a legal battle that would most likely be lost, he would strike a deal with Trade and Industry Secretary Cecil Parkinson. The resulting Parkinson–Goodison Accord, between the Thatcher government and the London Stock Exchange settled the restrictive practices dispute between the Office of Fair Trading and the London Stock Exchange. The Accord also presaged the introduction of electronic settlement systems.

The date for deregulation, 'Big Bang', was set for 26 October 1986. Big Bang abolished minimum fixed commissions on trades, thereby increasing competition; it ended single capacity, the separation between stocks and shares traders (jobbers) and those who advised investors (brokers), thereby opening the floodgates of mergers and take-overs; and it permitted foreign firms to own UK brokers, putting up a 'welcome' sign to international banks. Trading switched from the face-to-face, 'open outcry' on the Stock Exchange floor to the new quote-driven dealing system – the Stock Exchange Automated quotation System, or SEAQ which itself was modelled on the established American NASDAQ system. This would render the Stock Exchange floor redundant and generate a huge demand for (by London standards) very large, densely occupied trading floors in corporate buildings (explained more fully in the next chapter).

The changes ushered in an era of global financial trading in which markets became globally integrated. Investment banks from America, Europe and Asia poured into London, accompanied by the tech companies to provide their computers. In the run-up to Big Bang, international investment houses and merchant banks invested heavily in broking and jobbing firms. The feeding frenzy saw old City names absorbed into rapidly expanding financial services conglomerates. Examples included Hoare Govett (Security Pacific); James Capel (Hong Kong & Shanghai);

Quilter Goodison (Bruxells Lambert); Rowe & Pitman (SG Warburg) and Wood Mackenzie (Hill Samuel).

In 1983, there were 214 broking and 17 jobbing firms in the City (following a long period of consolidation), and these were quickly snapped up by UK merchant banks and international investment houses. At this point, banks could take a maximum 29.9% in any target business. Some were bought by UK clearing banks but many more were snapped up by much bigger US, European and Japanese banks. Clearly, new organisations were in the making: larger, more complex, more international and more technologically intensive.

The combination of reorganising and expanding businesses, together with rapidly evolving technology for international dealing systems, led directly to a sharply increased demand for premises: new, bigger, larger floorplates, more highly serviced. Not only did the larger conglomerates wish to bring their merged companies together under one roof, but they also needed more highly specified buildings to cope with expanding technological demands. Developments such as Broadgate and Canary Wharf were a direct response to these enormous changes in the office economy. Big Bang unleashed a transformation that would result in London cementing its role as a leading global city.

Summary. The post-war period saw the internationalisation of the rational office with the spread of large American corporates. Up to this point, London and European offices remained dominated by cultural mores from the previous century. The stark contrast between the two traditions would erode over the coming decades as the European tradition became ever more accepting of the rise of managerialism and company man.

Managerialism involved complex organisational structures with multiple layers of management and status modelled on the ranking system of the military or civil service. Management created an intermediary form of worker, the salaried class, who answered to the capital owners and shareholders. Their evolving role led them to rationalise office practices and procedures, and scientific management was one early symptom of this process in America. But managerialism was much more than the practice of management practices and procedures; it was about a broader culture among workers and managers and the rise of a new middle class for whom the office was more than simply a place of work.

Wright Mills was one of the first authors to directly describe the growing masses of office workers as ‘middle class’, suggesting that the emerging class could be described not only in terms of skill and function, but also in terms of income and in degrees of power over other people. In this way, occupations become tied to class, status and power as well as to skill and function. Wright Mills also posited that managerialism overwhelmed workers by turning them into a compliant, doleful and self-satisfied stratum of society, who were kept in their unchallenging position through a combination of coercion, authority and manipulation.

Whyte was concerned by a particular aspect of corporate conformity: the ceding of the Protestant ethic to a social ethic, which subjugated the individual to the group; a shift to a collectivist ethic rather than the frontier individualism of their grandparents. The new generation had sacrificed individualism on an altar of corporatism.

Writing nearly four decades after Wright Mills and Whyte, Sampson described how the global recession of the early-1990s brought about a wholesale shift in managerialism, whereby most organisations no longer felt paternal towards their employees. At the same time, digital technology began to radically alter the shape and structures of corporate organisations.

The launch of the IBM PC in 1981 and the Apple Mac in 1984, introduced desktop technology; and very soon the arrival of mobile phones, smart phones, the World Wide Web and email would have a profound impact on every aspect of life: everyone’s personal and business lives were radically changed through the ability to collect, receive, transform, analyse and communicate information.

The unfolding of the 'technological revolution' would ensure that there would be no return to the unwieldy corporate structures of earlier times. If for no other reason, the revolutionising and democratising impact of ICT would not allow that to happen. Instead, businesses began to explore ways of using technology to support the more intensive and more agile occupation of property. The so-called *new ways of working* were employed with the aim of introducing changes to the way in which workers would engage with employers and colleagues, by becoming more flexible and dynamic, and thereby reducing the amount of space that organisations required.

By the end of the 1980s, the digital revolution had begun. Old corporate structures and company man were consigned to history, and the digital office emerged with novel ways of working. As ever, Frank Duffy captured the moment perfectly.

> *We have come to the end of simple office landscaping and perhaps to the beginning of a more complex period in office design. A new series of challenges is being met: better understanding of the different styles of organisation; greater interest in providing a range of space types; exploiting a much wider range of furniture options; using building services in a more sophisticated manner ……*[70]

Notes

1 Black A; Muddiman D & Plant H (2007) *The Early Information Society: Information Management in Britain before the Computer* Ashgate, Aldershot p25.
2 Spiegelberg R (1973) *The City: Power Without Accountability* Blond & Briggs, London p110.
3 Kynaston D (2002) *The City of London Vol IV A Club No More 1945–2000* Pimlico, London p401.
4 Kynaston D (2002) *The City of London Vol IV A Club No More 1945–2000* Pimlico, London p324.
5 Spiegelberg R (1973) *The City: Power Without Accountability* Blond & Briggs, London p103.
6 *The Banker* October 1969 pp1063–71 Cited in Kynaston D (2002) *The City of London Vol IV A Club No More 1945–2000* Pimlico, London pp402–403.
7 Perkin H (1989) *The Rise of Professional Society: England since 1880* Routledge, London & New York p291.
8 Perkin H (1989) *The Rise of Professional Society: England since 1880* Routledge, London & New York p292.
9 Perkin H (1989) *The Rise of Professional Society: England since 1880* Routledge, London & New York p294.
10 Berle AA (1954) *The 20th Century Capitalist Revolution* Harcourt, Brace p39.
11 Micklethwait J & Wooldridge A (2003) *The Company: A Short History of a Revolutionary Idea* Orion Publishing, London p102.
12 Harris R (2021) *From Clerical Factory to Digital Hub: London's Global Office Economy* Routledge, London.
13 Davoudi L; McKenna C & Olegario R (2018) The Historical Role of the Corporation in Society *Journal of the British Academy*, Vol 6 No 1 pp17–47. DOI https://doi.org/10.5871/jba/006s1.017 Posted 17th December 2018.
14 Sampson A (1995) *Company Man: the Rise and Fall of Corporate Life* Harper Collins, London p98.
15 Sampson A (1995) *Company Man: the Rise and Fall of Corporate Life* Harper Collins, London p137.
16 Drucker PF (1954) The Practice of Management: A Study of the Most Important Function in American Society Harper & Brothers, New York p4.
17 Ward SC (2011) The Machinations of Managerialism: New Public Management and the Diminishing Power of Professionals *Journal of Cultural Economy* Vol 4 No 2 May 2011 pp205–215.
18 Burnham J (1941) *The Managerial Revolution: What is Happening in the World* John Day, New York.

19 Sampson A (1995) *Company Man: the Rise and Fall of Corporate Life* Harper Collins, London p95.
20 Sampson A (1995) *Company Man: the Rise and Fall of Corporate Life* Harper Collins, London p95.
21 Berle AA & Means GC (1932) *The Modern Corporation and Private Property* Macmillan.
22 Berle AA & Means GC (1932) The *Modern Corporation and Private Property* p313.
23 Smith A; Tennent KD & Russell J (2019) Berle and Means's The Modern Corporation and Private Property: the Military Roots of a Stakeholder Model of Corporate Governance *Seattle University Law Review* Vol 42 pp535–563.
24 Smith et al., Berle and Means's The Modern Corporation and Private Property.
25 Perkin H (1989) *The Rise of Professional Society: England since 1880* Routledge, London & New York p460.
26 Perkin H (1989) *The Rise of Professional Society: England since 1880* Routledge, London & New York p460.
27 Perkin H (1989) *The Rise of Professional Society: England since 1880* Routledge, London & New York p461.
28 Riesman D with Glazer N & Denney R (1953) *The Lonely Crowd: A Study of the Changing American Character* (2nd edition) Doubleday Anchor Books, New York.
29 Centers R (1962) An Examination of the Riesman Social Character Typology: A Metropolitan Survey *Sociometry* September Vol 25 No 3 pp231–240.
30 Centers, Examination of the Riesman Social Character.
31 Micklethwait J & Wooldridge A (2003) *The Company: A Short History of a Revolutionary Idea* Orion Publishing, London p103.
32 Micklethwait J & Wooldridge A (2003) *The Company: A Short History of a Revolutionary Idea* Orion Publishing, London p109.
33 Wright Mills C (1951) *White Collar: the American Middle Classes* OUP, New York.
34 Wright Mills C (1951) *White Collar: the American Middle Classes* p107.
35 Wright Mills C (1951) *White Collar: the American Middle Classes* p109.
36 Wright Mills C (1951) *White Collar: the American Middle Classes* p64.
37 Wright Mills C (1951) *White Collar: the American Middle Classes* p68.
38 Wright Mills *White Collar* p71.
39 Wright Mills C (1951) *White Collar: the American Middle Classes* pX.
40 Wright Mills C (1951) *White Collar* pXV.
41 Wright Mills C (1951) *White Collar: the American Middle Classes* p21.
42 Wright Mills C (1951) *White Collar: the American Middle Classes* p77.
43 Wright Mills C (1951) *White Collar: the American Middle Classes* p80.
44 Wright Mills C (1951) *White Collar: the American Middle Classes* p100.
45 Wright Mills C (1951) *White Collar: the American Middle Classes* pp68–69.
46 Wright Mills C (1951) *White Collar: the American Middle Classes* p71.
47 Wright Mills C (1951) *White Collar: the American Middle Classes* p74.
48 Wright Mills C (1951) *White Collar: the American Middle Classes* p70–71.
49 Wright Mills C (1951) *White Collar: the American Middle Classes* p75.
50 Whyte WH (1956) *The Organization Man* Simon & Schuster, New York.
51 Whyte WH (1956) *The Organization Man* Simon & Schuster, New York p97.
52 Whyte WH (1956) *The Organization Man* Simon & Schuster, New York p7.
53 Whyte WH (1956) *The Organization Man* Simon & Schuster, New York pp4–5.
54 Whyte WH (1956) *The Organization Man* Simon & Schuster, New York p5.
55 Whyte WH (1956) *The Organization Man* Simon & Schuster, New York p3.
56 Sampson A (1995) *Company Man: the Rise and Fall of Corporate Life* Harper Collins, London.
57 Sampson A (1995) *Company Man: the Rise and Fall of Corporate Life* Harper Collins, London pp168–169.
58 Sampson A (1995) *Company Man: the Rise and Fall of Corporate Life* Harper Collins, London p307.
59 Drucker PF (1992) The New Society Organisations *Harvard Business Review* September–October pp95–104.

60 Cited in: The Horizontal Corporation *Business Week* 20th December 1993 pp44–49.
61 Cited in Sampson A (1995) *Company Man: the Rise and Fall of Corporate Life* Harper Collins, London p140.
62 Sampson A (1995) *Company Man: the Rise and Fall of Corporate Life* Harper Collins, London p307.
63 Sampson A (1995) *Company Man: the Rise and Fall of Corporate Life* Harper Collins, London p307.
64 Sampson A (1995) *Company Man: the Rise and Fall of Corporate Life* Harper Collins, London p308.
65 Stelzer I (1993) Transition to Save the American Dream *The Times*, 15th April p25.
66 Sampson A (1995) *Company Man: the Rise and Fall of Corporate Life* Harper Collins, London p312.
67 Gregory J & Nussbaum K (1982) Race Against Time: Automation of the Office. An Analysis of the Trends in Office Automation and the Impact on the Office Workforce *Office: Technology and People* 1, 1982 pp197–236 Elsevier Scientific Publishing Company, Amsterdam.
68 See Friedman J and Wolff G (1982) World City Formation: an Agenda for Research and Action International *Journal of Urban & Regional Research* Vol 6 pp309–344.
69 Lowe GS (1984) The Enormous File: the Evolution of the Modern Office in Early Twentieth-Century Canada *Archivaria* 19 (January) pp137–151 https://archivaria.ca/index.php/archivaria/article/view/11139
70 Duffy F (1992) *The Changing Workplace* Phaidon Press, London p78.

12 New environments for working

The history of the office described in this book has demonstrated how the nature of office buildings has evolved along with changes in the economy and in office work. The general pattern is that building design has responded to economic and business changes rather than anticipated them, which is no surprise in a 'form-follows-function' sense. It would be foolhardy to design and build buildings for which no specific purpose currently existed.

While recognising almost infinite variability in individual buildings, particular periods have shown distinctive building formats over the past 250 years. The earliest phase was typified by the 'shop', where ground floor trading took place in a building that was often a converted house. The defining characteristic was the domestic architecture associated today with Georgian town houses. Vere and Asgill's bank of 1757 in Lombard Street (as discussed in Chapter 3) is the example given here. Owners typically lived on the upper floors ('living over the shop'), and as businesses became larger, owners moved to dedicated homes and accommodated their clerks in the living space.

The shop was followed by the buildings delivered for the new joint stock banks in the nineteenth century, which, as we saw in Chapter 6, were larger, and more grandiose and ostentatious than their predecessors. This was the age when the clerking workforce expanded rapidly and offices became single-use buildings. The example given here is National Provincial Bank's 1866 building on Bishopsgate (discussed in Chapter 6). As can be seen, there is a step change in scale and image.

Around the turn of the century, office buildings became altogether more refined and complex, with lifts, electricity, modern sanitation, air handling and heating and so on. They were responding to more complex organisations, that were departmentalised and layered in ways not seen before; and females fully entered the workforce. These were comparatively large buildings and heavily sub-divided. The example shown here is Africa House on Kingsway (1932). This was the age of office mechanisation and the introduction of high volume, repetitive and labour intensive work, as discussed in Chapter 10.

The post-war corporate offices featured cheap, poorly specified buildings (see 'Working in the pre-digital office' below), represented here by St Alphage House (1962), one of the 'Seven Sisters' on London Wall. This approach lasted through the 1950s, 1960s and 1970s. The buildings were designed to meet the needs of relatively unchanging organisations, where much work was sedentary; there were few technological demands on building infrastructure, and above all, they were built to minimum budget. Once Big Bang got going, the buildings were found to be entirely inadequate to meet the needs of the digital age.

As a result of the organisational and business demands posed by Big Bang companies a whole new building design agenda was ushered in at great speed. 'Big Bang buildings', as represented

DOI: 10.1201/9781003582205-16

here by Exchange House in Broadgate, dating from 1990, were altogether larger, sometimes with floorplates the size of football pitches. The buildings were deeper and accommodated services to cope with high intensity data and communications technology. And all this was wrapped within an architecturally high-quality exterior.

While the United States had its skyscrapers and very large buildings, Europe, Australasia and the Far East were different. Nowhere had replicated the rational office of the US, partly because countries outside the US had too much history and culture wrapped up in their built stock and partly because nowhere outside the US went through an economic transformation, on such a scale, and from a standing start, as did America in the latter years of the nineteenth century. When American companies expanded rapidly in the UK and Europe in the 1960s and 1970s they did so, largely, within the confines of the existing 'cultural' office building tradition. Big Bang, however, provided the rationale for breaking that tradition, and the hybrid cultural-rational building was born.

Working in the pre-digital office

The post-war London office remained very different to its US counterpart. In the three decades following the Second World War, office buildings remained relatively primitive environments. Building design was driven primarily by consideration of how cheaply buildings could be constructed. As Oliver Marriott explained in *The Office Boom*, what "*developers wanted from their architects was a commercial service. They needed functional buildings designed to a certain price, usually the lowest possible…*"; furthermore they "*were just not concerned with design*". To emphasise the point, Marriott presented some comparative costs. "*Whereas [speculative] developers put up office blocks for £4 10s 0d to £7 per square foot ... the owner occupier perhaps spent £8 to £10 a square foot on average*".[1] When speculative developers were spending half as much as owner occupiers, the impact on the quality of the resulting work environment needs little further illustration.

The supply industry was simply not organised to address the needs of the companies who occupied its buildings. There was no perceived need for intelligence, or research: the industry had a standard product (designed to institutional specification), available on standard terms (the Full Repairing and Insuring lease). Property's triumvirate of investor, developer and agent, in which the investor was the customer, the agent the intermediary and the developer the supplier, prescribed a passive process in which there was no impetus for innovation, and there was no reward for added value. The FRI lease secured a secure, long-term (25 year) income stream, and this more or less guaranteed income was the triumvirate's *raison d'etre*. This inertia helped to underpin the notion that property was a supply-driven process. The process was rigid and unchanging, and one of the side-effects was that the industry's products, namely its buildings were, at best, mediocre and more often than not, of a very poor quality.

Gone was the Gothic Revival, Italianate and Neoclassicism soup of architectural styles from the Victorian period. Instead we saw the emergence of an entirely new style – commercial architecture. The grandiosity and ostentation of earlier times were banished in favour of a stripped down facadism of a more or less uniform style. There was very little product differentiation, either in terms of physical attributes or tenure arrangements; and there was virtually no consideration of what infrastructure would be required within buildings to ensure that they performed to an acceptable standard through their life cycle.

The 1950s, 1960s and 1970s saw some of the most brutal and ugly commercial office buildings ever built in the UK. Design was driven primarily by consideration of how cheaply buildings could be constructed: the era of the 'lowest price wins' bidding process arrived.

Architectural firms found innumerable ways to cut the cost of construction, with consequent impacts on the quality and durability of both the façade and the interior work environment. Externally, a standard concrete structure with coloured glaze cladding of little architectural merit was standard. There was no meaningful understanding of how the use of the building and its management would interface.

The majority of speculative office development consisted of buildings designed and built with a rigid floor plan (typically, a central corridor with rooms either side), low capacity service provision and a low quality internal specification. Small, shallow floors were standard with a structural floor to ceiling height of 8ft to 9ft (2.4m–3.0m), and limited potential for horizontal distribution of power, data and telephony cabling. Floors were generally finished in woodblock or tiles, with no provision for under floor cabling; while suspended ceilings were largely unheard of – light fittings being fixed directly to the plastered ceiling. Single glazing, coupled with large window areas gave rise to problems of heat gain in many buildings. Office furnishings were 'industrial' in character: desks were large and bulky, and made of steel; and chairs were immobile and uncomfortable, with minimum adjustability. To let accommodation, the minimum of tenant facilities was required by the developer, and toilet areas and entrance halls were generally extremely basic.[2]

The life expectancy of the buildings built during this period (around 50 years) seems, in retrospect, to have been ill-judged. Though it might be argued that the buildings were functionally successful for the time, in so far as they responded to the immediate needs of their occupiers; it might also be argued that in planning and investment terms they were woefully inadequate. Neither is this verdict one reached with a long span of hindsight. As long ago as 1977, and while recognising the progress of 'corporatism' in its own terms, Frank Duffy and colleagues argued that the work environment had failed to modernise, and caustically wrote:

> *Older architectural magazines reveal with cruel clarity the prevailing standard of office design in the late 50s and early 60s. Advertisements display more vividly than editorial what seems to us now to be an astonishingly uncomfortable internal physical environment – hard floor finishes, lots of glazed partitioning, surface mounted light fittings, bulky steel desks… .*[3]

The authors went on to suggest that, in the UK, technical problems rather than user requirements or experiments in plan form were what obsessed architects.

Buildings with expected life spans that would leave them still in use in the early part of the twenty-first century were being demolished for redevelopment just 20 years after they were built. Clearly the buildings lacked the flexibility in their design to respond to the changing economic, business and technological circumstances.

By American standards, European offices were small. As late as the mid-1970s, only 15% of office units in the City of London were larger than 10,000 sq ft (929 sq m) and over 60% of units were smaller than 3,000 sq ft (c279 sq m); the buildings themselves were uncomplicated, often constructed for different purposes (including residential and light industrial) and the demands placed upon them by their occupiers were relatively light. London had to wait until 1967 for the 365ft (111m) Centre Point; while the eponymous 600ft (183m) Nat West Tower (later rebranded Tower 42) was not built until 1981.

The work environment: stable and grey. Office interiors thus remained relatively unchanged for an extended period after the war. Throughout the City and its offices, the work remained routine and process-driven. Kynaston cites the example of Dundas Hamilton, the son of a senior partner, who started out in 1946 at Carroll & Co, a small stockbroking firm a 5 Copthall Buildings:

The general office consisted of a girl who worked with a comptometer ... a man who kept the books, an office boy and a girl typist. We all sat in the outer office. There were gas fires, the lift went by hydraulics and not electricity.

Little changed during the greater part of the 1950s, and the words of the historian of the National Mutual Life Assurance Society were broadly applicable to many other firms.

If someone who had worked in the head office in 1900 had been able to return in 1955, he would have recognised, apart from pension schemes, all that was going on. He would have found the policy registers, still bound in half calf, still being written up by hand. He would have found himself, together with many others, working out each member's bonus by an individual calculation and after all had been checked, writing the answers with pen and ink on notices to be sent to members.

Kynaston noted that adding machines and mechanical calculators were starting to become more common, but the older, time-honoured skills persisted, such as bookkeeping, which remained thoroughly manual.[4] He then described the ambience of a few British banks, which underlined the 'cultural' characteristics attached to them in this work.

Each interior had its distinctive ambience. Martins Bank in Lombard Street was famous for its display of flintlock pistols, Barings in Bishopsgate had huge lead water tanks, the Smiths branch of the National Provincial feature the Smith family roll of honour. Other banks were noted for their furniture polish or disinfectant smells and rosewater sprays in summer. Blazing open fires in winter, the cool of Lazards marble in August. Several, such as Rothschilds, were baronial and impressive.[5]

Workplaces were remarkably stable because organisations during this time changed only very slowly: headcount might change either through organic growth or merger; but functionally businesses were relatively stable. There was little need to move people around buildings with any regularity; once a functional use was established it might remain in the same place, doing the same work, for many years, even decades. The great majority of the workforce arrived at more or less that same time in the morning, and the reverse happened in the evening.

Corporate offices were 'grey'. There was very little colour, no art, no planting and very little personalisation of space. Walls, carpets and furniture provided largely undifferentiated space. Little attention was paid to the environmental or ergonomic conditions on the occupied floor. David Kinsella summarises the primitive nature of offices at this time.

Office equipment and fittings were basic but of good quality. All the chairs and desks and cabinets were of wood, the larger desks often leather-topped with drawers on either side. Several rulers were of the round, black wood type, and all calculating machines were mechanical, simple in capacity and often of the hand crank kind. Pens and ink prevailed, my first biro being seen at the HQ of the Midland Bank.

In other areas of office technology 1960 seems to have been the key year. In the course of those twelve months Linklaters & Paines acquired its first Rank Xerox photocopier and first telex machine ... Ashurst's, had by then four electric typewriters.

That same year Durlachers marked its move to Austral Houe by inaugurating a much publicised, diamond-shaped dealing room, equipped with as many as forty or fifty telephone handsets.

Summer of 1964, for example, saw the arrival at Philips & Drew of a round-the-clock telephone answering service.[6]

But as well as their physical attributes, pre-digital offices also clung to many of the social mores of the previous era. For example, as office design introduced more uniform approaches to layout, so means had to found to reflect social status. This was achieved most simply by dividing space into open plan and cellular space according to job function and level of seniority. Increasingly, private offices claimed the outward-looking parts of office floors, pushing open plan clerical staff into the less well-lit, central areas. Even in the open plan there was hierarchy, expressed through desk size and pedestals – and filing and carpeting was used to designate areas and status.

The office thus remained riven by hierarchy and social class division. Senior managers and directors were disproportionately public school educated. Males and females undertook distinct tasks with the latter undertaking the more menial and manual and it was, of course, during the post-war era that the secretary became a defining feature of office workplaces. Managers were frequently allocated one-to-one secretaries. The manager–secretary relationship would often be close, with the secretary undertaking personal tasks for her manager as well as routine office work.

This gender divide in the office was reinforced in subtle ways. For example, desks were designed for different roles, but more often than not, reflecting the gender of their anticipated user. As observed by Amy Thomas, as the main subject of office planners,

the desk became a highly gendered object in the City, bound up with material signifiers of status and hierarchy. For City men, the typical freestanding mahogany desk was a luxurious rest stop in the busy movement of the day and part of a grand assemblage of domesticity in management offices.

By contrast, typists (predominantly female), were provided with the smallest desk, suited to their single-task role, and machine operators worked on moveable trolleys mounted on rollers, which simultaneously doubled up as storage. "*For the thousands of women employed as clerical workers, desks were little more than equipment – an efficient extension of the typewriter, which they operated for eight hours a day*". Thomas concludes that women were "*judged not by their social skills or by their reputations, as men were, but rather by their productivity and efficiency*".[7]

Saval underlines the continuing significance of social strata in the post-war office and its influence on office layout and social relations within the office.

Around the serried rows of L-shaped desks in the steno pool were offices partitioned by glass; these were filled with couches, desks made of mahogany and sometimes topped with marble, lounge chairs, and ottomans, all done up by the very best names in American design.[8]

The luxury never stopped. Take the elevator up to the executive suites on the top floors, and you would find an extraordinary lavishing of care and detail, snugly insulating the top brass from the ricocheting clacking of the typewriter pools below.[9]

Larger businesses would provide canteens, and even these reflected social division – many of the larger banks and insurance companies would have four or five different levels of catering for different ranks of workers; with the highest receiving silver service and waiters. Similarly meeting rooms had their own hierarchy – those that anyone could use; those reserved for senior executives and, that most mysterious of room for all but the cognoscenti, the Board Room. Few knew what went on in there, but it was obviously of the utmost import.

Offices were organised to reflect support functions as much as business units. Functional departments grew, developing their own hierarchies and social structures. Accounting, data processing, filing and records, legal, office supplies, printing, stock control, typing and myriad others spread. Again, given their 'support role', many of these functions were grouped in basements and lower floors of buildings – never was a typing pool allocated the top floor!

Social relations in the office. Typing pools, almost exclusively female, were consolidated to receive hand-written documents, which they would type onto standard paper, send back to the originator for checking and receive once again for alterations. Sometimes this circle could be completed dozens of times before the final version was signed off. The process would be overseen by a matronly character who enforced discipline, checked quality and interfaced with executives. Beyond the typing pool, female workers were also subject to being boxed into particular kinds of work. Wright Mills suggested that as office machinery was introduced, the number of routine jobs increased and consequently the proportion of positions requiring initiative decreased, leading to some managers preferring to "*hire girls who are content to remain simply clerks, who will attempt to rise no higher than their initial level*".[10]

The period of relative quiet in workplace layout and design, with notable examples, some of which are examined below, lasted through to the 1980s. Vincent Giuliano was able, in 1982, to paint a picture of the unchanging office during this time.

> *An anthropologist visiting an office today would see much that he would have seen 25 years ago. He would see people reading, writing on paper, handling mail, talking with one another face to face and on the telephone, typing, operating calculators, dictating, filing and retrieving files from metal cabinets. He would observe some new behavior too. He would see a surprising number of people working with devices that have a typewriter like keyboard but also have a video screen or an automatic printing element.*[11]

As well as the physical characteristics of the office, the social aspects were also slow to change. The nature of the work reinforced hierarchy through a complex web of roles and responsibilities, ranging from quasi-manual work (such as printing) through to the much sought after managerial roles. Wright Mills provided an American perspective, in which the larger, more bureaucratic offices had lost a good deal of the informality of times past and that the office was no longer "*an informal, friendly pace. The drag and beat of work, the production unit tempo, require that time consumed by anything but business at hand be explained and apologised for*".[12] He described the rise of the office manager from a chief clerk to a responsible executive reporting directly to the company treasurer or vice president as "*an obvious index to the enlargement of offices and to the rise of the office as a centralised service division of the entire enterprise*".[13]

In comparing the personnel of the new office with that of the old, Mills cited the mass of clerical machine operatives as particularly striking. "*They are the most factory-like operatives in the white collar worlds*", and these

> *interchangeable clerks often punch a time clock, are not allowed to talk during working hours and have no tenure of employment beyond a week or sometimes a month.*[14]

He went on

> *The new office is rationalised: machines are used, employees become machine attendants; the work, as in the factory, is collective, not individualised; it is standardised for interchangeable, quickly replaceable clerks ... . The employee group is transformed into a uniform mass in a soundless place, and the day itself is regulated by an impersonal time schedule.*[15]

Mills also describes the new office hierarchy, which he says is based "*upon the power and authority held by the managerial cadre, rather than upon levels of skill*".[16] Smaller hierarchies fit into larger ones and are interlinked, with the management cadre permeating the whole and forming the backbone of the hierarchy. He then links organisational status and social standing.

> *Where one stands depends, first upon the extent to which one participates in the cadre's authority, and second, the closeness of one's association with its members ... . Educational level and experience naturally lend status, but only secondarily. It is from the managerial cadre that esteem is derived and status borrowed.*

Finally, Mills suggests that if hierarchy was purely bureaucratic then it would operate purely on authority, as it would in the military. But this is not the case.

> *Within and between offices, there is usually a system of cliques, which often cut across the formal line of authority and work. Through them 'the man in the know' can cut red tape and secretaries of top men ... can call other secretaries to expedite matters that would take much longer through the regular channels.*[17]

Writing in 1977, Evelyn Glenn and Roslyn Feldberg cite a federal task force conclusion:

> *Secretaries, clerks and bureaucrats were once grateful for having been spared humanization of the factory ... they had higher status than blue collar (workers). But today the clerk ... is the typical American worker ... and such positions offer little in the way of prestige ... imparting to the clerical worker the same impersonality blue-collar workers experience in the factory.*

The authors suggest one of the most striking features of 'paper work' in large organisations in the 1970s was the elaborate sub-division of tasks and the extreme specialisation of workers. They blamed the use of electronic data processing and word processing for part of this. They note the (mainframe) computer's capacity and speed to store huge quantities of information which, while labour saving in their own right, increased the demand for data, "*on the assumption that more data results in more rational, better-informed decisions*", but this in turn called for more clerks to prepare the quantities of data involved. Similarly, they did not find that clerical workers had experienced increased autonomy.

> *As a result, work throughout the organisation is structured by the requirements of the computer. Although the clerk may be less directly supervised, she does not gain autonomy; the requirement of the machine replace the directives of an immediate supervisor. Her mental choices are limited to predetermined categories. She has little discretion to do the work as she sees fit or to set her own pace ...*[18]

The modern office worker was a slave to her machine in a similar way to the clerk of yesteryear being a slave to his ledger. Indeed, the post-war period led to a reprise of the 'deskilling' debate of Braverman and others by introducing yet another cycle of short-lived, but repetitive and manual roles. For example, Glen and Feldberg argued that 1970s' office workers were being proletarianised in the same manner as Klingender had argued in the 1930s.

But the whole century from the 1880s to the 1980s was one long story of the creation and extinction of roles and tasks in the office. Just as mechanisation introduced machines to automate human endeavour and create repetitive tasks, so electrification created its own form of manual labour: while often cleaner, quieter and less manual, tapes, keypunches and programmes rarely made work more interesting or challenging in areas such as the typing pool, data processing department or in central filing.

The process reached its apogee with the 'back office' – a separate, low-rise building, typically in a cheaper location, where manual and semi-manual functions would be gathered in spartan interiors. The front and back office arrangement was loaded with social symbolism in terms of where power and influence was concentrated across an organisation. The back office was to the front office what the nineteenth century factory floor was to the office: a place for doing rather than deciding. Many back offices were even arranged like factories with serried rows of workers arranged like production lines. Most back offices disappeared with the advent of the digital office in the nineties.

Landscapes for work

We saw in Chapter 10 how the rational office model expressed itself in skyscrapers, and in Chapter 11 the rise of corporate managerialism. On the other side of the Atlantic, northern European countries began to experiment with different approaches to the office, while remaining within the European, or cultural office tradition. Two such examples are highlighted here: Centraal Beheer and NMB.

Centraal Beheer, Apeldoorn, Amsterdam. In terms of the cultural, or European office model, for much of the twentieth century, the *modus operandi* had not been shaken fundamentally. In several respects, the workplace contained many of the rigidities of earlier times: hierarchical, spatially divided according to strata of seniority, process-based and uniformly dull. One famous example that sought to redefine the European work landscape was that created by Dutch architect Herman Hertzberger: the Centraal Beheer insurance building at Apeldoorn in the Netherlands.

The building was commissioned by insurance company Centraal Beheer (now part of Achmea group). Constructed in 1972, it was designed to accommodate around 1,000 people, consisting of 56 tower-like cubes connected on each floor by over-passes. Each cube comprised a 30ft by 30ft (9m by 9m) space, located around a core containing the lifts, stairs and bathrooms. In each block there are four of these elements per floor with a gap of three metres between them for human traffic. These larger structures are grouped into four quadrants of office space; north, south, east and west, and are held together by a central core containing the social centre of the building.

For both Hertzberger and his client Centraal Beheer, the key to success was the 'human scale' of the design, which individual workers could relate to. Indeed, the brief for the building called

Figure 12.1 Centraal Beheer Insurance, Netherlands. Licenced under the Creative Commons Attribution-Share Alike 3.0 Unported license.

for employees to have the sense that they were part of a working community rather than an anonymous mass of workers. This was what drove the village philosophy of small office units for only eight–ten people, each linked to raised walkways, atria and common spaces. The result was a mosaic of interior spaces, with many small break out areas where people could have a coffee, relax, or hold meetings.

It is fair to say that the building revolutionised contemporary thinking about office design.

NMB, Amsterdam. In 1978, banking group NMB (now ING) held a limited competition to build a new head office. The winner was a relatively unknown architect, Ton Alberts, whose designs to date were the antithesis of what was expected of a modern bank. Rather than the image-reinforcing architecture of solidity, convention and boring, what resulted was, to say the least, idiosyncratic. Alberts's previous work had been limited to housing and social projects.

The end result was a unique development comprising ten distinct office blocks connected by a 1,150ft (350m) walkway. Within each block, an office floor comprised five work units of 950 sq ft (88 sq m) and meeting space arranged around a void. Upstairs, the clusters are separated by lifts and emergency stairs.

Figure 12.2 NMB Banking Group, Amsterdam.

The general amenities are ranged along the internal street. Natural materials such as stone and wood are exploited in the interior. Each void has its own colour; here water features and plants dictate the ambience, with artworks fully integrated into the building. Beneath the NMB building, which boasts 366,000 sq ft (34,000 sq m) of office space for 2,500 bank officials, are two levels of parking space. Outside, the cladding is predominantly brick combined with brown precast concrete lintels and blue aluminium window frames.

In contrast to the north American model, these European examples followed the established cultural model. They are low level, the floor plan is predominantly shallow, they are inefficient (in space planning terms) and heavily reliant on natural light. There is a high degree of internal sub-division, with generous space standards and staff amenities.

Europe also experimented with a compromise between open plan and enclosure, in the form of the 'combi-office'. One of the oft-quoted examples is Canon's headquarters in Sweden. The building was designed by Tengbom Architects, an early advocate of office landscaping. The essence of the combi-office was that cellular offices were lined up along perimeter walls (to maximise natural daylight), while their inward facing walls were glazed to allow transparency and visual connection with the central areas. The latter accommodated both employees in open plan layout and also shared spaces for meeting and collaboration.

Contemporary with the northern European examples given here, the only UK example that came close to replicating the 'democratic workplace' was the Willis Faber Dumas building located in Ipswich, in Eastern England, which opened in 1975. This pioneering building was designed by Norman Foster. The three-storey building is constructed on a grid of concrete pillars 46ft (14m) apart, and is faced with a curving smoked glass curtain wall. The interior is arranged around an atrium with three pairs of escalators – highly innovative for the time – to bring people up to a third floor restaurant and rooftop garden. The building was provided with a private swimming pool. Another innovative aspect of the building design was the inclusion of raised floors for the flexible distribution of cabling.

The architects and client, an insurance company, consciously set out to design a work environment that was seen to be democratic and nurturing a sense of community. The floorplans provided for a range of open and flexible interior spaces. However, Willis Dumas was very much the exception in the UK and while many northern European offices were experimenting with different office landscapes, where workers had a very significant impact on layouts (making them more 'democratic'), the situation in London was rather different.

By the 1980s, it was clear that the landscaped office so favoured on mainland Europe had failed to gain traction in the UK. Part of the reason was that the UK supply chain was so much stronger than in Europe. In London, speculative developers and investors held sway and themselves determined the most favourable building form (driven by cost efficiency). Many more occupiers in Europe owned and built their own buildings than was the case in London. There were few opportunities in London for corporate occupiers to influence the form of the buildings they occupied: they simply took a 25-year lease on a standard, speculative building. Consequently building design in the UK for much of the post-war period was driven by cost, to formulaic layouts that emphasised regularity and efficiency, and with generic users in mind rather than specific requirements.

In important ways, office building design in the 1970s and early-1980s was little different to the pre-war period. This was particularly the case when the occupation was laid over the base building design. Offices here maintained the cultural tradition by doubling down on cultural features such as class and status, which in practical terms meant layer upon layer of entitlement in terms of enclosed offices, the sizes of those offices, the quality of the carpet and furnishings, space allowances and so on. By way of one example, in the 1970s and early-1980s, many of the large clearing banks and insurance companies had tiers of canteens to which different layers of staff were allowed to go.

As described in Chapter 10, from the late-nineteenth century through to the interwar years, American and European offices had evolved in parallel but different ways. A fundamental difference between the two traditions was scale: American high rise buildings were built to a very large scale, allowing a template form of building to evolve – large, orthogonal, contiguous and deep floorplans, quickly necessitating (or being enabled by) air conditioning, electric lighting, left technology and sanitation systems. European businesses and their buildings operated on an entirely different scale.

On the eve of the digital revolution that would radically change the design and occupation of all offices, it would seem that the separate cultural and rational traditions were being maintained in Europe and North America, respectively, with one important caveat: London. As suggested above, it was here that there emerged a hybrid of the two models, firstly at Broadgate and then across the market. In this context it is ironic that, on the eve of the digital revolution, London should for the first time embrace (albeit reluctantly) the north American tall building. Canary Wharf, conceived, promoted and delivered by north Americans, offered tall buildings with central cores and large simple floorplates composed of space up to 20m (65ft) deep from window to core. The irony lies in the fact that London had resisted tall buildings for close on a century but chose to do so at a time when the economy and technology were starting to drive fundamental change in the nature and form of organisations, and therefore in the manner in which they occupied space.

Post-war office landscapes in Europe and America largely continued with their respective cultural and rational traditions. However, in terms of their internal layout and functionality further distinctions became more evident. Post-war mainland Europe experienced less intensive speculative development, and boasted a tradition of tailor-made buildings, financed by their occupiers. Additionally, in several countries, including Germany, the Netherlands and Scandinavia, Worker Councils exercised their substantial rights to influence the design and quality of their workplaces. This combination of supply and demand factors ensured a European tradition of lower rise, less regular and shallower buildings, with natural ventilation, access to external views and great attention to workplace amenity.

By contrast, through to the Second World War and beyond, the rational office of the US remained 'industrial' in character. Desks and chairs were heavy and lacking any of the ergonomic features familiar today. Most were made of wood. Wooden bookshelves sat awkwardly alongside the steel filing cabinets that had by now become ubiquitous. The layouts were relatively 'fixed' as power outlets helped determine where machines and telephones could be located, and moving furniture was hard work, time consuming and expensive. With their large, regular, column-free floorplates, office interiors resembled factories

> *in their modular assembly of component parts. Desks were carefully laid out in rectilinear patterns. Clerical staff toiled inside large bull pens within the deep space while more senior executives were given, progressively, window access, perimeter offices, larger desks, more domesticated furniture (usually in wood), artworks and private dining rooms as they moved up through the command-and-control ranks.*[19]

It is important to emphasise that the two traditions were not as distinctly different as portrayed here. There were large 'grey' areas, and nowhere was this more the case than in London, which increasingly became a hybrid of the two traditions. More perhaps than any European city, London was influenced by the presence of large numbers of American corporations

The contrasting traditions of the post-war offices in Europe and the USA are captured in two approaches to layout – the cultural Bürolandschaft approach and the rational Action Office system approach. European buildings remained comparatively small by US standards, they were generally inefficient and interior layouts reflected social and cultural traditions. America, by contrast doubled down on its rational approach through the widespread adoption of systems furniture. The two approaches are described below.

Bürolandschaft. Eberhard and Wolfgang Schnelle also worked in the furniture industry, in a company owned by their father in Hamburg, Germany. The brothers were sceptical of the Taylorist scientific management approach to the workplace, with its tendency towards uniformity and serried rows of desks, rigid corridors, production line management and reinforcement of hierarchy. They were far more interested in the flow of information, and were amongst the earliest to recognise that the emerging office economy workplace was not a production line, one enormous 'in-pending-and-out tray', but a complex set of relationships, where interaction was more important than process.[20]

In 1958, the brothers established consulting group Quickborner Team to pursue their unique approach to space planning, which quickly became known as Bürolandschaft, or 'office landscape'. The basic concept was for a more natural, or organic, layout. Desks were arranged according to need and the floor was softly divided by plants and low level screens. The result was open plan, free flowing and conducive to better communications: Bürolandschaft.

Duffy and colleagues summarise three key features of Bürolandschaft. First, the approach addressed managerial and physical arguments in parallel. "*The apparently idiosyncratic layout was not an arbitrary design whim but, so it was argued, an inevitable physical concomitant of a certain managerial style*." Second, the arguments for Bürolandschaft "*had a distinct moral edge. Not only must you have office landscaping but you ought to adopt this managerial style*". Finally, approach complete: "*Once the decision had been made, the techniques for applying office landscaping were already codified (in 68 rules) and ready to be put into effect*".[21]

Bürolandschaft offered a 'third way' to the inefficiency and anonymity of cellular offices, or efficient sterility of the American open plan. With office landscaping, the choice was much richer. Whilst at first glance Bürolandschaft appears chaotic and random, the layout is in fact a carefully considered reflection of the needs of individual teams for interaction with others. There are break-out and informal meeting areas; there is planting and carpets, and there are no private areas. The layout, or rather the approach to layout, had many advantages, including its ultimate flexibility and the speed and economy with which change could be effected.[22]

One of the first Quickborner applications of Bürolandschaft was an office for Bertelsmann, a large German publishing house located in Gütersloh. It was a project for 270 employees that served as a pilot study for a new headquarters with 2,000 employees. The building consisted of a rectangular open floor with removable screens and lightweight furniture as the main structuring elements. Wall-to-wall carpets, ceiling treatment and screens with acoustical surfaces to control the noise. Workplace arrangements were based on the detailed study of communication patterns between the different groups of the organisation. One of the first large-scale projects outside Germany was Volvo's headquarters in Torslanda, Sweden, designed by Lund and Valentin in 1967.

Over the years, the concept became established in Europe (although not in the UK), leading to a good deal of innovation in building services (air flow, heating and so on). As Duffy observed the "*big open plan offices of the 1960s were the direct result of an attempt to rethink the design of the office building from first principles, through a fundamental understanding of office communications and processes*". Moreover they were an attempt to eliminate status and encourage interaction with softer elements such as carpets and decorative plants.[23] By contrast, America adopted a 'cookie cutter' approach with uniform space standards and furnishing and finishes, low-level partitions and cubicles offered privacy while maintaining a general sense of open plan.

The Action Office. Robert Propst (1921–2000) was born in Colorado, where he began his career in the late 1940s as a graphic artist, teacher, and sculptor. In 1953, he formed the Propst Co., a Denver-based firm specialising in speculative product development. His is said to have invented a vertical timber harvester, a quality control system for concrete, an electronic tagging system for livestock and a mobile office for a quadriplegic lawyer. However, it is for his innovation in the world of office furniture that Propst secured his place in history.

In 1958, Propst was enticed out of his part-time role as professor of art at the University of Colorado by Herman Miller president Hugh DePree, to head the firm's new Research Corporation at Ann Arbor in Michigan. Tellingly, Propst's interests were many and varied, but included anthropology and sociology. It was perhaps this unusual angle that gave him the ability to see office work not as a production line. He immediately set out to turn Taylorist thinking around with a vision led by an understanding of office work as a mental activity.

In 1964, Propst co-wrote a brochure, *The Action Office*, with his then boss, Design Director George Nelson.[24] Page 6 of the brochure contains the following quotes.

> *Today's office is a wasteland. It saps vitality, blocks talent, frustrates accomplishment. It is the daily scene of unfulfilled intentions and failed efforts. A place of fantasy and conjecture rather than accomplishment. It fosters physical and mental decline and depresses capacity to perform. It is the equivalent of doing business on clay tablets in an age of the computer and instant communication.*
>
> *The office is a place for dealing in symbols. It rarely traffics directly with the actual objects its company may make, the actual money it negotiates with, or the markets it serves.*
>
> *The office, then, primarily should be a mind-oriented living space.*[25]

By describing the plight of the common office worker thus, Nelson and Propst introduced the insight and aesthetics behind Action Office, including examples of how Action Office promoted health and productivity: by encouraging people to change postures throughout the day; giving them ways to store and display materials; and allowing for adaptation so furnishings can adjust to the ebb and flow of the workday.

Just as his anthropological reading taught him about the relationship between man and his environment, so it informed his thinking about the worker and the office. The system offered workers lightweight, colourful furniture that enhanced both privacy and movement.

Whether Action Office was 'over-designed', not sufficiently flexible, too expensive or simply ahead of its time, it was a commercial failure – sending Propst back to his drawing board. The result of much further thinking, in 1967, was *Action Office II* (AO-II). This time the egalitarian and flexible approach was more successful. The AO-II furniture system comprised two- and three-walled, fabric-covered hinged dividers, built for movement and flexibility. The kit was modular allowing many arrangements; it was standardised and elements were interchangeable. It could be simply and cheaply disassembled and rebuilt elsewhere at short notice in response to business change. It offered a blend of privacy and personalisation with shared space. Screens of varying height doubled as storage units, shelving and display space, whilst also allowing workers vistas across the office floor. Floor-to-ceiling panels with windows and doors could easily partition space and create discrete departmental and functional areas within the office.

The blended system of desk, storage and chair produced a 'workstation' fit for the emerging knowledge worker. Instead of tying the worker to an immovable desk to undertake repetitive and routine tasks on a factory line principle, the new system approach provided for flexibility, agility and differential tasks.

In 1968, Propst published *The Office – A Facility Based on Change*. Reflecting his hybrid academic-practice experience, the paper was part academic paper, part brochure and part operating manual. The paper establishes the concept for AO-II by describing change as a new constant; one that "*the office as an institution will no longer be able to ignore*". He described the "*lack of mobility in our physical facilities as the most stubborn laggard in offices*".[26]

Ironically, the simplicity and flexibility of the AO-II system was also an inherent problem. Because it was so simple and flexible, anyone could move it, and on many office floors, the system was used to create acres of sterile cubicles, as immortalised in Scott Adams' Dilbert cartoons. This angered Propst and he let rip during an interview for *Metropolis* magazine in 1998.

The dark side of this is that not all organizations are intelligent and progressive. Lots are run by crass people who can take the same kind of equipment and create hellholes. They make little bitty cubicles and stuff people in them. Barren, rathole places ... I never had any illusions that this is a perfect world.[27]

Propst died 2 years after the *Metropolis* interview.

While superficially rejecting the factory style of the early corporate office, the dull reality was that the new system furniture (Herman Miller and derivatives) somehow reinforced command and control management in the workplace. Instead of improving internal communications and interaction and enabling the faster and cheaper reconfiguration of space and people, in reality, it allowed offices to be planned at higher occupancy densities, and certainly did nothing to address hierarchy and isolationism; for most, the dull thud of office routine continued daily.[28]

Figure 12.3 illustrates the contrasting styles of Bürolandschaft and the Action Office. Figure 12.3a shows a typical Bürolandschaft office landscape, quite clearly in the cultural tradition of the European office. By contrast, Figure 12.3b shows the more formal layout of low-level partitioning of the corporate office in the United States in the 1950s–1970s, a derivative of the Action Office. Again, this is clearly adhering to the North American rational tradition of office planning.

In an attempt to balance the strengths and weaknesses of the Action Office and Bürolandschaft, the variant 'combi-office' was adopted in some offices. The combi office consists of a mix of enclosed offices and open plan space. The enclosed offices are generally either for a single person or a small group of people. They are typically arranged around a common area that includes break-out space, filing and storage, vending areas and so on.

Despite the different approaches to office layout described here, the fact is that offices changed remarkably little throughout much of the twentieth century. As late as 1993, Frank Duffy and Jack Tanis felt able to observe:

Taylor's ideas changed the face of the workplace 100 years ago and still dominate office design to this day. Compare such leading offices of the turn of the century as Frank Lloyd Wright's Larking Building of 1904 with those of corporate America of the last few years. The physical similarities are striking, the conservatism colossal.

Noting the resemblance between the Larkin Building and a factory, the authors go on

Taylor's ideas of the mechanical, top-down, inhuman, status-rich, invention-poor, alienated workplace live on, manifested in every workstation, ceiling tile and light fitting.[29]

Setting aside the reservations expressed in Chapter 7 about the real versus claimed impact of Frederick Taylor on the office, Duffy and Tanis were correct in principle. Corporate buildings in the US evolved remarkably slowly.

Despite different origins and traditions in office design and layout, both the rational and the cultural office succumbed to the imprint of corporatism. The death of the clerk and the rise of the office worker, the mechanisation of the office and the feminisation of the workforce were all powerful forces, but they failed to radically change the 'white-collar factory' image of the office workplace. Reflecting on his PhD research in the late-1960s, and upon his subsequent research and experience, Frank Duffy made two observations as the office economy was on the cusp of the digital revolution in 1992. First he suggested that organisations would do well to strip away the apparatus of status because "*such distinctions are dysfunctional and irrelevant in a modern*

(a)

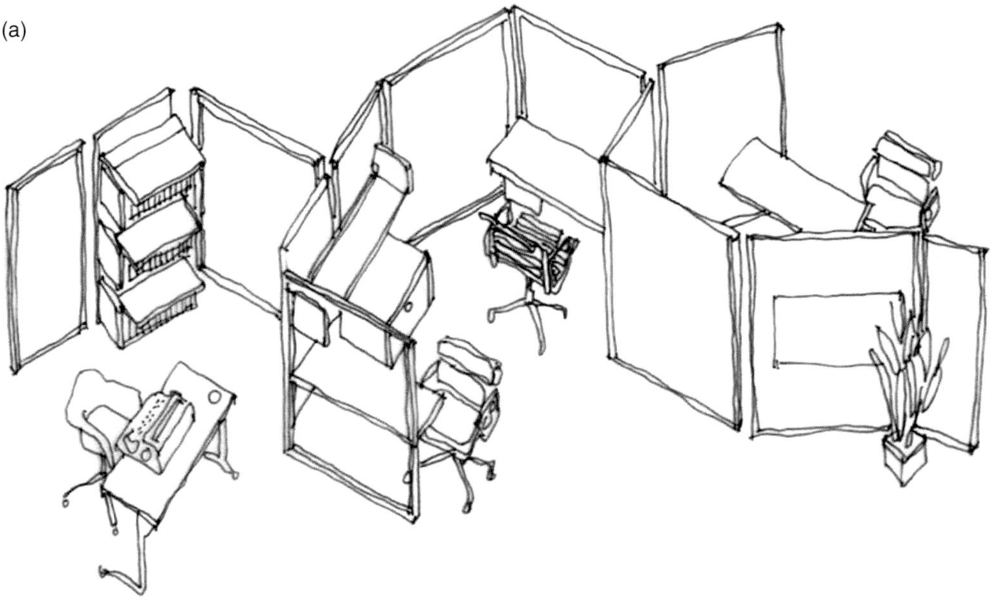

(b)

Figure 12.3 (a) Office landscapes: Bürolandschaft. (b) Office landscapes: Action Office. Images by kind permission of Michael Bedford.

Source: Harris (2021), p284.

climate of industrial democracy". Secondly he argued that, given the "*primary concern of the office is to improve interaction between its members*", then "*partitions and barriers should be torn down to maximise the openness achieved*".[30] Duffy's experience included his input to the Broadgate development in the City of London.

From paper files to digital files

A defining feature of the modern, post-war office was its role in collecting, storing, manipulating and disseminating information. Information was the oil that lubricated the whole corporation. In the same way that companies became more complex – with departmentalisation, hierarchies and multiple service lines – so the information required became more complex. As late as 1975, geographer Peter Daniels defined the office function in terms of five activities: receiving information, recording information, arranging information, giving information and safeguarding assets. Further, he defined the office environment as a place where "*people sit at desks or tables, equipped or supported in their work by paper and pens, typewriters, dictating machines, telephones, telex, calculating machines and more advanced data-processing machines*".[31]

The short period spanning 1945 to 1990 saw the traditional role of the office, as described by Daniels, transform into an altogether different role, that of providing a setting for the exchange of knowledge. This would have fundamental implications for the design and management of the office and for the skills profiles of those working there.

Information management. The single product that linked all of the collecting, storing, manipulating and disseminating functions of the office was paper. Copious amounts of it, everywhere. And there were two critical drivers of this proliferation of paper: first, the ability to store it efficiently and, secondly, the ability to generate more of it – limitless amounts with cheap means of production. It seems at the same time trite and obvious to point out that the revolutionary step to organising information was vertical filing – first widespread in the 1880s.

The filing cabinet's most familiar format is the unit with four pull-out drawers, stacked one on top of another format, wrapped in a metal case. Completely anonymous, understated and uninteresting. The provenance of the filing cabinet is something of a mystery, although its invention is usually traced to 1898 and credited to Edwin G Seibels (1866–1954). Seibels' central insight was that sheets of paper could be kept in large envelopes standing vertically in a drawer, obviating the need for folding and rolling. Not only that, records could be arranged according to subject, contract, customer, supplier or whatever instead of, as was widely practiced, by date of receipt or despatch. Cabinets and boxes could replace the older, bulkier bound ledgers and bundles of paper. Greatly excited by his insight, Seibels convinced the Globe–Wernicke Company of Cincinnati to make five wooden filing boxes to his specifications. Confident in his invention and with an eye to personal wealth, he applied for a patent, only to be brusquely informed that his invention was not a device, merely an idea, and ideas were not patentable!

Very quickly, the filing cabinet found its way into virtually all corporate offices as well as those of smaller businesses, hospitals, academia, indeed, anywhere reliant on the medium of the age – paper. Dozens of manufacturers sprung up to exploit a virtually insatiable market all rather unusually for a product with virtually no product differentiation. No matter how nondescript the filing cabinet was, the 'invention' of vertical storage of paper was a major breakthrough. The filing cabinet was a pioneering aspect of office work, providing a stepping stone towards modernity. Indeed, it signalled the arrival of 'the information office' – fundamentally changing the way in which people and organisations thought about the information they held – its purpose, its format and its management and retrieval.

Craig Robertson produced a fascinating history of the filing cabinet in 2021.[32] He noted that it arrived alongside the skyscraper, that symbol of American capitalism and modernity in the twentieth century. He describes how, as vertical structures, the skyscraper and the filing cabinet stacked discrete horizontal units (floors containing offices; drawers containing folders and papers). Such stacking was to serve two main purposes: to reduce the structure's overall footprint and to facilitate access to the units. The stacking was supposed to encourage order and efficiency. Just as the skyscraper brought order to an organisation's workflow and workforce via stacking and layout, so the filing cabinet would bring order to information.

And in concert, according to Roberston, the emergence of a flat-top desk and the vertical filing cabinet offered a coordinated rethinking of the relationships between paper and work and between storage and retrieval. He breaks the material qualities of the filing cabinet into three important areas: verticality; integrity; and 'cabinet logic'. Verticality is made most explicit in manufacturers' comparisons of filing cabinets and skyscrapers. Verticality was seen to symbolise efficiency and to make use of otherwise empty space. Integrity refers to the strength or robustness of the cabinets; their ability to withstand damage and fire. Finally, 'cabinet logic' was the sense of control and granular access that cabinets gave to their users; the filing cabinet promised to make information visually identifiable at a glance. Robertson suggested that, in partitioning space, companies could outsource thinking and memory *to* the cabinet.

Figure 12.4 Lateral filing cabinets, c1982. Licensed under the Creative Commons Attribution-Share Alike 3.0 Germany license.

By segregating papers, the filing cabinet provided epistemological support to information as a specific unit. This articulation of the archive and efficiency often took the form of claims that a well-organized cabinet created information that was alive, in contrast to mere storage, which was characterized as "dead" or at least dormant.[33]

The filing cabinet above all signified efficiency and modernity. It saved on space, by maximising density in use. It enhanced the visibility and usability of shared information. But it also had another, more subtle impact: it was used to reinforce the gender demarcation of work in the office. Robertson suggests that, as more women entered the workplace, they were channelled towards working with files; while male executives interpreted the information in those files.

Of course, looking back, whilst the early filing cabinets signified the coming of the modern office, and of information management, in later years they also came to signify the old office. Medium and tall cabinets came to be used to divide space; they separated departments, groups, teams. Often this could be related to personality clashes as much as office functionality. They also came to signify the fixity of space. A worker had to go to a cabinet or the place where the cabinets were. The very act of storage acted as an anchor; it fixed people in space.

Copying. Traditional clerks copied documents by hand. Clearly this was time consuming, mind numbingly boring and also prone to errors. The prospect of copying documents without rewriting them was a distant dream. By the early-twentieth century, carbon paper had been around for some time and the state-of-the-art copying machine was the mimeograph, which used ink to produce a small set of copies that got weaker with each duplication. It was imperfect and no long-term solution.

Carbon paper and mimeograph machines were improvements over hand copying, but neither could be used to reproduce documents that already existed. They were also very limiting in the quantity they could reproduce. Photostat machines, which were introduced in the early-1900s made copies photographically on sensitized paper, but were way too expensive for ordinary office use. They were also too big, too slow, and too hard to use.

In the early-1950s, the Minnesota Mining and Manufacturing Company (what became 3M) and Eastman Kodak both introduced office copiers: Thermo-Tax and Verifax, respectively. On the positive side, both were small and inexpensive; but on the negative side, their output was of a poor quality, and the Verifax produced a nasty odour. Neither machine made copies suitable for permanent storage and both required special paper. The first real breakthrough was not to come until the late-1950s. The following precis of this breakthrough is abridged from a fabulous article by David Owen in 1986.[34]

Chester Carlson (1906–1968) was born in Seattle where he grew up in poverty, but managed to find his way through junior college before joining the California Institute of Technology, from which he graduated with a degree in physics. After college he worked for a short time as a researcher at Bell Telephone Laboratories, before arriving in the New York patent office of a small electronics company called PR Mallory & Co. He frequently needed copies of patent specifications and drawings, and there was no really convenient way of getting them at that time. Specifications had to be retyped and then proofread for errors; technical drawings had to be sent out of the office to be copied, which was expensive, on a Photostat machine. Carlson often wished that there was a simpler method and in the late-1930s he started experimenting with a process that he called electrophotography.

Carlson spent the next 5 years refining his invention and seeking a buyer. Following many disappointing conversations, with the likes of IBM, RCA and General Electric, Carlson failed to win anyone over. Finally, in 1944, he managed to convince the Battelle Memorial Institute, a private research foundation in Columbus, Ohio of the merits of his prototype. They agreed to invest $3,000 in development in return for 75% of future royalties.

The project gained further momentum in the following year when photographic paper and related supplies' manufacturer, the Haloid Company, based in Rochester, New York approached Battelle and Carlson. President Joseph C. Wilson and Chief of Research John Dessauer set off for Columbus to see a demonstration. Early in 1947, Haloid agreed to support Battelle's research in exchange for the right to manufacture products based on it. The pieces were falling into place.

Haloid introduced its first xerographic copier in 1949: the Model A was a large, cumbersome machine, and customer feedback was not favourable. A series of other xerographic products followed, and sales were made, but the idea of an automatic, compact and desk-based copier remained elusive.

While Haloid and Carlson were working on the development, they decided that the phrase "electro photography" was too difficult for marketing purposes and consulted a classical language professor at Ohio State University. As a result they changed the name of their process to Xerography, which meant "dry writing" in Greek, and the name of the new machines to Xerox; while photocopying became popularised as Xeroxing. Eventually the first production model, the 914, rolled off the production line in 1959.

Having invested so much, the company took the further, risk-laden decision to pay for expensive television advertising. The launch was a huge success and Haloid was on its journey to becoming the Xerox Corporation. The company's revenues grew from $32m in 1959 to $60m in 1961 to more than $500m in 1966. A $10,000 investment in Haloid stock in 1960 was worth $1m by 1972.

All the way through to the 1980s, paper storage and reproduction was a major space user in offices: copiers similar to the one shown in Figure 12.5 proliferated. Filing cabinets, libraries, information centres, archive rooms and records' offices all proliferated along with post rooms, drawing rooms, photocopying rooms, printing rooms. Paper management could often account for 15% of a total space budget – second only to desk space. All this was to change dramatically with the advent of digital technology.

Digital technology. By far the most iconic technology of the 1980s was the desktop computer. IBM launched theirs in 1981, and the first Apple Mackintosh went on sale in 1984. From the later 1980s on, desktop computers spread rapidly throughout the white-collar workforce, radically changing the manner in which work was undertaken: from pen and paper to screen and keyboard, along with changes to social relations within the office. Secretaries and administrative staff suddenly found themselves increasingly surplus to need, as professional staff began to prepare their own reports and other output.

Similarly, professional workers began to experience the first chinks of daylight between their daily tasks and the oversight of their managers. While we had to wait a few more years for the internet and smart phones, the potential for independent work – away from the desk and away from the workplace – was clearly visible. The impact of personal technology is discussed more fully in Chapter 14.

Business, work and the workplace were co-evolving rapidly in response to rapid developments in technology. And the link between new technology and the changing workplace was highlighted in a number of early research reports.

One of the earliest was prepared by consultancy EOSYS in 1982. The report highlighted the difficulty in establishing in precise detail the full effect of office automation on offices "*because the technology itself is changing so rapidly*", but cautioned that it would be essential for all those concerned with planning office space "*to take note of the accelerated reality of technological*

Figure 12.5 Ineo 650 photocopier. Licensed under the Creative Commons Attribution-Share Alike 3.0 Germany license.

developments". It concluded that "*flexibility and a basic understanding of the requirements of Information Technology will be the keynotes to successful space management in the 1980s*".[35]

In 1983, the Centre for Advanced Land Use Studies (CALUS) published *Property and Information Technology – the Future for the Offices Market*.[36] The report was based on extensive consultation across the real estate industry and with occupier organisations, and set out to demonstrate how technology would call time on the entirely supply-driven approach to real estate.

> *The challenge facing architects, interior designers and space planners is to combine the conflicting requirements of machines and people to create an environment conducive to office activities.*[37]

CALUS found that investors would need to be concerned with providing a basic office shell, which offered occupiers the flexibility to arrange office areas in different layouts. They also argued investors would have to pay increased attention to the quality of their offices because the standard of development was likely to become increasingly dictated by occupiers. They pointed to higher expectations of office organisations who had

> *become more reliant upon a smaller number of more highly skilled individuals who are more concerned with the quality of the working environment. In response there has been a definite*

> *trend in design to 'humanise' the office environment. Technology may be adequately housed in bland boxes, but today's office workers are demanding higher quality designs.*[38]

Also from the same stable, and in the same year, appeared a book by Anthony Salata entitled *Offices Today and Tomorrow*. Referring to the growing numbers of occupiers dealing with 'sophisticated products', Salata rather quaintly observed that they will "*increasingly demand at least a modest degree of sophistication from the buildings that they are to occupy*" and that those "*in the property professions must be mindful of their needs*".[39]

These quotes from CALUS and Salata seem almost trite in today's context; but in the early-1980s there really was virtually no connection between building design – base build or interior – and the organisations who occupied the space. But this was a time when the implications of ICT in the workplace were just emerging. It was thought that workstations would be significantly larger in order to accommodate word processors and the CPU monitors of PCs. Lighting would have to be more sensitive to screen-based working. The spread of machines would lead to a steep growth in power loadings across buildings and a proliferation of power, data and telecoms cabling, which would need to be distributed vertically and horizontally. The same machines would generate a lot of heat that would need extraction. Acoustics would need to address the noise generated by printers, copiers and other equipment. But these were just the beginning of a developing understanding.

The real breakthrough came in a multiclient study undertaken by three firms: DEGW, Building Use Studies and EOSYS. Privately published in 1983, under the authorship of Frank Duffy, the study examined the growing impact of information and communications technology (ICT) on organisational change and the workplace. The report, *Organisations, Buildings and Information Technology,* became universally known as ORBIT.[40]

ORBIT recognised the direct effects of ICT on the workplace including, for example, that technology equipment would require novel forms of servicing, with an obvious impact on design in areas such as heating, cooling, ducting and so on. It also addressed indirect effects, such as how ICT would change organisations by changing the mix of skills required; introducing different forms of decision-making, and changing work patterns and locational imperatives. The report's key conclusions included the following.[41]

- Most technology expected to have the greatest impact already existed.
- ICT was not easy to assimilate into office buildings.
- Demands of technology on new buildings were more stringent than assumed.
- Problems in accommodating computer equipment were not short-term.
- ICT would change the way in which organisations used buildings.
- ICT changes patterns of space use in buildings.
- ICT was changing the rules of office choice.
- Buildings vary greatly in their ability to absorb technology.
- Many existing buildings were in danger of premature obsolescence.

The significance of ORBIT for our story here is that its explicit agenda was to understand the impact of ICT on the nexus of economy, organisation, work and workplace. Technology was changing business by opening up new markets and allowing work to be undertaken in radically different ways; and this in turn was feeding through to a reappraisal of how the workplace should be laid out, serviced and managed. ORBIT was the first evidence to suggest that white-collar factories were in the process of yielding to a fundamentally different approach to office design, management and occupation.

Figure 12.6 IBM 5150 personal computer, 1981. Licensed under the Creative Commons Attribution-Share Alike 3.0 Unported license.

Two years before ORBIT, IBM launched the first personal computer (PC), the 5150 model (Figure 12.6). Within a very short time, PCs were on most workstations, not only forcing a redesign of desks to accommodate the deep cathode ray tube (CAT) monitors but also adding enormous demands on cabling, ducting and power – fuelling the 'arms race' that was beginning to unfold in building specification. Suddenly workers were no longer dependent upon secretaries and typing pools as they began to learn typing skills and they were also less dependent upon data processing units and centralised computing services. Within a very short time, a further decisive step was taken with the launch of the laptop computer.

Figure 12.7 shows IBM's ThinkPad 300 from as early as 1992. As can be seen, it was a good deal 'chunkier' than later versions and, of course. It was yet to be hooked up to the internet and email but, crucially, it symbolised the cutting of the umbilical cord between worker and workstation. This was the birth of the agile worker.

Rational comes to London

Chapter 11 closed with a description of the impact of Big Bang on the City economy and on corporate structures; its impact on the built environment was also seismic, unleashing one of the greatest office development booms in London's history and perhaps the most significant in historic terms since the rebuilding of the City of London in the latter half of the nineteenth century. But the defining feature of the latest evolution of the office workplace was the arrival of

Figure 12.7 IBM ThinkPad 300, 1992. Licensed under the Creative Commons Attribution-Share Alike 4.0 International license.

'rational' buildings in London; in other words, large, deep, regular buildings built more closely to the US template than the traditional European 'cultural' template.

Suddenly it seemed that the whole financial services industry needed rehousing in purpose-built buildings, with dealing floors the size of football pitches. The property industry responded with an orgy of development. Not only were many old buildings demolished and replaced, but whole swathes of land were developed. Perhaps the best known of these was Broadgate at Liverpool Street Station, where railway land, air rights and new construction technologies were all exploited to permit the creation of a whole new district in record time. Other rail stations followed suit, with major developments at Ludgate; Cannon Street and Charing Cross. And, of course, there was Canary Wharf, a largely disused dock in east London.[42] As we saw in Chapter 10, this scheme saw North American towers come to London for the first time. The alien architecture was symbolic of the new financial order; almost a position statement on American economic and cultural hegemony in Europe. Canadian developer Olympia & York delivered regular, efficient, contiguous, deep floorplans and were primarily occupied by large investment banks (for example, Credit Suisse and Morgan Stanley).

The arrival of rational office buildings in London was not entirely coincident with Big Bang; it did in fact begin just a few years earlier. Development company Greycoat, founded in 1976 by Stuart Lipton (later of Stanhope), Ron Spinney (later of Hammerson) and Geoffrey Wilson (later of Equity Land) was highly innovative. Between 1979 and 1983, it delivered the redevelopment of Cutlers Gardens (previously warehouses once occupied by the East India Company), into modern office space on the eastern edge of the City. It then took a major gamble with the development of Victoria Plaza at Victoria Station, drawing heavily on Lipton's knowledge of US construction techniques.

Unconstrained by small plots and irregular street patterns, the building exploited 'air rights' over the tracks outside of Victoria railway station. The building was designed by local architect Elsom Pack & Roberts (EPR). The very term 'Plaza' indicated the American leaning of the building which provided over 200,000 sq ft across large, deep floorplates with what were to become trademark atria. The building was completed in 1986. The American theme was complete when, in an enormous shock for the City of London Corporation, American investment bank Salomon Brothers decided to leave the Square Mile and take the whole building. The move was unprecedented in scale and while it came as a shock to the property industry, it simply reflected the desperate lack of suitable space in the City.

While Victoria Plaza was very American, or rational, in many respects, it is No.1 Finsbury Avenue that can justifiably claim to be the first rational building in London. Delivered more-or-less simultaneously with Victoria Plaza, Finsbury Avenue was developed by Greycoat in joint venture with Godfrey Bradman's Rosehaugh. Peter Foggo of Arup Associates was selected to design the building in 1984 and it was completed in 1986, described by Lipton as

> *1 Finsbury was the first real definitive modern post war building. Large floors, solar shading, heated curtain wall, very rational grid, built to Cat A in 15 months, very disciplined architectural approach, very logical building.*[43]

Finsbury Avenue set a new benchmark in office building design. It was a magnificent addition to the City's office stock, and remains one of the most elegant commercial buildings in London. But its significance is much greater than its aesthetic: it fundamentally changed approaches to the design and development of office buildings by integrating research on occupier needs. It demonstrated that speculative buildings did not have to be cheap, of maximum density and of minimum quality. It was bold, exciting and of a very high quality.

A particular feature of Finsbury Avenue was the design of the structure (Figure 12.8). The building has a steel frame with metal decking as permanent formwork for the concrete floors. The use of structural steelwork with sprayed fire protection allowed extremely rapid construction. The building has bronze coloured facades with horizontal metal grille sun screens and exposed cross-bracing, a stepped section, planted roof terraces and, what became a Lipton trademark, a large atrium with sculpture and large-scale planting.

The finishes demonstrate remarkable craftsmanship and attention to detail with components such as balustrades and light fittings being made of fine metal sections. It is relatively low rise (at eight floors), with large 34,000 sq ft rectangular, 14m deep floor plates, and structural floor heights of 3.5m. More importantly, it was specifically designed so that the internal environment could be repeatedly redesigned without compromising or being constrained by the longer life span structural elements. Finsbury Avenue was listed Grade II in 2015.

Finsbury Avenue was, in fact, a pioneer building; it was the first building to be designed and built for the rapidly growing tech intensive financial services businesses. In some senses, it was a practice run for what happened in 1986. It was the first joint project of Bradman and Lipton, then of Rosehaugh and Greycoat, respectively. Sometimes referred to in the trade as 'Phase One Broadgate', the project was indeed the test bed for the development duo's more famous project next door.

Big Bang research. The impact of Big Bang was pre-empted by research commissioned by developer Rosehaugh Stanhope and undertaken by design practice DEGW. A report, *Accommodating the Changing City*[44] predicted that the internationalisation of financial trade, increasing competition from large US and Japanese integrated merchant banking and securities houses, and proposed changes in the Stock Exchange rules and in the Bank of England (all the result of Big Bang) would in combination have a profound effect on the organisational structure of the City

Figure 12.8 No1 Finsbury Avenue.

institutions.[45] The key conclusions of the report suggested the emergence of profound changes in the demands of occupier businesses:

- shifts in the financial structure of the City being compounded by advances in ICT;
- growth and restructuring resulting in size and servicing constraints of many buildings;
- larger floors, lower occupancy costs and flexibility to cope with change;
- getting the right space becoming more critical than location and image; and
- the conservation policies of the draft City plan forcing companies to consider a location beyond the City boundaries.

A number of specific building requirements also emerged:

- floors of at least 10,000 sq ft (929 sq m) and up to 30,000 sq ft (2,800 sq m), with uninterrupted space;
- structural floor heights of at least 12–13ft, (at least 3.9m);
- finished floor to ceiling heights of 9ft (2.7m);
- highly serviced buildings with adequate vertical ducts, raised floors and zoned lighting, power and air handling; and
- dealing floors of 60ft (18m) depth; very deep raised floors; support space to cope with data/telecommunications needs, and boosted air conditioning.

In 1985, these were radical demands that lay well outside the accepted specifications. Some of the mechanics and findings of the research might seem a little naïve in today's market, but at the time it reflected a wholesale shift, from a position where for three decades buildings had been designed and built to institutional standards in a kind of 'one size fits all' regime, to a position where product innovation through understanding and responding to customer needs became an acceptable means of developing buildings. The importance of the research also lay in the fact that it sought to discover what was happening within the businesses that occupied buildings and worked outwards to the implications for their occupancy needs and, ultimately, to the changing morphology of the city.

The building design agenda unleashed by Big Bang not only created the largest ever development boom in London, but it also set a template for building design for the next three decades and more. And this was not only in the case of London, it applied through Europe and beyond, and it extended from CBDs to the increasingly important out-of-town business park markets to which the technology companies flocked. Building design had made a step change and the architecture of the office was changed fundamentally. But the significance runs deeper than a mere change in form.

The new approach to office design was informed by the businesses and people who occupied the buildings. This was a fundamental leap away from the standard institutional specification of the 1960s and 1970s, which favoured low cost, low specification and low design, to a new era in which funders, developers and designers would explicitly address the changing needs of their customers. There emerged a widespread recognition that occupier businesses were not fixed and immutable, but rather were, particularly under the influence of emerging digital office technology, far more dynamic. They needed office space that allowed change in response to business change; space that could be easily reconfigured and retro-fitted to respond to new business imperatives.

And there were the first glimmerings that individuals within businesses were not all the same, stereotypical office work with similar needs and preferences. It came to be recognised that office interiors would increasingly have to respond to the needs of workers as well as their employing organisations.

The research from this time, and continuing into the 1990s, taught us four critical things about the co-evolution of business, work and the workplace. First, that different parts of buildings had different life cycles, creating a distinct tension between the time horizons of investors (long) and those of occupiers (short). In combination, these building features can evolve over time, allowing buildings to adapt to new uses and occupiers: "*a whole series of layers of materials with varying degrees of longevity*".[46] This crucial insight suggested that each element should be designed interdependently, to allow replacement and upgrade at different stages of the building life cycle. Never has this message been more important than today when we consider zero carbon growth, and the balance that must be struck between new build and reuse.

Secondly that buildings could be objectively compared for their suitability to meet modern occupier needs. Building appraisal is a technique that tests whether the spatial needs of a particular occupier, or an occupier type, can be met by an actual building, or a design for one. Thirdly, there was the impact of technology, particularly its role as an agent of change: technology was fundamentally changing the way in which office work was undertaken in a much shorter period than the mechanisation of the office in the later-nineteenth and early-twentieth centuries. This was eloquently summarised by Duffy:

> *Ways of working are changing radically. Information technology is seeing to that. Based on very new and very different assumptions about the use of time and space, new ways of*

> *working are emerging fast. They are inherently more interactive than old office routines and give people far more control over the timing, the content, the tools and the place of work.*[47]

He wrote of the "*possibilities inherent in information technology to allow staff to enjoy a new freedom in relating home, work, and leisure*".[48]

Finally, there was the relationship between dynamic organisations and the space they occupied, which lay at the core of Frank Duffy's thinking. He understood that organisations themselves were subject to a rapidly changing economy.

> *There will be a need to find dynamic new ways to accommodate ever-changing organisations that continually have to respond to an increasingly unstable and unpredictable business environment.*

And then he was fascinated by the relationship between the individual and the corporation, the constant negotiation between the two, and how this expressed itself in the nature of space.[49] His key insight was that organisations were coming to be managed less as huge corporate machines and more as systems or networks in which work was increasingly about short-term assignments, communication, mobility and connectivity. In such organisations, knowledge becomes the currency of exchange, and traditional approaches to command and control structures break down. He explored the need for 'interaction' and 'autonomy' in work to describe the need for four generic types of work setting.

Big Bang buildings. Developments such as Broadgate and Canary Wharf were a direct response to the demand created by Big Bang. The first Big Bang buildings emerged on land that had previously been Broad Street rail station and the adjacent operational Liverpool Street station in a scheme called Broadgate. The 12ha (29 acre) site lay immediately to the east of 1 Finsbury Avenue, and the developer was Rosehaugh Stanhope (Lipton had left Greycoat after Finsbury Avenue to set up his own development business, Stanhope). The first buildings at Broadgate, including its iconic Circle and Arena were designed by London-based Arup Associates, led by Peter Foggo while the later phases, five to fourteen, were designed by Chicago-based Skidmore, Owings & Merrill.

Phases One and Two Broadgate were large by City of London standards 150,000 sq ft (14,000 sq m) and 310,000 sq ft (28,800 sq m), respectively. Phase One had floorplates of up to 50,000 sq ft (4,600 sq m) – unheard of in a European context. They were the first two buildings of an integrated, masterplanned scheme totalling some four million square feet, across 14 buildings. The two buildings were built in an astounding (for the time) 12 months, reaching shell and core completion in June 1986. Both were pre-let to giant US financial services firms, with Phase One going to Security Pacific and Phase Two to Shearson Lehman.

Big Bang resulted in firms of unprecedented scale and complexity, dealing in UK and foreign equities, gilts, US Treasury bonds, Eurobonds foreign exchange, commodities and financial futures. The most iconic feature of their buildings was the dealing floor – hugely dense, large open plan floors, packed with new technology allowing 24-hour dealing. Some dealing floors were up to 45,000 sq ft (4,200 sq m) – equivalent to a full size international football pitch.

Almost overnight, demands on office buildings were transformed. There were fundamental changes in building form, with the sudden imperative to accommodate enormous amounts of technology hardware – physical space and voids to cope with cables and kit; together with mechanical and electrical systems to cope with high power demands and heat output. Buildings had to be designed with 'deeper' space to accommodate the large trading floors and higher concentrations of workers, along with raised floors and drop ceilings, and generally highly

Figure 12.9 The London Stock Exchange trading floor, early-1980s.

Source: Mary Evans Picture Library.

engineered systems and infrastructure.[50] Figures 12.9 and 12.10 show the contrast between the pre-Big Bang trading floor at the London Stock Exchange and a typical new trading floor, in this case Canadian Imperial Bank of Commerce's dealing floor at the Cottons Building in London Bridge City.

Technological capability and technology adoption proceeded in a kind of space race in which more capability served only to increase the appetite of occupiers for more and more technological capability. As the bulk of occupiers caught up on one innovation so there seemed to be a dozen others to succeed it. The knock-on effect for buildings was profound.

The amount of cabling, the capacity and demands on heating and cooling and power loadings all increased greatly, alongside a proliferation of kit (leading to server rooms, communications cabinets, computer rooms). Suddenly raised floors (sometimes a metre deep), drop ceilings and air conditioning all became *de rigeur*. Building depth, typically 40–50ft (12–15m) previously, grew until 65ft (20m) became quite normal. Such depth could accommodate much larger numbers of workers, as well as the new dealing floors.[51]

There followed an enormous building boom; a scramble to build the new generation of buildings required to meet the needs of the new international financial houses. The architecture of the buildings was entirely novel in London, breaking dramatically with the norm up to this

Figure 12.10 Canadian Imperial Bank of Commerce dealing floor, 1990s. © Historic England Archive/John Laing Photographic Collection/Mary Evans.

point. The buildings were specifically designed to meet the needs of international businesses operating in the global financial services market. This was possibly the first time that speculative development had used research to inform critical design issues such as floor size, planning grid, contiguity and servicing.

But while dealing floors were confined to the international banks, the explosive growth of office technology spread to every sector and it was not long before the traditionally staid environments of lawyers, accountants and consultants were changed forever when the leading firms began to intensify the use of their buildings and adopt desk-top technology throughout. And as this revolution in building design was taking place, the buildings themselves began to assume a less passive role in the lives of their occupiers.

Summary. For four decades following the Second World War, North American experience and European experience in office buildings was very different. While the American economy kicked on after the war, consolidating the growth of corporatism at a rapid rate, Europe was seeking to repair the ravages of another war. Economies were stretched and money was in short supply. In the 1960s, London saw the emergence of 'commercial architecture': out went the grandiosity and ostentation of earlier times; in came a stripped down facadism, a more or less uniform style with minimal product differentiation. The 1960s and 1970s saw some of the most brutal and ugly commercial office buildings ever built in the UK. Design was driven primarily by cost reduction with consequent impacts on the quality and durability of both the façade and the interior.

Workplaces were remarkably stable for an extended period after the war because organisations themselves changed only very slowly. There was little need to move people around buildings with any regularity; once a functional use was established it might remain in the same place for many years. At the same time, the office remained riven by hierarchy and social class division: the nature of the work reinforced hierarchy through a complex web of roles and responsibilities reflecting power, class and position.

While London continued with its cultural tradition of office design – small inefficient buildings with interior layouts that reflected social and cultural traditions – other northern European countries began to experiment with different approaches to the office landscape, which were more occupier focused. America, by contrast, doubled down on its rational approach through the widespread adoption of systems furniture.

As the digital age dawned in the early 1980s, the potential impact on office design and management came to be understood, quickly followed by the deregulation of financial services and the unleashing of an unprecedented development boom. The defining feature of this boom was the arrival of 'rational' buildings in London: large, deep, regular buildings built more closely to the US template than the traditional European 'cultural' template.

Developments such as Broadgate and Canary Wharf were a direct response to the demand created by Big Bang and signalled the arrival of American rationality in European real estate. The buildings challenged the traditional cultural model and replaced it with an identifiably modern, American rational approach. While not ready for skyscrapers, London and other European cities did change dramatically. Rabeneck has referred to the SOM buildings at Broadgate as "*representing a confrontation*" between "*British and US approaches to commercial architecture*",[52] a confrontation in which the latter triumphed. Rabeneck is blunt in his assessment of the contest

> *The cheaply built existing commercial office product, the arcane protocols of 'planning', the primitivism of the contracting community and the conceits of the architects, sat ill with a new clientele of international businesses in a hurry to house large populations in prestige buildings packed with modern technology.*
>
> *The combination of indifference to users and yearning for permanence, shared by British architects and owners alike, was deadly when facing an explosion of demand. Few were prepared.*[53]

The Big Bang, rational, set a building and workplace design agenda that would last for at least four decades.

Notes

1 Marriott O (1967) *The Property Boom* Pan Books London p41.
2 Harris R (2021) *From Clerical Factory to Digital Hub: London's Global Office Economy* Routledge, London p237.
3 Duffy F; Cave C & Worthington J (1977) *Planning Office Space* Architectural Press, London p68.
4 Kynaston D (2002) *The City of London Vol IV A Club No More 1945–2000* Pimlico, London pp138–139.
5 Kynaston D (2002) *The City of London Vol IV A Club No More 1945–2000* Pimlico, London p143.

6 David Kinsella The London Money Market – As Seen from Smith St Aubyn – 40 Years Ago Cited in: Kynaston, *The City of London Vol IV* pp144–146.
7 Thomas A (2024) *The City within the City: Architecture and Change in London's Financial District* The MIT Press, London pp273–274.
8 Saval N (2015) *Cubed: a Secret History of the Workplace* First Anchor Books, New York p143.
9 Saval N (2015) *Cubed: a Secret History of the Workplace* First Anchor Books, New York p144.
10 Wright Mills C (1951) *White Collar: the American Middle Classes* OUP, New York pp205–206.
11 Giuliano VE (1982) The Mechanisation of Office Work *Scientific American* Vol 247 No 3 September 1982 pp148–165.
12 Wright Mills C (1951) *White Collar: the American Middle Classes* OUP, New York p204.
13 Wright Mills C (1951) *White Collar: the American Middle Classes* OUP, New York p205.
14 Wright Mills C (1951) *White Collar: the American Middle Classes* OUP, New York p206.
15 WWright Mills, *White Collar* p209.
16 Wright Mills C (1951) *White Collar: the American Middle Classes* OUP, New York p209.
17 Wright Mills C (1951) *White Collar: the American Middle Classes* OUP, New York p210.
18 Glenn EN & Feldberg RL (1977) Degraded and Deskilled: the Proletarianisation of Clerical Work *Social Problems* Vol 25 No 1 pp52–64.
19 Myerson J & Ross P (2022) *Unworking: the Reinvention of the Modern Office* Reaktion Books, London p26.
20 Harris R (2021) *From Clerical Factory to Digital Hub: London's Global Office Economy* Routledge, London p282.
21 Duffy F; Cave C & Worthington J (1977) *Planning Office Space* Architectural Press, London pp68–69.
22 Duffy F; Cave C & Worthington J (1977) *Planning Office Space* Architectural Press, London p282.
23 Duffy F (1997) *The New Office* Conran Octopus, London p35.
24 Nelson G & Propst R (1962) *The Action Office* Herman Miller Inc, Zeeland, Michigan www.hermanmiller.com/stories/why-magazine/the-office-documented/
25 Propst R (1964) *The Action Office* Unpublished Brochure.
26 Propst R (1968) *The Office: A Facility Based on Change* Herman Miller Inc. Zeeland, Michigan p25.
27 Abraham Y (1998) The Man Behind the Cubicle *Metropolis* February pp76–115.
28 Harris R (2021) *From Clerical Factory to Digital Hub: London's Global Office Economy* Routledge, London, p281.
29 Duffy F & Tanis J (1993) A Vision of the New Workplace *Site Selection and Industrial Development* March/April Vol 162 No 2 pp1–6.
30 Duffy F (1992) *The Changing Workplace* Phaidon Press, London p95.
31 Daniels PW (1975) Office Location: An Urban and Regional Study G Bell & Sons, London pp4–5.
32 Robertson C (2021) *The Filing Cabinet: A Vertical History of Information* University of Minnesota Press.
33 Robertson C (2021) *The Filing Cabinet: A Vertical History of Information* University of Minnesota Press p158.
34 Owen D (1986) *Nobody Thought we Needed Xerography Until we had It: Copies in Seconds* The Atlantic Monthly magazine February 1986 pp64–72.
35 EOSYS (1982) *Office Structure and Design* (Managing Office Automation No9) Unpublished report p40.
36 Plumb C (1983) *Property and Information Technology – the Future for the Office Market* CALUS Research Report, College of Estate Management, Reading.
37 Plumb C (1983) *Property and Information Technology – the Future for the Office Market* CALUS Research Report, College of Estate Management, Reading p39.
38 Plumb C (1983) *Property and Information Technology – the Future for the Office Market* CALUS Research Report, College of Estate Management, Reading p42.
39 Salata A (1983) *Offices Today and Tomorrow* College of Estate Management, Reading p85.
40 Duffy F (1983) *The Orbit Study: Information Technology and Office Design* DEGW/EOSYS, London.
41 Duffy F (1983) *The Orbit Study: Information Technology and Office Design* DEGW/EOSYS, London. pp4–8.

42 Harris R (2021) *From Clerical Factory to Digital Hub: London's Global Office Economy* Routledge, London, p181.
43 Personal communication, 27 June 2024.
44 DEGW (1985) *Accommodating the Changing City* Unpublished report DEGW, London.
45 DEGW (1985) *Accommodating the Changing City* Unpublished report DEGW, London p6.
46 Duffy F (2008) *Work and the City* Black Dog Publishing, London p39.
47 Duffy F (2008) *Work and the City* Black Dog Publishing, London p46.
48 Duffy F (2008) *Work and the City* Black Dog Publishing, London p54.
49 Duffy F & Worthington J (1977) Organisational Design *Journal of Architectural Research* Vol 6 No 1 March 1977 pp4–9 Locke Science Publishing Company, Inc.
50 Harris R (2021) *From Clerical Factory to Digital Hub: London's Global Office Economy* Routledge, London, pp301–302.
51 Harris R (2021) *From Clerical Factory to Digital Hub: London's Global Office Economy* Routledge, London, pp303.
52 Rabeneck A (1990a) Broadgate and the Beaux Arts *Architects' Journal* 24th October 1990 pp37–48.
53 Rabeneck A (1990b) The American Invasion *Architects' Journal* 28th March 1990 pp37–54.

13 Three centuries of change

From clerk to nomad

The preceding chapters have traced nearly 300 years of history in the office economy, through to the millennium. The following chapter looks at work and office life in the twenty-first century, the years covering the transformative impact of the on-going digital revolution; while the final chapter looks ahead and speculates on the future of white-collar work. This chapter pauses the story in order to pull together the key themes emerging from the preceding chapters, summarising around three centuries of white-collar work.

Financial revolution and the arrival of the office business

The insurance companies that emerged from the great conflagration of 1666 were at the very root of the modern office economy. Vestiges of these companies survive to the present day (for example, the 'Sun' in Royal Sun Alliance). They were soon followed by the establishment of the Bank of England in 1694 and the growth of commodity exchanges; while the goldsmiths and merchants turned their attention to money-lending; their tentative steps leading to merchant banks in the City and private banks in the West End serving royalty, aristocracy and the landed classes, principally for deposit and personal loans.

Again, reminders of this time remain today in private banks such as Hoare and Drummonds. In combination, these developments started a financial revolution that was just as critical to the development of the wider economy as was the much more widely understood Industrial Revolution. They formed the crux of the nascent financial services sector.

It was during this time that the 'company', a concept that we take for granted today, was formalised into its modern form. Governments had, for many years, chartered companies to oversee trade links around the world, granting royal charters that gave the companies monopoly rights to trade in specified geographies. These were the first 'joint stock companies', in which individuals exchanged cash (normally) for a share in the business. This then entitled the 'shareholder' to a slice of the profits, proportionate to the size of their holding. Critically, individual shareholders were protected should business profit turn to loss: their liability being limited to the extent of their shareholding (limited liability).

The financial structure of joint stock companies allowed corporations to outlive the interest or life of individual shareholders, and it also created a legal body that shared certain rights and characteristics with individuals. For example, a company or corporation could enter into contracts, own assets, pay taxes and take out loans; it could invest and expand. In these important senses, the modern company was born.

From the beginning, insurance companies were well funded by private capital, often in the form of mutual and joint stock insurers, privately funded by local merchants and manufacturers.

DOI: 10.1201/9781003582205-17

As the market evolved, so the number of 'insurance products' expanded away from property into general insurance; for example, the diversification of shipping insurance and the introduction of life-insurance companies. Banks were not permitted to establish joint stock arrangements until the mid-nineteenth century.

As the eighteenth century drew to a close, the City's exchanges and coffee houses were buzzing, chaotic and crowded places where the exchange of gossip, knowledge and news was just as important as it has been in offices ever since. At the same time, the rise of the insurers generated a growing demand for platoons of black-coated workers – the clerks. The rapid growth in policies led to an enormous demand for writing, copying, bookkeeping and corresponding with an ever-growing customer base, placing physical strains on their premises. Similarly, while the banking houses themselves reflected the growing wealth and social stature of their owners, and perhaps even more importantly their reputations for thrift and probity, they were *banking houses*, they were not bank buildings. Their owners had personal wealth, but their businesses were not sufficiently capitalised to allocate funds to the construction of dedicated buildings.

Increasingly, the insurers and bankers were out-growing the domestic premises they had previously occupied – even allowing for creative solutions for adaptation and expansion. The first purpose-built bank building was the Bank of England, originally built in the 1730s. By 1813, the Bank's army of clerks grew to over 900 (the largest concentration of white-collar workers in the world). More generally, it was not until the brash new joint stock banks emerged in the mid-nineteenth century that it became acceptable within banking to build grand and ostentatious premises, both to house their various operations and to serve as architectural expressions of their wealth, respectability, and reliability.

The early-nineteenth century saw a major European influence on London banking. The turmoil of the Napoleonic and Revolutionary wars in mainland Europe during the latter part of the eighteenth and early-nineteenth centuries led many merchants to flee their home cities for a safe haven in Britain and London, where they could trade freely, often diversifying into banking. Many of the names of these banks echoed through the City for the following two centuries. Baring, Hambro, Kleinwort, Lazard, Rothschild, Schroder and Seligman were among the European arrivals, while others came from the USA, such as Peabody who emigrated from Baltimore in 1831 to found the firm that was to grow and split into JP Morgan and Morgan Grenfell.

Up to this point, virtually every bank occupied a converted house; the buildings were adapted for a mix of business and residence – a mix which had long-lasting impacts on work culture and organisation. The ground floor or 'shop', and basement, provided the customer space and ancillary space, with the upper floors allocated to the owner and his family or a manager, or a combination of these.

A very large proportion of premises had a narrow frontage, with one room at the front and one room at the rear, and extensive depth: overall a narrow, oblong layout. Adding to this layout rudimentary lighting in many premises, together with the domestic furniture and accoutrements, and one can easily garner an image of a quite primitive business environment. The Lombard Street buildings typified a period when banks were small and lacked capital for investment in property on a large scale. Their continuous adaptation through remodelling interiors and adding neighbouring properties was a practical response that allowed them to expand and grow their businesses in a piecemeal fashion.

The Lombard Street tradition continued through the eighteenth century and one such building was that for Child & Co in the West End at Temple Bar. In common with many private banks, Child & Co was conservative and discrete, much like the architecture of the Georgian townhouse it occupied. The larger windows and central doorway flanked by simple pilasters on the ground floor betrayed its commercial function.

Nevertheless, many such institutions were precarious because their small, family owned nature exposed them to difficult economic headwinds, or even just rumours. It was only a matter of time before joint stock status was made available to them. In May 1826, the Banking Co-partnership Act authorised the establishment of shareholder banks, or joint stock banks, in England and Wales (outside London). Joint stock banks represented the separation of ownership and management; as the Act directed that the new banks were to be led by boards of directors, who appointed professional managers. The agency of the manager, operating between owners and workers had been established. It was not until further legislation in 1833 that joint stock banks were allowed to operate in London, and even then for deposit only, not note issue, because the Bank of England retained its monopoly on note issuing within a 65 mile radius of Charing Cross. However, joint stock banks were not universally welcomed; indeed, they were subject to widespread suspicion and contempt.

The first joint stock bank to open in the capital following the legislation of 1833 was the London and Westminster Bank, which opened offices in both the City and West End in March 1834. The growth and progress of joint stock banks was largely inevitable and, by 1867 they outnumbered private banks in London.

Setting a pattern for the newly capitalised enterprises, the joint stock banks used their buildings more than the private bankers ever thought possible to directly amplify their presence in the City. The design and style of the architecture exuded a sense of financial and corporate strength, demonstrating the additional funds that they allocated to creating impressive premises, and conveying to potential customers a sense of solidity. The joint stock banks and insurance companies then indulged in a period of outdoing one another in terms of building ever more grandiose and ostentatious premises for themselves.

Legal and accounting firms were next on the scene and before long the 'financial and business services' sector emerged, with its complex system of supply chains and cross-business relationships. The stage was set for the emergence of the office economy as we understand it today.

Social organisation and work

The owner–clerk relationship was extremely varied. Sometimes feudal, sometimes paternalistic, sometimes formal and sometimes familial. However, there can be no doubt that as organisations grew in size, as the number of clerks in each business grew and as job specialisation grew, owner–clerk relations formalised. Mutual dependence – one for the security of his business; the other for the security of his income – yielded to something closer to the employer–employee relationship we understand today.

It is perhaps difficult to appreciate from the perspective of the twenty-first century that the growth of clerking as a profession represented a huge shift in the structure of the workforce; they not only formed a new economic function, but also in the twentieth century, the core of the emerging middle class. In the pre-Industrial Revolution eighteenth century, there was a gulf between the agricultural and labouring workforce and the gentry and aristocracy; a gulf that carried deep cultural and socio-economic meaning. The emergence of people whose role sat somewhere between these two worlds was in many senses revolutionary. This is important because it shaped so much of what was to follow in terms of the rise of the middle class, the professional class and the office economy.

In their adopted dress, clerks were making a clear sartorial statement that they were not part of the proletariat; that they were socially mobile and that they shared their cultural outlook with the owning class rather than that of the working class. After all, clerks could not only navigate through the clerking hierarchy, but could also rise from clerking into management, or even into ownership. Clerking was about social advancement.

Through to the mid-nineteenth century, there was no concept of a 'manager' class – just clerks to provide the owner with the means to execute their business. The closeness and interdependency of owner and clerk meant that the clerk got to know everything about a business and was often in a position of trust – having access to the firm's money and assets. They were frequently in personal contact all day, most days.

The notion of being 'managers in waiting' was also wrapped up in the Victorian concept of social mobility. In the middle of the nineteenth century, the development of 'Victorian values' led to a strong sense in the value of 'social improvement' – the notion that one could (and even should) progress up the social hierarchy through the application of hard work, diligence and deference. This was nowhere more the case than among the clerking profession.

Despite the potential for social mobility, for many clerks, office work was repetitious, mundane and unhealthy: the work environment was dry, dusty and dingy. Even many of the higher-level tasks, such as accounting and correspondence, were full of routine and monotony. All in all, offices were generally uncomfortable and unhealthy places. Phthisis, otherwise known as consumption, was of course symptomatic of the broader social conditions of the period (sanitation, diet and housing conditions), but its prevalence among clerks was notable. Early American offices were no less unhealthy.

It should also be noted, for balance, that working conditions in offices were almost universally better that in many manual and agricultural roles. The clerk worked fewer hours for his income than the wage-earner and enjoyed paid holidays much earlier than the mass of manual workers. Nevertheless, as offices grew in size and work processes were broken down into discrete parts, roles became more specialised. The 'power' of all-rounder clerks – many of whom had close relationships with their business owners; were trusted and relatively autonomous, and who enjoyed relatively elevated social status – started to wane. Even many of the higher-level tasks, such as accounting and correspondence, were full of routine and monotony.

In time, clerks evolved into lowly valued cogs in bureaucratic mazes, no longer with any overall picture of the businesses for which they toiled. Ultimately they morphed into simply office workers. The growing feminisation of the workforce also helped to further transform work and social relations in offices. A male-dominated managerial class quickly evolved alongside a female workforce that dealt largely with administrative matters – secretarial, typing, data entry, filing and so on.

Many European organisations were structured like the army, built-up from the bottom, comprising ever greater units of command: section, platoon, company, regiment, battalion, brigade, each run by an appropriate rank, with deference and discipline holding the whole together. The development of office organisational structures on this side of the Atlantic was thus influenced far more by culture and social structure than in the US. Here we sought to shoehorn age-old cultural structures into the rapidly changing economic structures and fairly unchanging physical structures: classic European muddling through. Not so in America. The integration of the US economy happened with very little historical baggage: new organisations and new buildings for a new age.

We have already covered many aspects of the cultural influence on the European office. The rise of private banking, catering for the needs of the aristocracy and wealthy merchants, in converted houses where clerks and managers lived 'above the shop', together with the faltering start of joint stock banks that were viewed with much suspicion. Sampson captured the essence of the situation: while the "*earlier social structures – the country estates, the regiment or the cathedral close – were beginning to fade ... the old patterns were re-emerging in new disguises within company headquarters*".

Between 1870 and the First World War, the world of office work was transformed. The role of the clerk became extinct, to be replaced by the manager class and the office worker. The office environment was provided with lighting, heating, ventilation, lifts and modern sanitation; mechanised equipment arrived; the telephone and telegraph revolutionised communications, and the workforce was feminised. America saw steel frame construction and the arrival of the skyscraper. All in little more than four decades.

The growth in office workers during the late-nineteenth century was accompanied by a step change in investment in machinery to improve efficiency and productivity. Innovations such as typewriters, bookkeeping machines and adding machines all spread with remarkable speed, allowing organisations to grow in scale and complexity. Banks were clearing more cheques, insurance companies were processing more policies and sales offices were meeting rising orders – all were struggling to keep up with the growing demands of a society that was growing accustomed to a huge increase in goods and services. Mechanisation not only allowed this growth to happen, but it also came with built-in savings in productivity: for example, an insurance business could double the number of policies it processed without doubling the number of clerks it employed.

Office mechanisation began in earnest in the final quarter of the nineteenth century. By the 1890s a range of mechanical devices were established in the office, notably Morse's telegraph, Bell's telephone, Edison's dictating machine, Remington's typewriter and a host of other gadgets including adding machines, addressographs, copiers and dictating machines. The early decades of the twentieth century saw further mechanical innovations, including the teletypewriter, ticker tape, the electric typewriter, duplicating machines, calculators, tape recorders for dictation, offset printing presses small enough for office use and data-processing equipment operated with punched paper cards.

The machines brought with them mixed fortunes for office workers. In some ways, they relieved the tedium of mundane jobs; on the other hand, they created additional mundane jobs. What cannot be denied is that they created *more* jobs. They allowed businesses to increase the volume of their output, while at the same time, improving quality in terms of error reduction. They also accompanied a dramatically changing office demography, particularly with women entering the workforce.

All manner of writers, concerned in one way or another with social and economic relations, formed a barely contested consensus that the later years of the nineteenth century and the early years of the twentieth witnessed the transformation of clerks from respected, skilled and responsible gentlemen craftsmen who had their eyes across whole businesses on behalf of an appreciative owner, into an under-paid and over-worked proletariat, undertaking routine, mundane and repetitive tasks, having been marginalised by mechanisation, technology and female workers.

The later Victorian period saw one of the greatest upheavals in the history of the office: the arrival of female workers. Up to this point the concept of women (universally 'ladies' in all contemporary literature) in the office workplace was entirely alien: counting houses were the exclusive preserve of men, or rather, gentlemen. So when it did happen there had to occur a massive shift not only in social relations in the office but also in the workplace environment itself.

The sheer scale of the change is clear in the raw data. Females grew from just 2,000 in 1851 to over half a million by 1921. This saw female clerks rise from less than one percent of all female jobs to 10% over the same period. Within 50 years from 1851, a more or less wholly male preserve had been transformed. In the USA, a similar picture emerges. All clerical jobs rose from just 91,000 in 1870, to 1.8m in 1910, growing from under 1% to over 5% of all jobs.

Female jobs grew from just 2,000 to 677,000 over the same period – rising to nearly 36% of all clerical jobs.

The feminisation of office work was driven by a combination of factors, the following among them. First the white-collar economy was growing at such a rate that there were not enough males to fill all the posts available; and the growing female workforce helped avert a shortage of clerical workers. Secondly, women were excluded from professional work by the expense and length of training. And they were excluded from more muscular work due to both physical limitations and social convention. Both meant that office work became a mainstream activity for them.

Thirdly, for their part, females were attracted to clerical work because it paid better than most traditional areas of the economy and it conferred upon them higher status, certainly compared to factory work, retail work and domestic service. Fourthly, even in this role, their chances of promotion to management were slim: as the number and range of managerial roles expanded, so the male workforce filled them, thereby gaining themselves greater work security and prospects of further advancement.

Thus it was that women became associated with the tedium of the office rather than with anything of a more senior nature.

Office work transformed. The economy expanded rapidly in the later-nineteenth century, internationalising and becoming more complex with an ever-growing range of products and services to make, sell, distribute, trade and so on. Inevitably, the businesses undertaking these activities themselves became more complex and, at every stage of the journey adopted new tools, techniques and processes to enable them to meet the new challenges. There was never a particular slice of time when clerks were deskilled. Adding machines [and other primitive office tools] were the start of a long process of mechanisation, which was then followed by electrification and, ultimately, digitalisation. And the same discussions are being had today about the new technologies and their potential to wipe out huge swathes of jobs.

Mechanisation, feminisation and new corporate structures led to increased social stratification in the office, leading to a split between highly qualified employees charged with handling matters demanding judgement, experience and responsibility, and a mass of unskilled employees assigned a series of simple unchanging operations. The previous patriarchal relationship between owner and clerk was broken forever, and the status of the clerk diminished.

Some have argued that the demise of clerking was part of the rise of an oppressed proletariat. However, it is also the case that they enjoyed the privileges of middle class comfort. There is no clear, single answer to this question, because the experiences of clerks were so profoundly varied. No doubt, some were part of an exploited proletariat. But others enjoyed close and positive relationships with business owners. And right through to the end of the nineteenth century, clerking remained an aspirational vocation: one which would offer social and economic advancement.

However, the number of office jobs and roles was expanding rapidly, and while clerking quietly disappeared, the broader notion of 'office work' came to dominate. As 'office work' replaced 'clerking', the office workforce became synonymous with the bulging middle class in most economies.

More efficient processes, mechanisation and the feminisation of the workforce have all been identified as the causes of the demise of the traditional clerk. As we have seen, there has been both some misplaced nostalgia around this account. The office workforce did not become oppressed, under-paid, generally exploited or poorly regarded in society at large. Instead office work came to dominate the workforce, producing countless new roles and opportunities and creating career ladders and social mobility for millions. Of course, along the way, repetitive and mundane jobs came into being, but these were generally superseded as technology evolved.

The period 1870–1945 saw the transformation of the office workplace. Utterly. Completely. Economic transformation and the birth of the modern corporation led to fundamentally new work and new ways of working, and the role of the clerk – especially in the context of his relationship with mainly owner-managers – changed the nature of social relations within the office. As office organisations underwent rapid expansion and increased complexity, so they inevitably introduced hierarchies of responsibility – particularly for male workers. At the same time office mechanisation was allowing businesses to handle huge increases in paperwork. Again, perhaps inevitably, this led to more division of labour, or more specialisation?

The many articles that argued that clerks were simply being deskilled by technology before and after the turn of the century can thus be challenged. More complex business led to more complex organisations led to more complex tools to complete the job – and all this in turn led to higher skilled, more specialised and better paid workers. Mechanisation of the office was a sign of advancement, innovation and modernity; it was an essential pre-requisite for growing business and a growing economy. In turn, the transformation of work necessitated the transformation of the physical spaces in which it took place.

Human filing cases

In the nineteenth century a growing number of banks let out the upper floors of their grandiose buildings in what we would call today 'office suites' – self-contained spaces for smaller firms looking to gain a foothold in the City. And by the 1850s and 1860s, most new office buildings contained a proportion of lettable space. But just as owner-occupier buildings with sub-let space were taking shape another trend was emerging, one that would become the normal pattern of development and occupation through to the current day. This was the development of speculative buildings.

Speculative buildings are designed, funded and developed for leasing in their entirety on the open market, for any occupier. It is speculative because the occupier and the financial return carries a good deal of uncertainty. The first speculative building appeared in the City in 1823, and by the end of the nineteenth century were starting to give the City of London a look and feel more akin to its image today. Quite quickly, the development industry evolved – the coalescing of architect, developer, constructor, funder and advisor. Development companies began to proliferate with the single end of buying up sites and covering them with office space.

Over the course of the later-nineteenth century the City was rebuilt. The level of construction between the 1850s and the 1880s was, in comparative terms, similar to the post-war rebuilding of the 1950s and the 'Big Bang' boom of the 1980s. More particularly, the physical environment was transformed to cope with new, dynamic, expanding businesses. Between 1855 and the turn of the century, around 80% of the City's physical fabric was rebuilt.

Between the 1850s and 1870s, the office building emerged in its modern form: large, single-use structures, dedicated to management and administration. Ever larger clerical workforces gathered daily in these new structures to benefit from cheap and reliable electricity, elevators, lighting, heating and sanitation systems. And in both Europe and America, burgeoning clerical workforces experienced rapidly developing workplace equipment and technologies. Desks, chairs, table lamps, meeting tables and filing cabinets all evolved; adding machines, copying machines and typewriters provided a mechanical dimension; telegraphs and telephones added a sense of the exotic. And then there was stationery: ball point pens, carbon paper, card indexes, envelopes, loose clip files, paper, paperclips, pencils, notebooks and waste baskets – all manner of equipment to transform office work.

More and more banks, insurance companies and related businesses were seeking offices with a prestigious address and a decorative façade. From the mid-1850s to the late-1860s, most joint stock banks moved to new premises that, by comparison to the norm up to this point, were grand and ornamented.

America joins in. America in the 1860s remained a largely agrarian economy and there were very few ventures set up to service national markets – just a few banks and insurance businesses, merchants and shippers. Functional relationships between owners and clerks were similar to those in Europe. Owners controlled everything, issuing orders directly to their clerks; each developed their own style and approach, which could be good news, or not, for the clerk. There was generally a high level of trust, and interpersonal relationships were generally more patriarchal than master–servant in character. However, in the US the owner–clerk relation was less impacted by the class backdrop found in the UK: American clerks saw their role in functional terms rather than as the outcome of social status.

All of this changed fundamentally, and at speed, with the spread of the railroad across the entire continent. The 1880s and 1890s saw the transformation of the US economy: it came to be dominated by large, highly capitalised, complex 'vertically and horizontally integrated' businesses, which gained huge economies of scale by stripping out the add-on costs of intermediaries, brokers, merchants and distributors.

This led to the spread and eventual hegemony of large, integrated industrial corporations that sought control of as much of the production and distribution process as possible in order to impose greater efficiency. In all cases, the result was the same: larger, more complex, centrally managed corporations. From this distance in time it is difficult to appreciate the socio-cultural impact of the emergence of large, faceless organisations, that seemed to operate to an alien set of rules, and which operated behind the impenetrable walls and foyers of largely anonymous office structures.

This was the dawn of the faceless bureaucracy, the corporation. It was here that the template was laid down for hierarchy, departmentalisation, communication systems and internal processes. Various levels of managers were established, clear lines of authority were devised, and formal rules were created to govern the company's operations. The managerial revolution helped to create a 'new' middle class. Unlike the older middle class, which consisted of farmers, shopkeepers and independent professionals, the new middle class was made up of white-collar managers in the corporations.

The twentieth century saw the transformation of offices and office buildings into enormous, single-use structures, dedicated to the handling, processing and management of paper and information. One of the severest critics of the new additions to the urban landscape was American sociologist Lewis Mumford (1895–1990), who described:

> *... files, vaults, places for live storage and dead storage, parade grounds of documents, where the records of business were alphabetically kept, with an eye to the possibility of future exploitation, future reference, future lawsuits, future contracts. This age found its form in a new type of office building: a sort of human filing case, whose occupants spent their days in the circumspect care of paper: numbering, ticketing, assorting, routing, recording, manifolding, filing, to the end that the commodities and services thus controlled could be sold to the profit of the absentee owners of the corporation.*[1]

Born in 1895, Mumford was one of the greatest observers of urbanism that ever lived, his long working life coinciding with the maximal period of growth in the office economy. He was also a great humanist. Leaving college early due to tuberculosis, Mumford spent enormous amounts of time studying, sketching and simply watching New York. What he saw was the dehumanising effect of modern urban life, its stultifying impact on the human soul, ultimately its denuding of human purpose. The term '*human filing cases*' to describe office buildings, and the caustic '*circumspect care of paper*' highlight Mumford's severe misgivings about the concentration of white-collar work in downtown.

Two traditions: cultural and rational. The fact that by the later-nineteenth century European offices had already been evolving for over a century, and that American offices really only got going at this point, led to European and American offices evolving in very different ways. By the turn of the century, the European office – its physical features, corporate structures and the social relations among those who worked in these early modern office buildings, was a world away from the burgeoning office blocks in the US. These two traditions are described here as the cultural office and the rational office, respectively. The model is a helpful tool in highlighting certain key traits, which help explain differences in the design, occupation and management of office workplaces through to the late-twentieth century

The European building was cultural, reflecting the values of the family businesses that occupied the spaces; it was domestic in scale and design; it was inefficient and personal. The US building was rational, built on a large scale, with repetitive design features and standardised components; it was organised internally for efficiency and to maximise output. The distinction between the European cultural and American rational office traditions was magnified during the early years of the twentieth century.

An important point to make is that the American business influence on London was such that Europe was not homogenous in terms of the office economy and London, in particular, became something of a 'cultural-rational hybrid'. Nevertheless, the model is a helpful tool in highlighting certain key traits, which help explain differences in the design, occupation and management of office workplaces. For a start, the model suggests that scientific management (see Chapter 10) was not the single most important factor determining modern office workplace designs and cultures that is generally accepted.

From corporate to digital

The rise of the bureaucratic behemoths in the post-war era was a central feature of the economic reorganisation that saw the coming of age of the office economy. The number of large bureaucratic organisations mushroomed as increasing specialisation in work led to the rapid expansion of workers to manage, supervise, co-ordinate and administer the new routines of colleagues. Workplaces rapidly ossified into rigid (social) hierarchies, graded according to the authority granted to each level to plan and execute their own work while planning and supervising the work of others who did not plan work or make decisions, but delivered assigned work.

Indeed, office workforces resembled military regimentation, with officers, non-commissioned officers and 'other ranks', each respecting of the others' rank and duties. The other ranks were assembled in large platoons, undertaking tiny, abstract, aspects of the overall operation and without sight of the greater plan. They were corralled and directed by NCOs who received daily operational orders from the officer class, some distance away on another, usually higher, floor of the building.

Layer upon layer of management evolved, and companies divided like amoeba into complex departmental structures. This was the age of corporatism, and 'corporate man'. In the US, workers abandoned the 'can-do' frontier individualism of their grandparents and moved instead towards a collectivist ethic. The new generation sacrificed individualism on an altar of corporatism: organisation knows best. In Europe, the post-war years saw the economy transition from predominantly manufacturing to predominantly service. And within this service element, office activity was the central growth pole. Banking and finance, legal and accounting, advertising and media, real estate, information and communications, along with many others, all grew rapidly. This was over and above the growth of office activity in the offices of organisations in education, energy, government, health and manufacturing.

The post-war London office remained very different to its US counterpart. In the three decades following the Second World War, office buildings remained relatively primitive environments. Building design was driven primarily by consideration of how cheaply buildings could be constructed. Developers commissioned architects to design functional buildings to the lowest possible price. The 1950s, 1960s and 1970s saw some of the most brutal and ugly commercial office buildings ever built in the UK, with clear consequences for the quality and durability of both the façade and the interior work environment. There was no meaningful understanding of how the use of the building and its management would interface.

Office furnishings were 'industrial' in character: desks were large and bulky, and made of steel; and chairs were immobile and uncomfortable, with minimum adjustability. Corporate offices were 'grey'. There was very little colour, no art, no planting and very little personalisation of space. Walls, carpets and furniture provided largely undifferentiated space. Little attention was paid to the environmental or ergonomic conditions on the occupied floor.

Workplaces were remarkably stable. Organisations during this time changed only very slowly: headcount might change either through organic growth or merger; but functionally businesses were relatively stable. There was little need to move people around buildings with any regularity; once a functional use was established it might remain in the same place, doing the same work, for many years, even decades. The great majority of the workforce arrived at more or less that same time in the morning, and the reverse happened in the evening.

Post-war office landscapes in Europe and America largely continued with their respective cultural and rational traditions. However, in terms of their internal layout and functionality further distinctions became more evident. For example, post-war mainland Europe experienced less intensive speculative development, and boasted a tradition of tailor-made buildings, financed by their occupiers. Additionally, in several countries, including Germany, the Netherlands and Scandinavia, Works Councils exercised their substantial rights to influence the design and quality of their workplaces. This combination of supply and demand factors ensured a European tradition of lower rise, less regular and shallower buildings, with natural ventilation, access to external views and great attention to workplace amenity.

By contrast, through to the Second World War and beyond, the rational office of the US remained 'industrial' in character. Desks and chairs were heavy and lacking any of the ergonomic features familiar today. Most were made of wood. Heavy wooden bookshelves sat awkwardly alongside the steel filing cabinets that had by now become ubiquitous. The layouts were relatively 'fixed' as power outlets helped determine where machines and telephones could be located, and moving furniture was hard work, time consuming and expensive.

By the 1980s, it was clear that the landscaped office so favoured on mainland Europe had failed to gain traction in the UK. Part of the reason was that the UK supply chain was so much stronger than in Europe. In London, speculative developers and investors held sway and themselves determined the dominant building form (driven by cost efficiency). Many more occupiers in Europe owned and built their own buildings than was the case in London. There were few opportunities in London for corporate occupiers to influence the form of the buildings they occupied: they simply took a 25-year lease on a standard, speculative building. Consequently building design in the UK for much of the post-war period was driven by cost, to formulaic layouts that emphasised regularity and efficiency, and with generic users in mind rather than specific requirements.

In important ways, office building design in the 1970s and early-1980s was little different to the pre-war period. This was particularly the case when the occupation was laid over the base building design. Offices here maintained the cultural tradition by doubling down on cultural features such as class and status, which in practical terms meant layer upon layer of entitlement in term of enclosed offices, the sizes of those offices, the quality of the carpet and furnishings, space allowance, access to canteens and so on. By way of one example, in the 1970s and early-1980s, many of the large clearing banks and insurance companies had tiers of canteens to which different layers of staff were allowed to go.

It is clear that, through to the 1980s, the separate cultural and rational traditions were being maintained in Europe and North America, respectively, with one important caveat. London, influenced strongly as it was by US businesses, had begun to carve out something of an unflattering hybrid of the two models. In London, the influence of the speculative development industry was over-powering of any consumer input and so buildings followed the American model of efficiency and cost reduction. But, unlike the rational model, the UK hybrid grasped hold of its socio-cultural baggage and continued to allocate and manage space to reflect class and status.

In this context it is a little ironic that, on the eve of the digital revolution, London should for the first time embrace (albeit reluctantly) the north American tall building. Canary Wharf, conceived, promoted and delivered by north Americans, offered tall buildings with central cores and large simple floorplates composed of deep space. The irony lies in the fact that London had resisted tall buildings for close on a century but chose to do so at a time when the economy and technology were starting to drive fundamental change in the nature and form of organisations, and therefore in the manner in which they occupied space.

Big Bang', unleashed one of the greatest office development booms in London's history and perhaps the most significant in historic terms since the rebuilding of the City of London in the latter half of the nineteenth century. The late-1980s set a design and development template that was to dominate for at least four decades. But the defining feature of the period was the arrival of 'digital work'.

Digital technologies changed the relationship not only between employer and employee, but also between employee and the workplace. Back offices disappeared as computers replaced administrative staff. Professional staff came to rely less on secretarial support as they increasingly wrote their own reports and managed their own diaries. New technologies based on knowledge and information production and dissemination, were increasingly driving change in the economies of many countries. The proliferation of desk top technology in the form of PCs, together with the quickly ubiquitous take-up of email and the internet combined to redefine the economy, office work, the office worker and the office workplace.

Personal computers and local area networks, quickly followed by WiFi, laptops, smartphones, the internet, email and social media. Office work, office workplaces and office organisations were about to enter a new era. The umbilical cord that connected the individual to a desk was to be broken forever. The 'fixed' nature of office workplaces would change and become more flexible to respond to rapidly changing economic, business and technological developments. The notion of digital nomads emerged – people who conducted their work from wherever they happened to be around the world. While such extreme examples were relatively few in number, the great majority could increasingly exercise choice around their workstyle, blending the office with home and ‘third places’ (hotels, cafes, airport lounges, trains and so on). The implications for office design and management, as well as corporate management human resource departments were profound.

Note

1 Mumford L (1940) *The Culture of Cities* (2nd Edition) Secker and Warburg, London p227.

Part V

1990 onwards

The digital office and beyond

First generation 8GB iPhone (2000s)
https://creativecommons.org/licenses/by-sa/2.0/deed.en

Office Era	Time period	Work & Place Themes	Office Technology
Market	**Part I: 1670-1770**	**Trade**. Goldsmiths as bankers. Central banking. Coffee Houses. Joint Stock Companies. Exchanges. Insurance.	Coal fires. Candles. Copying. Ledgers. Paper. Bills of Exchange. Quill pen. Cash book. Shop counter.
	Part II: 1770-1870	**Organisation.** Stock Exchange. Clearing House. Private and merchant bankers. Joint Stock Banks. Office buildings. Speculative development.	Gas lighting. Copying. Ink Pen. Ledgers. Paper. Stenography. Polygraph. Correspondence book.
Factory	**Part III: 1870-1945**	**Production**. Rise of corporation. Division of labour. Work: fixed, manual, repetitive and mechanical. Female workforce. Rise of professions.	Mechanisation. Adding machine. Electricity. Telegraph. Telephone. Typewriter. Copiers. Elevator. Air conditioning. Fluorescent light.
Corporate	**Part IV: 1945-1990**	**Process**. Globalisation. Scale. Bureaucracy. Command & Control. 'Corporate man'. Work landscapes. Birth of digital office.	Photocopier. Mainframe. System Furniture. Golf Ball Typewriter. PC. MS Office. Mobile phone. Internet. Email.
Digital	**Part V: 2000 onward**	**Knowledge**. Agile work. Data. Power.	Smart phone. The Cloud. Social Media. Augmented Reality. Automation. Artificial Intelligence. Robotics. Virtual Reality.

14 The digital office

During the later-nineteenth century, the world of the clerk shifted from that of the small, simple business, where he and the owner had oversight of the whole enterprise, and where he was the trusted and skilled employee whose reward and esteem placed him above the average manual worker in social terms. His domain had survived a degree of growth in the size of businesses in mid-nineteenth century, and was diminished with the rise of corporate businesses employing large numbers of clerks. But his domain was finally swept away as businesses scaled up, bureaucratised feminised and mechanised as the twentieth century got underway.

During the interwar years and those following 1945, particularly in the USA, large corporations reached previously undreamt of scale and complexity. These were the business behemoths driving America's expanding economy: laying railroads, drilling oil, making cars, casting steel, producing food and selling goods. And then there were those that helped make all these things happen – accountants, advertisers, bankers, lawyers, publishers, realtors and countless others. And all these great bureaucracies coalesced in city centres: the central business district or downtown. Company man was born – the identikit office worker who sold his life and soul to the corporation in return for a comfortable, suburban lifestyle and the gradual climb through the countless layers of corporate seniority. Europe followed the broad trends: concentration and bureaucratisation – but on an altogether smaller scale, and carrying with it much cultural tradition in workplace practices.

The post-war years witnessed a period of relative calm in the office economy as multinational corporations grew and consolidated. As we saw in Chapter 7, the work of Wright Mills, Whyte and Sampson in different ways described this process in great detail, and it is somewhat ironic that, contemporaneous with the first two authors, Peter Drucker was proposing an altogether different approach to management. In his *Practice of Management*, he dismantled the Taylorist notion that planning and doing should be done in isolation – a primary factor in the rise of the demiurge. While recognising Taylor's insight that planning and doing are different tasks, Drucker argued that they are "*separate parts of the same job; they are not separate jobs. There is no work that can be performed effectively unless it contains elements of both*".[1]

Drucker set out a prospectus for an approach to management, which embraced the concepts of decentralised decision-making (managers empowering employees to make decisions); management by objectives (including setting goals and defining outputs), and management with social responsibility. Planning did not need to be carried out by some designated manager; it could be achieved by workers themselves if they were given the necessary information. Drucker was reimagining management 70 years ago in terms that we are just beginning to witness happening today.

Nearly half a century after *Practice of Management*, Drucker was on the case again, this time in the digital era. In the early months of the new millenium, he was musing about the fact

DOI: 10.1201/9781003582205-19

that more and more workers had far greater choice than earlier generations.[2] He was referring specifically to those workers who were coming to be known as knowledge workers, who he predicted would become "*the single biggest factor for competitive advantage in the next 25 years*". Such workers, Drucker argued, are attracted to US companies "*because they are treated as colleagues, not as subordinates*" and that they do not "*believe they are paid to work 9 to 5; they believe they're paid to be effective*".

> *Organisations that understand this – and strip away everything that gets in their knowledge workers' way – will be able to attract, hold, and motivate the best performers.*

Whilst it was ahead of its time somewhat, Drucker's key point lies at the epicentre of the changes in the office economy in the early-twenty-first century. The digital era of the office in the 1990s and 2000s, combined with a new approach to corporate management focused on efficiency and productivity, brought forward the greatest revolution in the office economy since the mechanisation and feminisation of the office at the turn of the previous century. The digital era is also associated with a fundamental shift in the relationship between business, workplace and worker. While up to the 1990s, the office was generally an inert backdrop to the process of work, technology-enabled knowledge work brought about a much sharper focus on the design and management of office space; on the office and wellbeing; on the impact of the office on individual and corporate productivity, and on the role of the office in attracting and retaining expensive workers.

So where does all of this change leave the place of work, the office? In the pre-digital era, businesses and buildings evolved more or less in tandem. The pace of change was slower and the buildings were comparatively simple structures. However, the 1990s exposed this as being a redundant notion: the rate of corporate and technological change now exceeded the rate at which buildings were able to respond. In 1997, Frank Duffy suggested that a yawning gap had opened up between the "*open-ended, fertile managerial aspirations and closed, sterile physical reality of conventional offices*", and proposed that as a result, office design, shaped as it was by "*inherited and outmoded assumptions, must also be re-engineered*".[3]

Then, in 2020, the office entered what appeared to be little short of an existential crisis. Suddenly, overnight, everyone was working from home as the first COVID-19 lockdowns came into force. Attitudes to office work, particularly around the how and where of work, changed across the board. Up to this point, most senior management teams remained wedded to the traditional, corporatist model of management, which involved daily attendance at the office and line-of-sight management. What followed was a period of universal exploration of the possibilities of organising work and workstyles differently.

This penultimate chapter summarises the main features of the office in the digital era, which is centred on the end of the demiurge, of corporate man and of traditional corporate structures; and the rise of the knowledge economy and knowledge work. The final chapter then turns to the future, speculating that the office economy is on the cusp of entering a fifth era – the network office.

Economy

For many years now, observers have been anticipating the arrival of the 'post-industrial' economy. One of the first to frame this in the context of the knowledge economy was Austrian-American Fritz Machlup (1902–1983). Born in Weiner Neustadt, Austria, Machlup studied economics under Ludwig von Mises and Friedrich Hayek at the University of Vienna in the 1920s.

He moved to the United States as a Rockefeller research fellow in 1933, holding successive appointments at Harvard University, the University of Buffalo and John Hopkins University. He became an American citizen in 1940. He published many influential books, of which the most relevant here was *The Production and Distribution of Knowledge in the United States* in 1962. Machlup suggested that as, what he termed the scientific revolution played through, so knowledge would become a critical economic resource.

Services activities generally came to dominate western economies in the 1990s, with up to three-quarters of all jobs being in non-productive sectors; and this was swiftly followed by rapid economic transformation by digital technologies. In 1992, Peter Drucker predicted that the traditional factors of production – land, labour and capital – would become secondary to knowledge.[4] He was describing the ascendancy of the knowledge economy, the knowledge-based company and the knowledge worker. Drucker was correctly interpreting the implications of the propensity within business, across all sectors, to collect, analyse, manipulate and extrapolate information and data. He was also watching the impact of computers and telecommunications technologies, and observing the symbiotic relationship between the outsourcing of non-core activities (e.g. advertising, marketing, payroll and recruitment) and the emergence of specialist firms to fulfil those functions. This, in turn, helped transform demand for offices; but the change in collective understanding was slow for some.

> *The economic, industrial and social landscape is changing, often in ways that are poorly understood and documented. Many of our statistics and frameworks are still based on a world of accounting conventions where the only thing that really matters is machines and physical infrastructure.*[5]

The early stages of the transition to the knowledge economy saw small, tentative steps. Office equipment became leaner and more efficient. Electric typewriters improved the quality of office documents with proportional character spacing to mimic printed documents. International Business Machines (IBM) introduced their Selectric 'golf ball' typewriter, with its interchangeable typing head permitting a variety of fonts in the early-1960s. Similarly, in 1964, Xerox introduced the first commercial version of the modern fax machine, the LDX, for Long Distance Xerography.

USA telephone monopoly Bell System introduced its Bellboy radio paging system in 1962. It was the first commercial system for personal paging. The pocket-sized device used solid-state circuitry, 'beeping' when someone called. The user then found a conventional phone to call the service centre, which informed the user of the caller's message. Motorola launched its Pageboy in 1974. And people soon began to realise the benefits of these devices, to help them to remain in some form of contact whilst out and about. Thus by 1980, there were 3.2m pager users worldwide. And by 1990 there were over 22m in use, just as mobile phones were beginning to appear.

The launch of the IBM PC in 1981 spelt the death knell not only for the large corporate mainframe with its desktop terminal linkages, but also for a range of other office tools, including the typewriter and word processor. IBM was quickly joined by a plethora of competitors in PC production, such as the Commodore 64 and Apple Macintosh (with a 6 MHz processor, 128 kilobytes of RAM and a black-and-white screen with 512 by 342 pixels, all dressed in a beige shell and weighing 7.5kg or 16.5 lbs).

These PCs were primitive and quickly superseded. During the 1980s, Intel released a series of processors, the 80286, the 80386 and the 80486, the last of which had 32-bit processor which had more than a million transistors on a single chip, a clock speed of 25 MHz and a 4-gigabyte memory space. And it was not just under the bonnet improvements. Apple's revolutionary 1984

Macintosh came with a Graphical User Interface (GUI). The screen of the 'Mac' was clear, mimicking paper with its black characters on a white background, and came with the iconic 'mouse' for clicking and dragging. It is difficult to conceive, four decades later, how novel and exciting such developments were. The GUI did not gain universal application, but by the end of the 1980s, PCs were everywhere in the office economy.

During the late-1980s, desktop PCs with their large CPUs and CRT monitors spread rapidly throughout the office, leading to a new generation of purpose-designed workstations, with sufficient depth to cope with the CRT monitor. Suddenly every employee had access to the corporate system. The PC was also a catalyst for a revolution in software, spawning countless new firms developing products for word processing and spreadsheets, as well as applications with the prefix 'computer aided', as in computer aided design, or CAD.

Of course, one of the most iconic innovations of the 1980s was, the mobile phone, beginning with Motorola's DynaTAC 8000X in 1983, complete with a $4,000 price tag. This was followed in 1986 by the 4500X which, together with its battery pack, weighed in at c3.5kg. But innovation was rapid. In 1989, Motorola released the MicroTAC (the flip phone) and in 1996 Nokia released its 8110 (the slider phone); which it followed up in 2000 with the 3310 which gave Nokia market domination, not least due to its introduction of SMS (Short Message Service) a kind of precursor to Tweets.

The first 'smart' phones included Ericsson's R380 in 2000 (the first to use the new Symbian OS) and in the same year, Samsung's Uproar (the first with MP3 music capability). The Palm Treo (2002) could dial from a list of contacts and send emails; while the Sanyo SCP-5300 (2002) was a camera phone in which photos could be viewed on screen. The Blackberry and the iconic Apple iPhone were launched in 2003 and 2007, respectively.

Personal computers were a major step forward, but their potential use remained relatively untapped until the advent of the Internet and the World Wide Web. While email was invented in 1971 by Ray Tomlinson, using the internet's predecessor, Arpanet, the more familiar formats of AOL and Outlook were released in 1993, with free email becoming available in 1996. Email gave companies instant communication, anywhere, allowing people to share information, messages and files in a truly revolutionary manner. It also transformed how employees interacted with clients and colleagues.

In 1990, the internet arrived, quickly followed by the release of MS Windows 3.1 in 1992. The latter came with major user interface improvements, including a True Type font system, thereby introducing desktop publishing capability for the first time. It also came with a new system allowing icons to be dragged and dropped.

This cocktail of technological innovations began to fundamentally question how office work was undertaken. Large, inflexible, hierarchical and slowly moving businesses dominated by clerical, process-based work and with clear demarcations between grades and functions of work dominated economic life through to the late-1980s. But the new technologies were highly disruptive to the stability of corporatism: they were creating the new infrastructure for an economy almost entirely reliant upon digital technology. Within a period of only ten years or so, almost all knowledge workers were equipped with a laptop, a smart phone and a broadband connection into their home. Suddenly, office workers were no longer tied to a desk to get their work done: they could work from a variety of different settings.

From the perspective of 2024, the dawn of digital work and the digital office was, at one and the same time, both a very long time ago and also quite a recent event! The spread of office technology has helped to define a new era, in which technology-enabled work has utterly changed how office workers engage with the office place, with their employers and with their co-workers.

Table 14.1 The old economy and the new economy

Feature	*Old economy*	*New economy*
Economy		
Markets	Stable	Dynamic
Scope of competition	National	Global
Organisational form	Hierarchical, bureaucratic	Networked
Structure	Manufacturing core	Services core
Sources of value	Physical capital	Human and social
Business		
Organisation of production	Mass production	Flexible production
Key drivers of growth	Capital and labour	Innovation & knowledge
Key technology driver	Mechanisation	Digitisation
Competitive advantage	Lower costs	Innovation, quality, speed
Importance of research	Low-moderate	High
Relations to other firms	Go it alone	Alliances, outsourcing
Consumers		
Tastes	Stable	Changing rapidly
Skills	Job-specific	Broad and adaptable
Education needs	Craft or degree, one off	Lifelong learning
Workplace relations	Adversarial	Collaborative
Nature of employment	Stable	Risk and opportunity
Government		
Business relations	Impose regulations	Encourage growth
Regulation	Command and control	Market tools, flexibility
Government services	Nanny state	Enabling state

Source: Reproduced from Coyle & Quah (2002).

While using the 'old economy-new economy' terminology, Diane Coyle and Danny Quah provided a very helpful overview of the transition in the economy from the corporate era to the digital era (Table 14.1). The old economy features stability, hierarchy, mechanisation and regulation, while the new economy features networks, digitisation, collaboration, change and flexibility.[6] The model helps describe the new work landscape, highlighting that 'fixed and linear' has been replaced by 'dynamic and complex'.

The concept of the knowledge economy recognises the fundamental shift from a focus on making goods and investing in physical assets towards an economy dominated instead by investment in intangible assets and activities. As Ian Brinkley argued, the knowledge economy is a story of how

> *new general purpose technologies have combined with intellectual and knowledge assets – the 'intangibles' of research, design, development, creativity, education, science, brand equity and human capital – to transform our economy.*[7]

He also noted that it is "*a universal process*", operating across all sectors, domestic and international, public and private, large corporation and SME.

> *Consequently traditional boundaries between sectors such as manufacturing and services are disappearing and previously unnoticed industries – such as the creative sector – have emerged as major employers, generators of value-added, and exporters.*

The critical point here is the universality of change, which has meant the rise of the office economy in more traditional sectors as well as emerging industries. The shift from investment in physical assets towards intangibles as the basis of economic activity is central to the knowledge economy story as "*firms and organisations rely increasingly on the exploitation of knowledge to secure competitive advantage and better performance*".

It is widely anticipated that most future economic growth will derive from 'things' that have no weight: computer software is perhaps the most obvious example; branding is another of increasing importance. And accounting for the economy, or quantifying it, becomes more problematic, as eloquently described by Danny Quah:

> *International trade becomes not a matter of shipping wine and textiles from one country to the next, but of bouncing bits off satellites. With economic value having no clear points of physical entry and exit, international trade statistics become that much murkier and ambiguous. Keeping track of trade is no longer just counting the bottles and bales that pile up on the loading docks in a port.*[8]

Haskell and Westlake described intangible investment in *Capitalism Without Capital*, noting along the way that intangible investment overtook tangible investment at the time of the global financial crisis in 2008.

> *Investment used to be mostly physical or tangible, that is, in machinery, vehicles and buildings and, in the case of government, in infrastructure. Now, much investment is intangible, that is, in knowledge-related products like software, R&D, design artistic originals, market research, training and new business processes.*[9]

The implications of the shift to a knowledge economy for office (as real estate and as workplaces) is profound. For example, the weightless economy sits uncomfortably with the 'heavy' real estate process. For most of the twentieth century, owner occupation and long leaseholds of offices were common, as buildings provided financial meaning in the form of assets and security; operational meaning in the form of bespoke environments and symbolic meaning in the form of architecture. But as the knowledge economy spread, the meanings of buildings as described above has largely disappeared. Many 'weightless' firms do not benefit from controlling physical assets; their operational needs are generic, and they attach little symbolic meaning to the exterior of their building. Yet, they are looking out to a property supply process that weighs a great deal indeed.

Business

In 1990, quite suddenly, the economy slipped into recession; a long, deep and highly disruptive recession. And this time it was not confined to 'old economy' jobs; suddenly, white-collar work was under attack. The pressure was on to reduce costs, wherever possible. This was a reasonable but narrowly focused reaction that had lasting effects because, while the short-term focus was on cost cutting, the long-term under-current of drivers that had been pushing the restructuring of the office economy continued their work.

The chill winds of change gathered pace and jobs for company man were severely culled. And the whole ethos was underpinned by a plethora of management books, all in various ways proclaiming to have found an angle on the emerging economic landscape: the management guru reached 'popular' status and in some areas, possibly, cult status. New wave management theory and the names of Drucker, Hammer, Handy, Porter, Peters and Senge, became very familiar to anyone with even a passing interest in what was happening to business culture. Each had their particular theses and approaches, but among the many books and articles that emerged, the consistent messages were about downsizing, outsourcing, delayering, reengineering, mobility and virtual work: in short, methods for becoming leaner, more efficient, more flexible and more competitive. Corporate management joined scientific management on the scrapheap of business thinking: new wave management had arrived.

Big business responded in kind. Outsourcing spread rapidly as companies sought to shrink to a newly defined core of activities, relieving themselves of the overhead burdens associated with support activities such as catering, cleaning, building maintenance, security, IT and even accounting and personnel. Demergers multiplied as the fashion for breaking down the lumbering leviathans took hold: BAT, BTR, Courtaulds, Forte, Hanson, ICI, Racal and Thorn EMI all got in on the act. Others, such as AT&T, Eastman Kodak and General Electric, delayered; while yet others underwent full-scale business process re-engineering, including Asea Brown Boveri, AT&T, Rank Xerox and Shell. By the early-1990s, ABB was able to report that it had 215,000 employees world-wide, with a central staff of just 200; 5,000 autonomous business units and only three layers of management.[10] In 1995, Europe's largest company, Shell announced the results of its own re-engineering process, the nub of which was redundancy for 1,200 head office staff and the release of half its famous headquarters on the South Bank of the Thames in London.

There emerged a general consensus that the nature of post-war corporatism was changing, rapidly and profoundly. Suddenly, it began to look like corporatism had reached its zenith and that Balzac's pygmies would begin a full retreat. And one of the most profound shifts to enable this was that from 'corporate technology' to 'individual technology' – mainframe to laptop; switchboard to cell phone; corporate intelligence to knowledge worker. Complex chains of command, not to mention arcane social structures, began to look somewhat antiquated. Similarly, we were told that large, complex and hierarchical organisations would yield to horizontal networks, with highly skilled, group-based knowledge workers replacing clerical, departmental information processors.

Robert Heller picked up the theme, arguing that since "*horizontal business processes are all multi-functional and multi-disciplinary, the manager is being led in similar directions. Cross-functional, synergistic and inter-departmental working is unavoidable now: so are task specific teams*".[11] And *Business Week* argued that horizontal corporations should organise around processes rather than tasks; flatten hierarchies; use teams to manage everything; let customers drive performance, and reward team performance.[12]

To some extent, the early evangelists for the new economy over-sold their case. The new wave management theories of the 1990s exhibited that familiar trait of dramatically over-estimating the short-term impact of innovation, while under-estimating its medium- and long-term impacts. Their proclamations began to look a little fragile as the promised economic results failed to materialise, particularly in the areas of 'downsizing'. Doubts over its efficacy began to surface: for example, one study found that fewer than half the American firms that had downsized between 1990 and 1997 had seen any improvements in quality, profitability or productivity.[13]

Then in a dramatic *volte-face*, one of the leading exponents of downsizing – Stephen Roache from US investment bank Morgan Stanley – declared, "*If you compete by building, you have a future. If you compete by cutting, you don't*".[14] More subtly, it came to be recognised that accountancy practices that recorded assets such as plant and machinery failed to recognise the inherent value of the intellectual capital of the workforce. In other words, the slashing of wage bills, which pushed up earnings and share value in the short-term, actually led to a reduction in shareholder value over the longer-term. Downsizing quickly took on a new persona: dumbsizing.

But despite reservations over the role of cost cutting, the broader message about the changing nature of the organisation in the digital era had longer term relevance. For example, Charles Handy expounded his concept of the 'Shamrock Organisation', an organisational structure with three parts, or leaves. The top shamrock leaf represents the core staff of the organisation, those who are highly trained, fully committed and those who make up the senior management. The second leaf comprises a contractual fringe, including both individual workers and other organisations, who provide the core with specialist services. The third leaf includes a flexible labour force who provide part-time and intermittent resource, who feel sufficiently engaged to ensure a high standard of work.[15]

There also grew a recognition that the digital technologies were opening up opportunities not only of business itself, but as the means of undertaking business. The 1990s recession was the cause of a massive rethink about the nature of the corporation – its scale, structure, complexity and purpose – and the advent of digital technology served to reinforce the new thinking. What we were seeing at the time was the death of management and the emergence of the knowledge businesses that were less reliant on internal hierarchy and more reliant on external networks and relationships. Manuel Castells and the late, great Peter Hall captured the process wonderfully:

> *Major multinational corporations continue to be the strategic nerve centers of the economy, and some of its most innovative actors. What is changing is the organisational form, both for large corporations and for small businesses. Networks are the critical form for the flexible process of production. They include networks between large firms, between small and large firms, between small firms themselves, and within large firms that are decentralising their internal structure [into] quasi-independent units.*[16]

Castells and Hall were describing what is referred to here as the knowledge economy, in which networks of organisations, individuals and specialists replace monolithic corporate structures; where small companies compete and collaborate with large companies; corporate islands become more permeable, specialist works with specialist, and trust and respect replace command and control. In this new world of work, companies and employees become flexible and adaptive and forsake career and job security for and choice and self-determination. Organisations are altogether leaner and more agile, they are outward facing; they are dependent upon team-based and collaborative work in which social skills are as important as technical skills.

Harris described four defining features of the knowledge economy that would, in turn, help shape the nature of workplaces.[17] Figure 14.1 highlights the critical nature of connectivity, changing corporate structures, the priorities of knowledge workers and the reduced importance of the 'corporate island' in favour of a more complex web of supply chain relationships. These four features help knowledge businesses to operate within short-term planning horizons, respond to an ever-changing economic landscape and seek to maximize their flexibility to adapt and gain competitive advantage.

The power of networks, involving collaborative production and multidisciplinary skills, is coming to be realised. As foreseen by Handy, more commoditised and non-core activities are being undertaken by specialists; more work is being undertaken collaboratively, and by small companies. This provides opportunities for individuals and small and nimble firms to use the ever expanding range of cheap and ubiquitous technology "*to go global without being big themselves*".

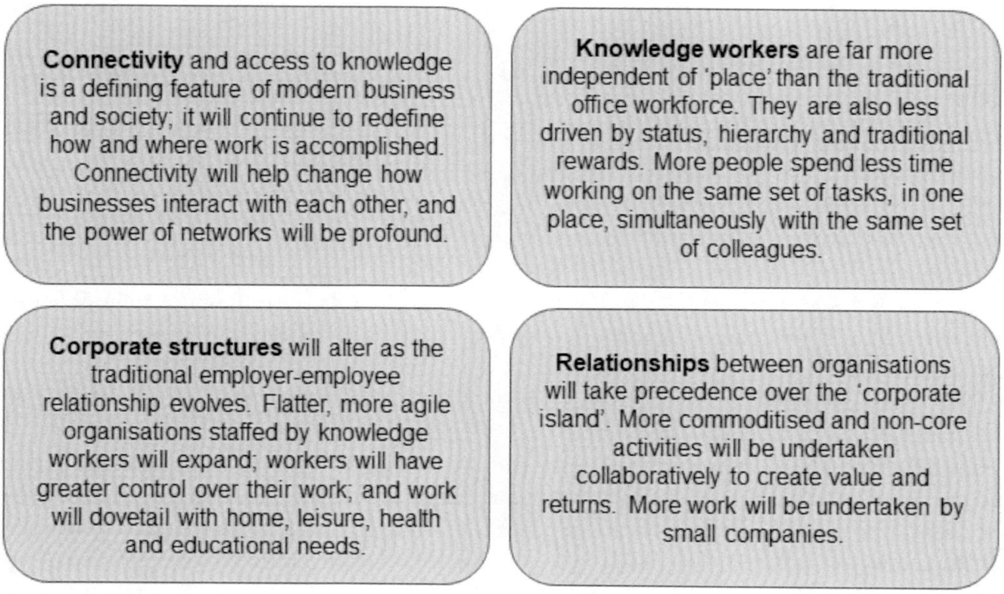

Figure 14.1 The knowledge economy business landscape.

Source: Adapted from Harris (2015).

> *Today they can expand very fast by buying in services as and when they need them. They can incorporate online for a few hundred dollars, raise money from crowdsourcing ... hire programmers from Upwork, rent computer processing power from Amazon, find manufacturers on Alibaba, arrange payments at Square, and immediately set about conquering the world.*[18]

Thus far, the switch to knowledge business has been described in largely benign terms. But, of course, there were far-reaching implications, which are still playing through today. The combined impact of the 1990s' recession and the emerging technologies was a more-or-less complete overhaul of the company–worker relationship. For example, whereas previously workers tended to receive security, benefits and a regular salary 'for life', and employers had a stable workforce in which they could make long-term investment, in the knowledge business this changed in a fundamental way. The increasing adaptability and flexibility of organisations led to the rapid growth of a 'contingent' workforce; an increasing proportion of the workforce who are not directly employed; they are consultants, interims, contractors, part-timers and supply chain partners.

As more and more non-core functions such as payroll, marketing, accounting, technology and others were outsourced to third party suppliers; as corporations 'delayered' and 'rightsized', and as technology replaced more routine jobs, the unwritten contract of loyalty between corporation and company, one that had lasted for two centuries, disappeared seemingly overnight. Just as mechanisation disrupted the world of the clerk, so digital technology disrupted the world of company man: suddenly he was as dead as the trusted clerk. As changes in the economy and work played through during the 1990s, particularly with the growth and spread of the knowledge economy, so office environments and working styles also began to adapt.

Most obviously a breakdown in traditional corporate structures and emerging workstyles mean different approaches to office design and management.

Work

A recurring theme through this book has been the co-evolution of work and workplace. As the nature of business and work has changed, so the physical environment in which it occurs has changed. Sometimes (such as the late-nineteenth century), business change is rapid and profound, and it takes a little time for buildings to catch-up. Other times (such as the post-War corporatist era), business change is slow, and the workplace can keep up. In both cases, as businesses respond to changes in their markets, they require adjustments to the form of their workplace; similarly as new innovations occur, so work processes and workplace layouts respond in kind.

This relationship between work and setting took on a new significance with the emerging knowledge economy because, for the first time, people were no longer tied to a single desk, or even a single building. Their work no longer (or at least rarely) required dedicated machinery; and the notion of working with the same small number of people each day began to break down. The 'fixed workplace' morphed into the '*new ways of working*', implying a more dynamic relationship between work and workplace.

One of the earliest pioneers in understanding the relationship between organisational structure, work activity and work setting was Frank Duffy, in his 1960s' doctoral research at Princeton. He later pioneered the field of 'space planning' and 'facilities management' in Europe, while co-founding innovative design practice DEGW in 1977. Duffy's thesis was that organisational types (for example, research, headquarters, design, clerical) could be mapped onto a grid showing on the one hand, features such as bureaucracy and differentiation and, on the other, requirements for space (for example, interaction and subdivision).[19] Duffy understood that an organisation and an individual will have different perspectives on what is a 'good' workplace, and that the design imperative is to provide settings, which accommodate a balance between the corporate and individual perspectives. Duffy was one of the earliest to link organisational ecology with physical form.

Impact of office technology on work. Reference was made in the previous chapter to Duffy's seminal study on the emerging impact of technology on organisational change and the workplace, the ORBIT study of 1983 (see From paper files to digital files, in Chapter 12). Duffy later reflected on this study, stating it gave him the insight that information technology would have far-reaching implications in terms of the need to accommodate the expanding aspirations of new kinds of office workers – more highly paid and more demanding – not just for a better quality of life in the office, but for more personal control over their own environments. This, he argued, was a sociological issue also.

> *The demography of the office would change from the old hierarchical pyramids in which a handful of self-important managers controlled a vast subservient labour force, to more complex organisational structures weighted, if anything toward a preponderance of professional and senior managerial staff.*[20]

A follow-up study, ORBIT 2, was carried out in North America, where Duffy teamed up with Gerald Davis, Franklin Becker and William Sims. The study argued that a principal impact of information technology (IT) was that all organisations would tend to move, at varying rates and in differing degrees, away from routinised working practices towards less predictable patterns

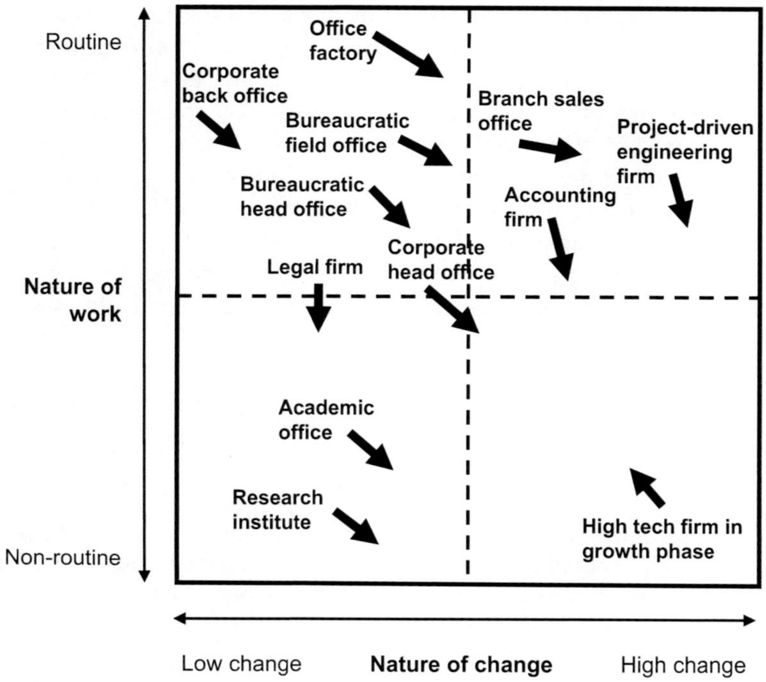

Figure 14.2 The impact of information technology on office work.

Source: Adapted from Davis, Becker, Duffy & Sims (1985).

of work, as illustrated in Figure 14.2. The study also predicted that, sooner or later, IT would accelerate organisational change in all organisations.[21]

Increasingly work was seen to be about enabling people to interact and collaborate, underlying the wider macro-economic shift from process labour to knowledge work. In doing so, the office was expected to have to provide a much richer palette of settings in which individuals and groups could work in a far more dynamic fashion compared with much of the sedentary work of the past. The technology would have to cope with a multitasking, mobile and itinerant work force.

Between the two ORBIT studies, a paper appeared in the *Harvard Business Review* that was particularly prescient in its claims for the unfolding impact of technology on the nature of work. Titled *Your Office is Where You Are,* the paper was written by two academics, Philip Stone and Robert Luchetti[22] from Harvard's Faculty of Arts and Sciences. The paper sought to

> *challenge the customary ways of thinking about offices and show how managers can gain the advantages and avoid the disadvantages of the new technologies. Managers can integrate physical layout, design, and communications to support organizational objectives that:*
>
> - *Emphasize informal exchange.*
> - *Reassign people to different work teams and study groups.*
> - *Provide many employees access to specialized equipment.*
> - *Value individual initiative and mobility.*
> - *Derive payoffs from serendipity.*

- *Attract talented employees. Increase productivity while reducing office costs.*

The authors went on to suggest that "*Any coherent rethinking of an office plan requires that management integrate facilities, communications, and computers in accord with company objectives*". The article was, possibly, the first formulation of the knowledge workplace.

The notion that 'your office is where you are' encapsulated as well as any phrase the essence of new ways of working: the knowledge worker was no longer shackled to a single desk, but released to work in a variety of settings, both within the office and beyond – at home, clients' premises, coffee shops, in transit via rail and air nodes, and so on.

As the impact of technology on the workplace became ever clearer in the 1990s in terms of supporting workstyles that were no longer tied to a specific space, so it became apparent that the workplace would become critical in maintaining business performance. Again Duffy was quick to capture the process:

> *The office has become, in effect, an adjunct to the computer and, as a result, the importance of the building as a means of achieving organisational success has been enormously enhanced.*[23]

This point cannot be over-emphasised, and deserves amplification here. The interplay between digital technology and changing corporate structures in the 1990s completely and utterly altered the relationship between occupier business and the physical environment. Up to this point, the office workplace was seen by senior management simply as a leaden, inert, inflexible and costly backdrop to doing business. But it was now coming to be recognised as a strategic resource that could be actively managed to improve the productivity and wellbeing of the valuable and costly assets which it housed: the knowledge workers.

The growing technology-enabled new ways of working brought about a much sharper focus on the design and management of the office workplace, on the office and wellbeing, on the impact of the office on individual and corporate productivity, and on the role of the office in attracting and retaining expensive workers. Indeed it led to the emergence of 'facility management' or 'corporate real estate management', as distinct functions in the property process, with a greater emphasis on the delivery of support services to enable new organisational structures and new patterns of work. Real estate managers began to explore ways of using technology to support the more dynamic occupation of property (see below); they started to experiment with new workstyles and the aim of introducing changes to the way in which workers would engage with employers and colleagues, by becoming more mobile and thereby reducing the amount of space that was required.

In short, businesses began to realise that organisational performance was linked to individual performance and that individuals were changing in terms of their priorities and demands on the workplace. As well as responding to the leaner times with cost cutting and efficiency measures and by using technology to support the more dynamic occupation of property, managements also started to experiment with new workstyles with the aim introducing changes to the way in which workers would engage with employers and colleagues, by helping them to become both more mobile and more productive.

New landscapes for work. New ways of working meant that the office was increasingly becoming less a place to go to work in a largely sedentary manner on a set of prescribed tasks, and more a place to visit and interact with colleagues and use support services. Frances Cairncross argued that the "*office will become a place for the social aspects of work, such as celebrating, networking, lunching and gossiping*"[24]. And she was right. The ubiquitous impact of mobile phones, laptops, the internet and email presaged an era in which work itself has been transformed, conducted in ways entirely different to even the recent past.

Working from home, hotels, trains and indeed just about anywhere was increasingly possible. Not only had the technologies released workers from the 'where' issue, but they had also been released from the 'when' by introducing the 'anytime' option. Work, as many commentators noted, was quickly being defined as an activity rather than a place. To cope with this shift in emphasis, office environments would have to be designed that enable increasingly complex relationships to flourish. As a result, the term 'new ways of working' came to be replaced by 'activity-based working' (to stress the different work settings in the office) and 'agile working' (to stress the detachment of work from a single place).

In 1995, Dutch consultant Eric Veldhoen was appointed by insurance group Interpolis to implement new working styles at its offices in Amsterdam. Veldhoen, who published *The Demise of the Office* in 1995[25] is usually accredited with creating the term activity-based working (ABW). His book argued that information and communications technology would make the office redundant.

Originally Interpolis was planning to co-locate four different buildings into a single, twin-tower building. Veldhoen + Company advised their client that such a solution would immediately create two cultures within the organisation. Instead, they proposed an activity-based approach that would reduce demand for space sufficiently to occupy a single tower. The work began with detailed consultation to uncover corporate priorities, high among which was retention of customers and employees. This crystallised into a design approach that encouraged a workplace based on trust and transparency, or *Helder Werken*.

Veldhoen's basic approach was to create a palette of work (activity) settings with different design and layout features across the building, including private offices, partially open spaces and fully open spaces. Dedicated personal desks were dispensed with completely as c2,500 employees occupied only c1,500 workstations. Each worker was provided with a locker in their 'home zone', their team area where they were based in the office. They were also given mobile phones (radical for the time) and allowed to work from home as they felt the need to do so.

The underlying Veldhoen philosophy was that ABW allowed workers to structure their days, using a variety of different work settings according to need, thereby leading to productivity and wellbeing benefits. It is worth emphasising that the new Interpolis workplace preceded mobile phone, laptops and email.

In 1998, DEGW published a study looking at the implications of knowledge work for modern working practices for the specification and performance of building form and environmental systems. The research led to the development of a model for "design logic in the new office", comprising four alternative office layout typologies (Figure 14.3) to express both the physical and behavioural characteristics of the different kinds of work they supported environments – den, club, hive and cell.[26]

The den environment supported group work where teamwork was high and autonomy low. The club environment supported creative and knowledge-based work where concentration was important and where interaction and autonomy were important. The hive supported process work where both interaction and autonomy were low. And the cell described individual work where interaction was low and autonomy high. This categorisation has largely stood the test of time, and remains a useful way of gaining a rapid understanding of how an organisation's work processes translate into types of workspace.

DEGW had captured the essence of technology-enabled agile working, as outlined by Stone and Luchetti, and provided an intellectual underpinning for those organisations seeking to

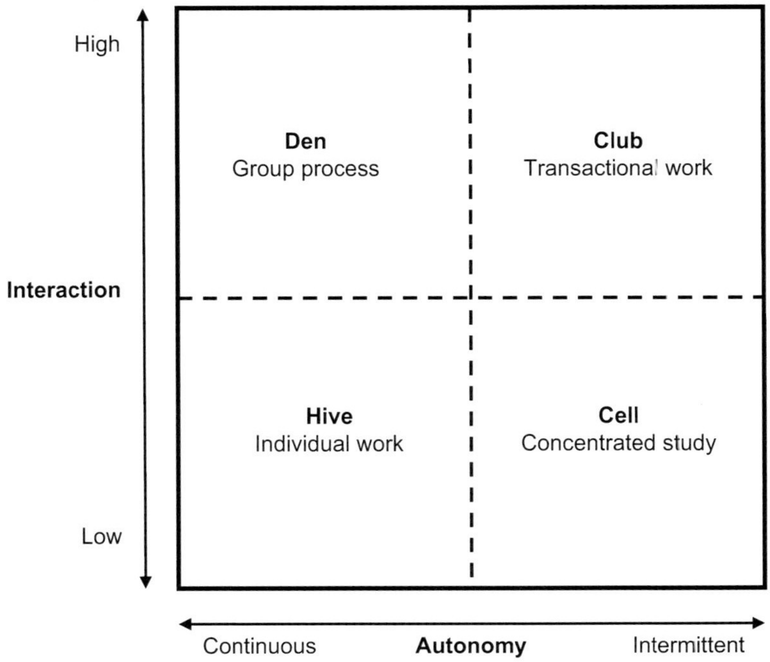

Figure 14.3 Workstyles and work settings.

Source: Reproduced from Laing, Duffy, Jaunzens & Willis (1998).

implement their own, specific projects. As the twenty-first century dawned, organisations were coming to be managed less as huge corporate machines and more as systems or networks in which work was increasingly about short-term assignments, communication, mobility and connectivity. In such organisations, knowledge became the currency of exchange, and traditional approaches to command and control structures broke down.

Increasingly the workplace was coming to be seen to be about enabling people to interact and collaborate, underlying the wider macro-economic shift from process labour to knowledge work. In doing so, the office was expected to provide a much richer palette of settings in which individuals and groups could work in a far more dynamic fashion compared with much of the sedentary work of the past. Such thinking was a world away from the command and control systems that had been so dominant for so long. It was holistic, it was purposeful and it recognised technology as an enabler, not a driver.

As agile working spread, a new language in office design emerged. Design responded to the potential of emerging office technologies and the fashion in management texts with a new language in office design. Enclosed offices, seen to reinforce hierarchies and individuals, as well as to consume space, were replaced by areas for group interaction, co-operation and innovation: *hot desking, hoteling, virtual offices, touchdown areas, lagoons, oases* and *docking stations* proliferated. And new types of workers were identified: *nomads, guests, teleworkers* and *core/periphery workers*, among countless others. Some reception areas became 'concierge' services where desks could be booked on an as-needed basis, and where personal belongings could be collected for the duration of the stay in the office.

Dynamic workplaces. In 1984, Scandinavian Airlines, SAS, held an architectural competition for the design of a new headquarters building for its 2,000 staff, having secured land in Frösundavik, just north of Stockholm. The competition was won by Norwegian architect Niels Torp, who then delivered the building between 1985 and 1988. What emerged on the site was little short of revolutionary (Figure 14.4). Rather than a single, regular, square-shaped building, Torp delivered a 590,000 sq ft (50,000 sq m) complex of seven separate building blocks with a street, covered by a glass roof, connecting the corridors.

The very concept of the SAS building was the antithesis to the oppressive corporate buildings of even the recent past. It was conceived more as a community rather than a single function workplace. The main office areas were bifurcated with a central boulevard along which were sited shops, cafes and restaurants, as well as various services, including a medical centre and leisure in the form of swimming pool and gym. There was plentiful planting and water features. Staff were encouraged to leave their desks periodically.

As well as the architect, SAS CEO Jan Carlzon was heavily invested in the project as he sought to use the new building to bring about organisational and cultural change. He was for example instrumental in creating the central street to encourage interaction amongst staff and to create a shared culture. The building is 'democratic' in its treatment of space.

The floorplates of the individual, low-rise buildings and the spacious central street defied normal property market conventions in terms of space efficiency and construction costs; definitely more cultural than rational.

Figure 14.4 SAS headquarters, Frösundavik, Stockholm.

The building was deemed to be so successful that Torp (this time supported by London practice RHWL) was subsequently commissioned by British Airways (BA) to design its new headquarters, Waterside, at Harmondsworth, just north west of Heathrow Airport. BA was another organisation that was seeking to re-invent itself for a new age. Under the leadership of Bob Ayling, the business saw the opportunity of using its office to assist the process. Indeed part of the aim with the new office was to reflect the airline's new inclusive 'citizen of the world' corporate identity.

Construction started in September 1995, and the building was progressively opened between December 1997 and May 1998. Waterside comprised six, four-storey, limestone-clad buildings. They were integrated by a central, enclosed 'street', 575ft (175m) long. The street included cafes, bank, supermarket, health centre, travel centre, newsagent and so on.

The project involved the consolidation of over 3,000 people from a large number of disparate sites. The 100,000 sq ft (9,000 sq m) office was laid out in open plan, with no assigned personal space, the aim being to create a community based on openness and team working. While departments coalesced in specific areas, within these areas there was hot desking and staff were encouraged to make use of shared spaces. The design unashamedly adopted and adapted the Scandinavian model for a London project.

The external design of the project was one thing that distinguished the project, but it was the internal environment that really set the project apart. Bob Ayling clearly set a brief for a building that would be a catalyst for organisational and cultural change for the airline, transforming the way it conducted its business and how workers interacted with each other. He wanted to ensure a smoother more efficient flow of information and to encourage collaboration and the exchanging of ideas in less formal environments. Staff were encouraged to undergo training in agile working, including minimising the use of paper.

Ten years apart, both the SAS and BA headquarters delivered tailor-made offices for their own occupation; both had strong patrons, in their chief executives. The two schemes shared a humanistic architect and a design philosophy, which had two key related elements. The first was that the new office buildings should help bring about organisational change by allowing the organisations to literally organise themselves around their preferred business model and to introduce cultural changes. Secondly, both projects involved the adoption of what were then called new ways of working. People were encouraged to use the space in a far more dynamic manner – sharing, collaborating, meeting and socialising. The two buildings were among the earliest expressions of how knowledge businesses would accommodate knowledge work in more dynamic workplaces.

With their design philosophy, the buildings rejected the north American rational approach (large, regular deep floors) and produced a twentieth century version of the cultural office. It is also worth noting that the two buildings continued the heritage of Centraal Beheer (1972) and ING Amsterdam (1978) – both discussed in Chapter 12 – with their emphasis on people and creating community rather than a monochrome office environment.

Agile working: time and place, fixed and fluid. In some cases, approaches to agile working were taken up with a kind of missionary zeal by managements who forgot the needs of individuals in their search for balance sheet savings. Managers saw hot desking and telecommuting as panaceas for space and cost saving. But they ignored the significant cultural and social difficulties that people would experience. Some offices took on a closer resemblance to production lines: clinical, depersonalised spaces, where family photos were banned for their association with personal space, and the trappings of office life such as stationery, files and 'bits and pieces' were pulled around the office in portable pedestals to the pre-booked desk, before being returned on leaving the office. Personal filing was commonly limited to one linear metre, where before several metres might have been the norm. Instead of liberating workers and allowing them to structure a work-style that suited their needs, for some agile working often led to a feeling in which the lack of personal space created a sense of not belonging, of insecurity and a feeling of being a production unit rather than a valued member of staff.

For many, the positive aspects of home working (such as more control over hours worked, less travelling time and more family time) were also outweighed by the practical disadvantages of structuring output, getting management input, missing opportunities for training, not feeling part of the corporate culture and a sense that not being seen in the office could be detrimental to career opportunities. Such issues underscored the importance of treating agile working initiatives as part of a broader change management process rather than simply a property initiative to reduce costs.

Of course the whole agile working debate and practice was given a turbo boost following the onset of the COVID-19 pandemic in 2020, with its enforced 'lockdowns'. Early on in the process, the term 'hybrid' work gained common currency well beyond the arcane world of workplace professionals. Mainstream media adopted the easy-on-the-eye term, and it has now become *the* term to describe post-pandemic working.

The media need iconic handles to use as a form of shorthand in their printed and broadcast pieces; and superficially, most people know that the term hybrid means: a work combo of office and home (or nearby). But the fact is that it introduces a major problem because there is an implication that we are moving from one 'old workstyle' to another 'new workstyle'. No variation, no nuance, no diversity, no complexity. But the fact is that there is complexity. We are not moving from one dominant model to another. As we have seen here, new ways of working, activity-based working and agile working all, in their specific ways, preceded hybrid working. There is context, there is continuity, there is evolution. We are adding to the richness of the work experience, not replacing one with another.

Figure 14.5 summarises the main forms agile working as practiced to around 2020. The figure offers a model based on both time and place; it offers a continuum of choice in both, from fixed to fluid, rather than a binary choice; and it allows for 'traditional', or fixed office work (because no matter what some of the more evangelical commentators say, there will remain a role for this type of work and worker).

The model suggests that 'hybrid' is simply one style of work that is added to 'fixed', 'flexible', and 'activity' working; and all are captured under the generic heading of 'agile', including fixed as one of four extremes.

The key aspect to the model is any organisation or workplace might accommodate all four workstyles. Equally, there might be just one workstyle. But the fact is that there is choice – dependent upon corporate and individual priorities.

Fixed working is self-evidently the previously almost universal approach in which everyone turned up at the office at more or less the same time, and left at more or less the same time, 5 days a week, for roughly 8 hours per day. There are some people and functions for whom this remains either a necessity or a preference.

Flexible working is a term that gained most widespread usage in the public sector. It rarely questioned the issue of location, which was assumed to be 'the place of work' as defined in the employment contract. The 'flexible' aspect referred to time, and was granted to part-timers, job sharers, those with caring responsibilities and so on. Today, it simply means the 'when' rather than the 'where'.

As the computer left the building and mobile technology became ubiquitous, so workplaces became 'activity-based' with a tapestry of settings to accommodate different work

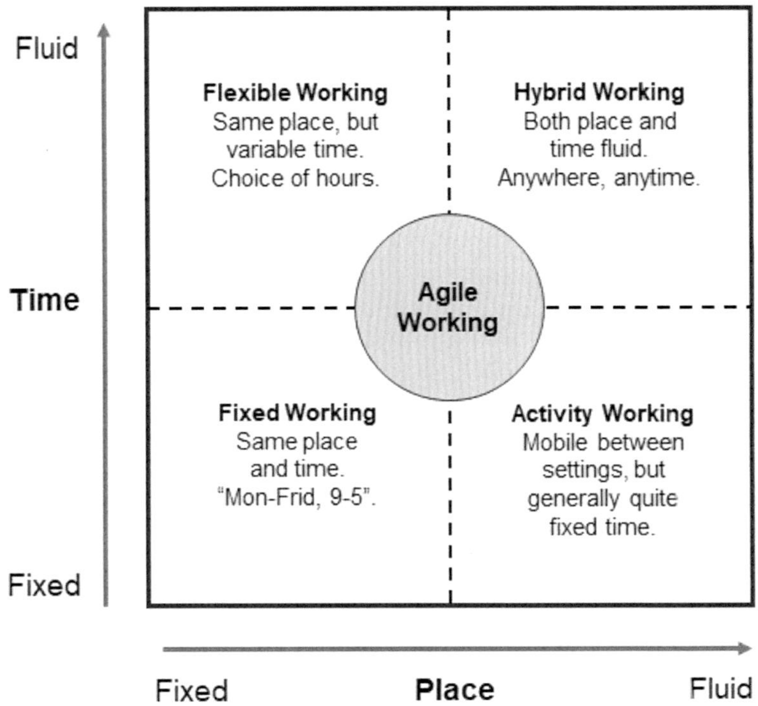

Figure 14.5 Agile working: time and place; fixed to fluid.

modes – concentration, collaboration, socialising, and so on. Even so, the office remained the anchor for most activities, with an element of 'third place' working.

Hybrid working, of course, is the label attached to the now widely anticipated practice of blending working in the office with working from home (or nearby). Both time and place are fluid. In this sense, we do not have hybrid workplaces or flexible time, just hybrid work. Hybrid work forms a bridge between flexible time and fluid place. Some people might adopt a blend of flexible and hybrid; others a blend of activity and hybrid.

The office today

The Big Bang design agenda of the 1990s established a dominant building form that was enormously complex internally. The hermetically sealed buildings had centrally controlled environments – to control air, temperature and light – and were packed with information and communications technology infrastructure – cabling, ducting, risers, raised floors and so on – all laid out across very deep and cavernous floors. Yet despite their modernity, and the fact that they were a direct response to new economic circumstances, there were hiding an irony.

One of the underlying tenets of the new design agenda was flexibility, the notion that – in sharp contrast to the buildings of the corporate era – buildings should be able to respond to change. Hence the large open floorplans, the lack of sub-division allowed different layout permutations, and the capital expenditure on specification to ensure that no demand on power, air or heat would be too great. The irony is that, just as all the capacity and technical enablement was being installed, digital technology was quietly simplifying work, lowering the demands that workers placed on buildings and reducing the need for floors the size of soccer pitches.

In this way, as the implications of knowledge work began to emerge in the 1990s and 2000s, so it became clear that buildings were changing more slowly than work, organisations and workers. The coming of knowledge work, with fundamental changes to the economy, to business and to work as outlined here, brutally exposed the very different rates at which buildings and their occupiers were evolving. In earlier times, when buildings were smaller, simpler and less capital intensive, refurbishment and replacement could be effected quickly in response to changes in business processes. This was no longer the case.

As long ago as 1993, Duffy *et al.* began *The Responsible Workplace* with the statement that something "*is very wrong with office buildings today*". They based this statement on the belief that there had opened up a divergence, which they were certain would increase,

> *between what people, individually and collectively wish office work and the office environment to be and what the majority of offices are actually like – highly conventionalised stereotypes, the meaning of which had hardly been challenged for decades.*[27]

The authors suggested a number of ways in which the office environment would have to evolve.

- Workplaces will have to be designed to accommodate much greater flexibility in the use of space and time, to provide more rapid responses to business and operational needs, and to allow employers to adjust to employees' rising expectations of workplace quality.
- New workplaces and systems for managing the work environment will have to be designed, which constrain users much less, and which enhance the quality of working life as well as productivity.
- Workstations will be freed from the locational constraints of the traditional office; more kinds of work will be possible in more places; the free-address environment will become viable for many.[28]

It is worth emphasising that these observations were made at a time, around 1993, before the impact of email, the World Wide Web and smart phones was even partially understood.

Duffy also observed that the changing nature of demand would affect all offices. "*With far higher stakes at risk, neither conventional North American nor northern European office buildings will be adequate to meet the number and complexity of the demands being made on office space.*"[29] Twelve years later Duffy would lament, in a vivid manner, that the situation had not improved:

> *It is hardly an exaggeration to say that the collective failure to understand the relationship between the working environment and business purpose puts us in the position of early 19th century physicians, with their limited and erroneous notions about the transmission of disease before the science of epidemiology had been firmly established.*[30]

Despite its apparent modernity the 'new office' buildings continued to be a static backdrop to work rather than an active agent of its execution. But despite such reservations, office work and the office workplace continued to evolve and were engaged in an irreversible process of change driven by emerging technologies. The role of the office was becoming one of enabling people to interact and collaborate; providing a palette of settings in which to work. In other words, the office was becoming less a place to go to work on a set of prescribed tasks, and more somewhere to visit and interact with colleagues.

Just as the economy could be described as weightless and dealing in intangibles, so office work became weightless and intangible. Workers were making fewer demands on the physical environment and the work process was digitised rather than arranged around static equipment. Networks, relationships and insight formed the new infrastructure of work rather than buildings, services and furniture.

As knowledge work spread, and as organisations evolved from traditional corporate structures into digitally enable organisations, so the role of the workplace, and its design, in supporting organisational effectiveness, or productivity, came under the spotlight of senior management teams, who began to view work differently. Whereas in the past the workplace was seen by senior management simply as a leaden, inert and inflexible cost of doing business, it was now coming to be recognised as a strategic resource that could be actively managed to improve the productivity and wellbeing of the valuable and costly assets which it houses: the knowledge workers.

The rise of the knowledge economy and knowledge work fundamentally changed our relationship to the office workplace. Previously, people commuted there to access the paraphernalia for work – phone, personal computer, filing and stationery – increasingly they were doing so for different reasons. The office workplace evolved from being a static backdrop for process-based, largely routine and solitary work, to an increasingly actively curated environment, managed more like a hotel than a traditional office, with a high level of service and experience for 'guests'.[31] It began to provide for novel activities, including socialising and collaboration, as well as offering management a mechanism for nurturing a common purpose. The workplace began to integrate work processes, technology and people in a more seamless manner. There was a growing appreciation that the workplace could contribute to corporate performance: it could be managed to support the technology-enabled, agile, digital workplace.

While the relationship between work and the workplace continued to evolve through the digital era, with a growing sophistication in terms of how managers and designers sought to manage the interface, there was emerging an even stronger force for change in the nature of the office economy. This was a social change rather than a technological one, but one with equally profound implications for the role and function of the office workplace. Indeed, Drucker argued that when the history of our time is written from a long-term perspective, it is likely that the most important event historians will see is not technology, it will instead be

> *an unprecedented change in the human condition. For the first time – literally – substantial and rapidly growing numbers of people have choices. For the first time, they will have to manage themselves.*

As we saw earlier, at the turn of the century, Peter Drucker was highlighting the degree of choice available to workers compared with pre-digital times. Knowledge workers are distinct from office workers of earlier times because they are – to a greater or lesser extent, dependent upon role – independent of place. Of course, as has already been noted, the new freedoms come at a price: the comfort and security of corporate employment are sacrificed, while the relationship between worker and employer is more transactional than in the past.

Knowledge workers are as light as the weightless economy in which they work – able to work anywhere because they use mental processes rather than physical processes to complete their tasks. They offer their skills to an employer who will engage them on far more flexible terms than, say, the corporate era. The notion of working for a single employer, through a career, is anathema. Knowledge workers take more control of their engagement with work and the workplace.

The flip side to these trends is, of course, further disruption of the corporate model of management. Knowledge workers are the antithesis of Balzac's pygmies and Wright Mills' demiurge.

Notes

1 Drucker PF (1954) *The Practice of Management: a Study of the Most Important Function in American Society* New York, Harper & Brothers p284.
2 Drucker PF (2000) Managing Knowledge Means Managing Oneself *Leader to Leader* No. 16 Spring 2000 The Peter F Drucker Foundation for Nonprofit Management.
3 Duffy F (1997) *The New Office* Conran Octopus Limited, London p49.
4 Drucker PF (1992) The New Society of Organisations *Harvard Business Review* September-October pp95–104.
5 Brinkley I (2008) *The Knowledge Economy: How Knowledge is Reshaping the Economic Life of Nations* The Work Foundation p10.
6 Coyle D & Quah D (2002) *Getting the Measure of the New Economy* The Work Foundation, London.
7 Brinkley I (2008) *The Knowledge Economy: How Knowledge is Reshaping the Economic Life of Nations* The Work Foundation p9.
8 Quah D (2019) The Invisible Hand and the Weightless Economy In: Fouquet R (editor) *Handbook on Green Growth* Edward Elgar, Cheltenham p468.
9 Haskel J & Westlake S (2018) *Capitalism without Capital: the Rise of the Intangible Economy* Princeton University Press p239.
10 Peters T (1992) ABB's Prophet of Smallness Points Way to Profit in the Nineties *The Times* 11th November p25.
11 Heller R (1994) The Managers' Dilemma *Management Today* January pp42–48.
12 Cited in: The Horizontal Corporation *Business Week* 20th December 1993 pp44–49.
13 American Management Association, cited in: Instant Coffee Management Theory *The Economist* 25th January 1997 p83.
14 Carlin J (1996) Guru of Downsizing Admits he Got it all Wrong Reported in the *Independent on Sunday* 12th May p20.
15 Handy C (1990) *The Age of Unreason* London, Arrow Books Limited.
16 Castells M & Hall P (1994) *Technopoles of the World: The Making of 21st Century Industrial Complexes* London, Routledge pp3–4.
17 Harris R (2015) The Changing Nature of the Workplace and the Future of Office Space *Journal of Property Investment & Finance* Vol 33 Issue 5 pp424–435.
18 *The Economist* (2015) Reinventing the Company *The Economist* 24th October.
19 Duffy F (1974) Office Design and Organisations: 2: the Testing of a Hypothetical Model *Environment & Planning B* Vol 1 No 2 pp217–235.
20 Duffy F (1997) *The New Office* Conran Octopus Limited, London p53.
21 Davis G; Becker F; Duffy F & Sims W (1985) *ORBIT 2: Organisation, Buildings and Information Technology* Harbinger, Norwalk, Conn Cited in Duffy F; Laing A & Crisp V (1993) *The Responsible Workplace* Butterworth-Heinemann Oxford.
22 Stone PJ & Luchetti R (1985) Your Office is Where You Are *Harvard Business Review* March-April pp102–117 p102.
23 Duffy F (1992) *The Changing Workplace* Phaidon Press, London p180.
24 Cairncross F (1997) *The Death of Distance* Orion Business Books.
25 Veldhoen E (1995) *The Demise of the Office* Uitgeverij, Rotterdam.
26 Laing A; Duffy F; Jaunzens D & Willis S (1998) *New Environments for Working* Routledge, London.

27 Duffy F; Laing A & Crisp V (1993) *The Responsible Workplace* (1993) Butterworth-Heinemann, Oxford p1.
28 Duffy F; Laing A & Crisp V (1993) *The Responsible Workplace* (1993) Butterworth-Heinemann, Oxford p1. pp2–3.
29 Duffy F (1997) *The New Office* Conran Octopus Limited, London p43.
30 Cited in CABE & BCO (2005) *The Impact of Office Design on Business Performance* CABE & BCO London p2.
31 Brunia S; De Been I & van der Voordt T (2016) Accommodating New Ways of Working: Lessons from Best Practices and Worst Cases *Journal of Corporate Real Estate* Vol 18 No 1 pp30–47.

15 The office in the network era

The office economy has passed through four broad eras, as was shown in Figure 1.2. The market era involved coffee houses and exchanges, as well as nascent banks and insurance firms whose owners and clerks often lived 'over the shop'. The clerical factory era saw the rapid growth, mechanisation and feminisation of offices and the emergence of the grandiose and ostentatious buildings of the joint stock companies in the mid- and late-nineteenth century.

This was followed by the corporate hegemony of the twentieth century, the rise of pyramidal hierarchies and the fealty of 'company man' in a relatively brief period of stability for the office economy. Finally, the 1990s saw the arrival of the digital era, when the office economy embarked upon a phase of transformation defined by technology-enabled knowledge work. The precise boundaries between these eras are less important than the fact that each can be seen to have a distinct set of social and economic circumstances.

It is the contention here that the office economy is on the cusp of entering a fifth era, that of the network era. This final chapter switches from an historical account of the first four eras of the office economy, office work and the office workplace to an assessment of the implications of recent trends and what these mean for the emerging network era. The chapter begins by summarising economic and business features of the network office, before turning to the nature of work and the form of the workplace. The chapter closes with some thoughts on how the traditional real estate supply process – involving owners, developers, designers and brokers – will need to adapt to the rapidly changing office economy.

Business: death of the demiurge and the rise of networks

The UK Nobel prize-winning economist, Ronald Coase (1910–2013), wrote in 1937 that the raison d'etre of the company is that costs attached to doing business are more effectively and efficiently dealt with by a single decision-making structure. His paper, *The Nature of the Firm,* provided an academic rationale for corporate organisations, which vertically integrated supply chains to reduce frictional costs.[1] But what happens in a weightless economy, where trade in intangible goods and services dominate? Where are the frictional costs in transactions involving bytes rather physical processes? What is the rationale for corporate structures in a digital economy?

While the concept of the company is something that is taken as a given today, this study has shown how its form has changed over time. What is now recognised as a company emerged in the seventeenth century, going through several evolutionary phases, including the creation of joint stock companies in the nineteenth century, and the rise of corporatism in the twentieth. The latter, with its labyrinthine organisational structures and the concept of 'company man' reached its maximal phase in the 1980s. The lesson is that the company is not a static

DOI: 10.1201/9781003582205-20

phenomenon; nor is it of a single form. Organisational structures, management philosophies and worker expectations have displayed significant cultural differences in different countries at different times. The past is not a guide to the future.

The corporate era saw organisations become ever larger and more complex. Sprawling international organisations brought together thousands of sedentary, clerical workers into single headquarter buildings. However, these 'corporate islands' were looking somewhat redundant in the digital era. Lower capital costs, growing connectivity, artificial intelligence and social changes (such as the aspirations of the digital generation) were all starting to negate the relevance of rigid corporate structures. This fact was thrown into sharp relief in 2020, with the onset of the COVID-19 pandemic.

Impact of COVID-19. The post-COVID period threw into sharp relief a previously hidden struggle between workers and what Wright Mills described with some disdain as the management demiurge. Suddenly, in 2020, workers were instructed to work from home. A workstyle that had previously been resisted by countless senior managements who relied instead on line-of-sight management to ensure that everyone was working effectively. But what the pandemic experience demonstrated was that, at least in the short-term, most office organisations could continue to function effectively without everyone working synchronously in the same office each day.

In 2024, the 'working from home' debate is continuing to rage and there remains widespread resistance to returning to the *status quo ante*, with mainstream media outlets regularly carrying articles highlighting instances of managements' attempts to introduce 'return to office' mandates and workers' refusals to fall into line. What seems most likely at this stage is that, for most people, a 5-day week in the office is a thing of the past. Hybrid working, in which workers and managers agree a balance of time between home and office, appropriate to role and responsibilities, looks set to become the new normal.

Fully remote working is rarely satisfactory for either employee or employer. While there are some roles where such an arrangement works, for most employees, it can lead to detachment from the corporate organisation, lost training and development opportunities and even to loneliness and overwork. For the employer, it can mean a weakening of corporate culture and direction, or purpose, as well as, possibly, a loss of productivity. While the COVID-19 experience demonstrated that there is no longer a need for everyone to attend the office 5 days a week, it did not suggest that we no longer need offices.

While synchronous work is no longer mandated by most employers, there is a natural tendency to co-locate, at least some of the time, if not most. Being alongside colleagues enhances socialising, sharing knowledge and experience, mentoring, training, collaboration, corporate opportunity, meeting and so on. Workers like to socialise and employers wish to create a common purpose. Similarly, the home is often not set up for work for one person, let alone two or more. For many, the quality of public realm, access to shops, leisure and services, and office support in the form of nutrition and wellbeing, are often far superior in the office and its locale to that which is available in the home environment.

It is likely that a highly variegated pattern of workstyles will emerge, crafted to suit, and importantly, balance specific company and individual needs. Most workers will find a blend of working from home, working in activity-based settings in the office and working from 'third places' that suits their individual and team preferences. It is unlikely that more than a handful of companies will go entirely remote.

Hybrid working can lead to much greater flexibility for the worker alongside cost and time savings associated with travel to work; while the employer should have a more satisfied and therefore more productive workforce. All of these benefits and disbenefits are hard to prove decisively but, intuitively, hybrid working is widely regarded as a good compromise.

Death of the demiurge. Given that hybrid work becomes the norm, businesses will need to adapt their approaches to management. While the shortcomings of traditional corporate management were evident before the pandemic, in many organisations the corporate structures and management cultures of yesteryear live on today, particularly in the largest organisations and through much of the public sector. Many workplaces remain ossified in rigid socio-economic hierarchies, graded according to the authority granted to each level to plan and execute their own work while planning and supervising the work of others who do not plan or make decisions. In such organisations, management seeks to instil a common corporate culture, almost a sense of identity, often using the language of the sporting arena: all aligned to a shared set of goals. In the worst cases, old style management techniques are being shoehorned into the new style of work.

As hybrid working evolves such styles of management will look increasingly anachronistic. For example, in extremis, workers are having their keyboard activity monitored and face-to-face management has simply transferred to the screen. Such approaches will evolve as management becomes less a function centred on supervision, discipline and control, and more one that focuses instead on mentoring and support, training and development and on creating communities of shared interest and a positive work experience. Fairness and inclusion, agreeing practices and protocols and communication, and engaging and co-ordinating will become critical management skills.

In the network era there will be fewer workers justifying layers of management; organisations will be flatter and workers, many of whom are specialists, will be more accustomed to self-management. In the network era, the demiurge will lose much of its meaning. Workers do not need to be corralled and closely supervised; many can work independently. Harry Braverman's distinction between conception (thinking, planning) and execution (doing) of work no longer holds sway because, in the network era, the difference between the two functions has faded.

But management is not dying altogether; it is evolving. Its role is shifting from one of controlling workers to managing relationships. Richard Donkin summarised the notion elegantly, arguing that while the co-ordinating role of management remained important, other functions such as planning and organisation are increasingly being handed over to employees.

> *But there is another significant role for managers working closely with professional employees who by the nature of their work will exercise high degrees of self-management. Advertising, education, the media, marketing, finance and research and development are all sectors where this relationship is evolving. This is the concept of the customer-manager, whose relationship with the professional is transactional rather than commanding.*[2]

Fewer workers feel part of fixed hierarchies, with rigid reporting lines, where they are treated according to their role which determines their perceived status. Such organisational features assume, implicitly, that workers are generally lazy and need to be managed to perform. There is, in short, a growing mismatch between the traditional imperatives of management and the new aspirations of workers.

One of the most powerful outcomes of the digital era of the office has been the democratisation of work. The historic relationship between 'boss and clerk' and then the social structure of corporate organisations with layers of deference being expressed between different classes and roles was replaced with a workforce that increasingly exercised choice and managed itself. Growing numbers of technology-enabled workers have been exercising choice in how, when and for whom they work, and many have chosen to control their own destinies by setting up their own small businesses.

This process will extend in the network era as more self-managing specialists replace hierarchies and task-driven workers. Self-managing individuals and specialists within organisations, alongside contingent specialist workers, all bound together by a common sense of purpose and shared interest will predominate (see 'Career versus contract' below). Traditional notions of company structures and career paths will be swept away. Networks will be the defining feature of the office economy, underlining the need for the CBD to offer more than single-use office blocks (see below).

Self-managing specialists will demand a different approach from management. Instead of a one-size-fits-all style, organisations will need to offer personalised support and development. In some senses, the emphasis will shift from managing people to suit the organisation's objectives to managing people to suit individual objectives.

Corporate islands to networks. A defining feature of corporatism was pyramidal departmentalisation – the tiered separation of functions into distinct areas of responsibility (finance, legal, payroll, human resources, marketing, training and so on), within which workers 'progressed' their careers through gossamer-thin layers of seniority. The corporate era saw organisations become ever larger and more complex to meet the demands of the rapidly evolving, global trading environment, and the office economy emerged as a distinct facet of the city – no longer an ancillary function to industry, but an activity in its own right. This was the time when sprawling international organisations brought together thousands of sedentary, clerical workers into single headquarter buildings – Mumford's '*human filing cases*' for the '*circumspect care of paper*'[3] – *where* command-and-control management oversaw unchanging and predictable corporate organisations, which planned ahead with a comparatively high degree of certainty, providing jobs for life. Corporate organisations were, in many respects, inward looking, impermeable and self-sufficient.

The digital economy questioned traditional assumptions about the role of the firm, and as the digital office evolves into the network office era, there is likely to be a decline in 'corporate islands' in favour of interdependent businesses, knowledge-based work and increasing specialisation. Business ecosystems will involve large and small companies working together, with contingent workers and specialists across extensive and collaborative relationships. The critical nature of connectivity, changing structures and the priorities of knowledge workers are all altering the role and form of the corporation.

These trends have huge implications for the real estate industry. Traditional notions of company structures and career paths are being swept away. Just as the previous office era was defined by technology, so the coming era will be defined by the management of networks, relationships and knowledge. The impact on the workplace is potentially profound: instead of managing a fairly uniform, compliant and sedentary work force, managers find themselves managing a mobile, demanding, disruptive and discerning workforce. The approach to the design and management of the workplace cannot remain the same.

While in the past new businesses had to invest in expensive and long-term assets, including ICT systems and real estate, today's technological landscape enables them to compete globally with relatively little capital outlay, using cheap and ubiquitous technology – *without creating vast hierarchical armies* and departmental structures. The network era will allow individuals and micro-businesses to trade their intellectual capital using cheap and ubiquitous technology. They can incorporate online and raise capital through crowdfunding. They can find skills around the globe. In larger organisations, working relationships will form and unform as required, driven by project and task needs; and competitors will collaborate.

The company continues to need its legal protection, but for growing numbers, there is no risk and reward issue, and few transaction costs. The raison d'etre of the firm is shifting to something more ephemeral – recognising the shifting priorities of workers: the desire for a community of relationships. Whether working in a micro-business or as part of a larger one, more and more people want to share purpose, values and experience rather than office politics and a bonus pot. To accommodate this, companies will be more fluid, porous and adaptive, encouraging networks within and between businesses.

One of the defining characteristics of the network era will be change – constant and unpredictable change. Organisations will need to be capable of responding and adapting, and networks make this more achievable than traditional pyramidal hierarchies. Networks contain far less inertia and nodes can be manoeuvred to suit changing business priorities.

Place versus culture. Despite the great leaps that were made in the design and management of offices during the digital era, there is an on-going debate about worker dissatisfaction, which in turn feeds a never-ending appetite among workplace professionals to reinvent the office. Complaints about the workplace are most commonly reported as the outcome of workplace satisfaction surveys, where workers are encouraged to self-report the quality of their workplace experience. The outcome is almost always the same – people are dissatisfied! But, as we have seen, this has been the case for over two centuries. Both Dickens and Lamb referred to the relentlessness of work on wooden stools at high benches, which hardly compare to today's latest Herman Miller's Aeron chairs.

Over the past three decades, the quality of the typical office workplace has improved greatly. Far more resources are expended on the design, the layout and the management of the workplace than ever in the past. The dynamic workplaces described earlier, with the examples of the SAS and BA headquarters, were early examples of the more user-focused approach to offices. And yet, a theme running through countless articles, papers and commentary over the past 4 years is that satisfaction with offices has not increased at all. There is clearly something more than the physical workplace that is failing.

It is suggested here that today's workplace dissatisfaction arises not so much from workplace failings as from management, or work culture failings. The fact is that workplace satisfaction surveys utterly fail to question workers about management regimes, team dynamics, career paths and workloads (the 'work culture') – reflecting the gap that often exists between workplace planning and other corporate resources areas such as IT and HR. Such surveys measure the workplace 'hardware', but they do not apply the same rigour to the workplace 'software'. Yet the influence of the latter over the former can be overwhelming.

As we saw with Reisman's thesis in Chapter 10, corporate office bureaucracy entails individuality being sacrificed to the corporate body as goals and direction were provided by 'others', the 'it' that is management, rather than being driven by the inner self. Moreover, people felt corralled into a fixed pattern of work where they were tightly managed, or controlled. But the fact is that traditional corporate management continues to live on in many organisations through the digital era. Work has moved on; management has not.

In wider society, notions of conformity, uniformity, docility and deference have long gone, and status and social stratification are much less important today. So if none of these features applies in wider society, why should they do so in the office? The answer is that before the pandemic, there was no widespread questioning of the typical work pattern; corporate managements knew how to run their organisations, and that was along the corporatist model. The pandemic lockdowns demonstrated beyond doubt that working from home was practical and, for many, preferred; and there has ensued a battle of wills between management and the managed, with 'return to office' mandates being widely ignored. The genie is out of the bottle. People are not refusing to return to the *status quo ante* because they are dissatisfied with the air conditioning, or the lighting or the distractions. These are surrogate complaints for the real issues which are a dissatisfaction with work itself and work culture.

Many businesses have a singular focus on profit (topic for another discussion) but then camouflage this in a saccharine covering of corporate culture – hard and soft benefits designed to 'attract and retain' the 'best talent', while nurturing a shared sense of purpose. But no amount of designed 'workplace experience' will encourage people to return to the Monday-to-Friday, 9-to-5 routine, because it will not overcome a deep sense of antipathy toward traditional management structures and corporate culture. Office workers have had a taste of freedom, enjoyed the comfort and security of their homes; benefited from the reduced time and cost of commuting and enjoyed opportunities to socialise and exercise in ways not previously possible. Unless there is a shift in management culture, there is unlikely to be any profound change in satisfaction or in office attendance patterns.

Work: why, what and how?

This book has traced office work over a 350-year time span. This has been characterised by periods of intense change, such as the closing decades of both the nineteenth and twentieth centuries, and periods of less intense change, such as the post-war years. Mechanisation, feminisation and digitalisation of the office were three signature symptoms of rapid development. Economic and social context has also been a defining factor; the most notable example of which has been the parallel development of the European cultural office and the North American rational office. What is abundantly clear is that throughout the 350-year time span office work has continually evolved. The dawn of the network era suggests a further period of rapid development, involving a reappraisal of why we work and how it will be undertaken.

The work ethic: why work? This is a big question, and one that extends beyond the office economy. But there are emerging trends, which make the question relevant to our enquiry here. Studs Terkel opened his book *Working* with a memorable statement: "*This book, being about work, is, by its very nature about violence – to the sprit as well as the body*". And he was not finished there: he suggested that to "*survive the day is triumph enough for the walking wounded among the great many of us*" imploring that work should be

> *about a search for daily meaning as well as daily bread, for recognition as well as cash, for astonishment rather than torpor, in short, for a sort of life rather than a Monday through Friday sort of dying.*[4]

This quote from a half century ago was relevant a century *before* it was written and appears to be equally relevant half a century *later*. We saw how the clerks of the eighteenth and nineteenth centuries laboured in poor conditions, in no small measure because they were early subscribers to the Protestant work ethic, a belief that work was a rewarding activity in spiritual rather than financial terms. Sociologist, philosopher and political economist Max Weber (1864–1920) is widely known for his book *The Protestant Ethic and the Spirit of Capitalism*, in which he described work as a calling: "*labour came to be considered in itself the end of life, ordained as such by God... .Unwillingness to work is symptomatic of the lack of grace*".[5] More recently, Richard Donkin wrote in *Blood, Sweat and Tears: the Evolution of Work*:

> *... .most of western society has become immersed in the Protestant work ethic – an ethos that has defined work for many people for hundreds of years, creating the belief among most of us that work is toil, that it is actually something we would rather not be doing but that we know we must do, nevertheless, because therein lies salvation; there is virtue in its accomplishment.*[6]

Donkin separates the ideology of the Biblical guidance from the practice: "*the example, the determination to pursue a moral code, to find salvation through work*".[7] From the clerks of the eighteenth and nineteenth century through to the office workers of the corporate world, work was imbued with the Protestant work ethic. But does this remain the case?

According to the Government's statisticians, around 22% of working-age adults in the UK are not actively looking for work. Put differently, in 2024, 9.4m working aged people are neither in work nor looking for a job (while 32m are in work). While the numbers include students, carers and the long-term sick, the total is still more than 859,000 higher than before the pandemic. Of significance, the data suggest that there has been an increase in people aged 16 to 34 ceasing work, suggesting that people in their early-20s were more likely to be not working than those in their early-40s.[8] Similar trends exist in mainland Europe and America.

The question here is whether the growing number of people electing not to work – including both those retiring early and those at the beginning of their working lives – are a symptom of a growing disillusionment with the concept and practice of work – a rejection of the work ethic. In earlier times, a career in the office was an aspirational opportunity, one that could offer job security, development and social recognition; and workers would accept varying levels of privation in exchange. If the office is no longer a magnet to people who wish to build a career and their social standing, or if they are no longer prepared to accept the privations, then tomorrow's office workplace might need to appeal to different motivations.

What will work look like? As quoted earlier, as late as the mid-1970s, the role of the office was described as a combination of receiving, recording, arranging and giving information; and a place where "*people sit at desks or tables, equipped or supported in their work by paper and pens, typewriters, dictating machines, telephones, telex, calculating machines and more advanced data-processing machines*".[9] With only the lightest of editing, this statement by Pete Daniels could have made about the 1870s. And the same holds true for more recent times. While the digital era has ushered in new work landscapes and new styles of work, the experience has not been the same for everyone: many people continue to be sat at desks equipped with machines (laptops) and performing a small function within a large bureaucracy.

While workplace professionals wax lyrical about designing workspaces for creativity, collaboration, innovation and serendipity at the water cooler, the plain fact is that much office work remains dull, repetitive, reactive and routine and often accompanied by unrealistic targets and unsympathetic management, just as it has always been. Often times, work is solitary, sedentary and screen-based: it encourages (in an echo of Charles Lamb) a sense of weariness, even of confinement. Even for relatively well-paid jobs undertaken by qualified workers, office work is often lacking in meaning. This phenomenon was captured recently by David Graeber in his book *Bullshit Jobs*. It is an uncomfortable situation to recognise, let alone to admit to being part of, but many jobs in the modern office economy simply don't matter; if they ceased to exist, no-one would really notice.

> *Those who work in bullshit jobs are often surrounded by honour and prestige; they are respected as professionals, well paid and treated as high achievers – as the sort of people who can be justly proud of what they do. Yet secretly they are aware that they have achieved nothing to earn the consumer toys with which they fill their lives; they feel it's all based on a lie – as, indeed, it is.*[10]

A little bleak? Perhaps. But not all office work is a daily journey of joyous innovation and creativity. The modern office environment, the work experience, despite its technological advancement, has singularly "*failed to address the growing bundle of problems that are arising from flattened organisations, greater numbers of mobile workers, information overload and time-based stress.*"[11] For many, work resembles Studs Terkel's daily grind within corporate anonymity – the sense of being a very small cog in a very large, impersonal machine.

There is little doubt that a potentially large number of jobs exist in the long shadow of AI, and will disappear as suddenly as clerks, file managers and secretaries did. Others are just pointless, giving the worker little satisfaction or sense of purpose. Either way, we have a rump of workers who gain little from their daily interaction with the workplace or their co-workers. They don't need a new workplace design; they need new work.

In addition, while technology was widely anticipated to have a liberating effect on workstyles, for many the reverse has been true. Rather than lessening the hours worked and yielding more leisure time, laptops, the internet and mobile phones have all served to increase the time spent working (if not productivity).

If the reasons for going to an office have changed, then what the office has to offer, in terms of amenities, services, relationships and so on, will itself have to change. Work in the network era must nurture socialisation, wellbeing and positive experience; there must be a sense that work is helping individuals achieve their goals – business and personal. The office and work within it will become a more personalised experience. Artificial intelligence, smart building technology and Apps-based user interfaces will combine to allow workers to plan their visits to the office: to find appropriate work settings at set times, to set up meetings with co-workers, to access concierge and technical services, to access information and to diarise time for wellbeing and nutrition (either in the building or in the locale).

Access to knowledge. As jobs are reorganised and reinvented, the manner in which knowledge and information is accessed, transmitted and analysed is shifting radically. Again this has profound effects on traditional relationships in the office (peer-to-peer and manager-to-managed) as well as where, when and how work is undertaken. Take, as an example, the armies of professional workers who make up a large proportion of the office economy workforce. In *The Future of the Professions*, Richard and Daniel Susskind provide a comprehensive analysis of how technology is likely to have a profound impact on how professional services are organised and accessed. The authors state as their main thesis that "*we are on the brink of a period of fundamental and irreversible change in the way that the expertise of these specialists is made available in society*". Furthermore, in the long-term, "*we will neither need, nor want professionals to work in the way that they did in the twentieth century and before*".[12]

While noting that the "*professions have been the main channel through which individuals and organizations have gained access to certain kinds of knowledge and expertise*", the authors argue that "*increasingly capable machines, operating on their own or with non-specialist users, will take on many of the tasks that have been the historic preserve of the professions*", which will eventually lead "*to a dismantling of the traditional professions*".[13] The process is expected to comprise both automation in which "*traditional ways of working will be streamlined and optimized through the application of technology*" and innovation, whereby "*systems will transform the work of professionals, giving birth to new ways of sharing practical expertise*".[14]

While the Susskinds focus on traditional professions (including law, medicine, teaching and divinity) the processes and outcomes they describe, of course, broadly apply to the wider business services aspects of the office economy, such as business consulting, finance, insurance and marketing. While not forecasting a widescale decimation of white-collar work jobs, the Susskinds are certainly pointing to widescale rearrangement of jobs, and that "*even as certain professions may wane, others may indeed emerge – so the longstanding framework whereby groups of people have to reskill from one era to another may apply*".

Career versus contract. It has long been the case that 'job security' is in decline. Certainly, in white-collar work since the recession of the early-1990s, the search for corporate efficiency has been relentless and job turnover has risen. It was here that the fashion for corporate re-engineering, championed by management gurus such as Michael Hammer first emerged. The overriding messages were about 'delayering' (dispensing with hierarchies) and 'outsourcing' (contracting out non-core functions to specialist providers). The impact on the office workforce was huge; up to this point, mass lay-offs were something that happened in manufacturing due to automation and so on, not something that happened to office workers. Yet this was simply the first step in a dramatic reorganisation of white-collar work that is still unfolding.

A key outcome has been a shift in the relationship between employer and employee: from one based on a desire for long-term commitment by both sides, towards a recognition that job interviews are about agreeing medium-term contracts. Increasingly the traditional career profile, of graded progression through labyrinthine layers and structures, all with perhaps just one or two employers over the course of a career, is now gone. Office workers now serially change jobs to secure greater opportunity and reward for themselves.

Many companies now engage a growing contingent workforce including contactors, part-timers, consultants, freelancers and so on. What were once jobs, are now projects, contracts or programmes. For these workers, the employer–employee relationship has shifted from symbiotic to transactional: they have no emotional investment in the corporate body; the office is no longer their second home. Many more workers now decide how many days they work each week, where the work is done, for what reward and how much holiday they take. Many of these workers are self-employed, either working in isolation or as part of peer networks, or virtual communities, where individuals coalesce to form project teams and then disperse on project completion.

It is to be guessed that most of the self-employed choose that particular employment status and are therefore reasonably content. But they will lose the benefits they previously enjoyed in terms of steady employment and pay, pension contributions, paid holiday, a career structure and wellbeing and other support. It must also be recognised that while for many such workers, these arrangements are a more flexible, self-determined and rewarding experience, for others it is a more precarious, marginal and unpredictable existence. Either way, mutual commitment, trust and loyalty are harder to develop, investment in career development shifts from a corporate to a personal responsibility and workers have to reskill more frequently.

Workplace: clustering, condominiums and landscapes for work

While the nature of office work and the manner in which it is conducted is changing, the fact of office work looks set to continue. Furthermore, it is proposed here that offices will continue to provide the main platform for work – recognising that the role and use of the place will evolve. Myerson and Ross call for a reimagining and re-engineering of the workplace – not just an iteration of the past, but a radical reset "*free from the vestiges of our industrial past*". The authors go on

> *A new approach to defining work and the workplace is needed; one that is based on the freedom to create an experience based around what people are trying to achieve. By putting people at the heart of the way we work, by recognising their preferences ... we can create a new vision for work more suited and aligned to the digital age.*[15]

Clustering for collaboration. The early-eighteenth century saw the birth of the company, the spread of coffee houses and growth of merchants and their associated exchanges. A financial revolution occurred, featuring the Bank of England, the Stock Exchange, merchant banks and insurance companies. A set of interdependencies developed, and the early office economy became physically concentrated. By the twentieth century, these physical concentrations had become much larger and infinitely more complex. The standard 9-to-5 working day and 5-day working week took root and became an unquestioned feature of office working life – until the digital economy introduced the potential for agile working, which the pandemic made the new normal.

The invisible glue holding the office economy together in dense clusters is known in economics as 'agglomeration'. Whether in accounting, advertising, banking, consulting, finance, legal, marketing and media; or in coding, developing, graphic designing, programming and software engineering; or in architecture, construction, design, engineering, planning and surveying, the office economy is bound together in enormously complex webs of relationships, involving partnering, collaborating, networking and socialising, and which drive innovation and higher output. At their most basic, the webs form vast job markets and supply chains; at their most sophisticated they form sustainable and competitive business ecosystems, or networks.

As we know, cities have maintained their centripetal pull on business and people for at least three centuries. Indeed, before the pandemic there was evidence of companies being attracted to move *closer* to the centres of cities where the labour pool is infinitely deeper, more highly skilled and more mobile than in suburban centres and smaller regional cities. Despite this, the arrival of digital technology has, since the 1970s and longer, spawned a rich seam of predictions about the dispersal of economic activity, notably the 'death of distance' discussion: if information is equally available everywhere, at all times, then why be in the city centre when we can be telecommuting from a bucolic idyll of our choosing?

As early as 1970, James Martin and Adrian Norman suggested that we might see a return to cottage industry, with the computer terminal replacing the spinning wheel, perhaps leading to a future where "*some companies may have no offices*".[16] In Daniel Bell's *Post-industrial City*[17] and Jean Gottman's *Transactional City*[18], the morphology of cities was expected to change radically with technological 'advancement'. In 1980, Alvin Toffler foresaw the evolution of the "*electronic cottage*" as part of "*a new emphasis on the home as the centre of society*".[19] Frances Cairncross' *Death of Distance* suggested that any activity that relied on a screen or a telephone can be carried out anywhere in the world.[20]

The fact that this has not happened on a widespread basis (apart from in some transactional environments) suggests more subtle forces are at work. One of the attractions of the 'death of distance' narrative is that it overcomes many of the issues around commuting and concentration, with city centres commanding high rents and labour costs. And of course this is attractive: why commute for an hour when tech allows you to work in the suburbs? Why pay high rents downtown, when tech enables a move to a regional town where the rent bill is 50% less? However, networks are not replicable in a fully dispersed world. It is something of a paradox that, since the dawning of the digital age, the pulling power of some city centres has strengthened rather than weakened.

Mass transit makes cities more sustainable in terms of bringing white-collar workers to their workplace, and the critical mass of jobs helps support what economists call 'externalities' – additional benefits in terms of recreation, culture, shops, restaurants and so on. Business networks utilise physical networks (e.g. infrastructure) and build social networks based on support networks and communicating via virtual networks. The higher the city's density the more intense the networks (compare US cities with Asian cities).

Urbanist Edward Glaeser is positive about the future of cities: "*a few decades of high technology can't trump millions of years of evolution*", going on to suggest that connecting in cyberspace "*will never be the same as sharing a meal or a smile or a kiss. Our species learns primarily from the aural, visual and olfactory clues given off by our fellow humans.*"[21] More seriously, Glaeser points to the paradox in which "*the declining cost of connecting over long distances has only increased the returns to clustering close together*", while humanity's ability to rise to new challenges "*needs all the strength it can muster, and that strength resides in the connecting corridors of dense urban areas*".

While agglomeration is likely to continue as a feature of the office economy in the network era, it is also clear that workers will be exercising greater control and choice over where they work. We are likely to see agile working become embedded as the normal pattern of work, with people spending more time working from home and at other locations. The net result will be that most firms reduce their physical footprint while maintaining a presence in their business cluster, but they are unlikely to dramatically reduce their demand for real estate because the extent to which agile working is practiced by individuals will be limited by some basic needs:

> *The sheer efficiency and economy of direct personal contact is always being undervalued – until people try to go without it, because electronic communication is not a like-for-like substitute, and may never be.*[22]

The centripetal glue that has held business together in tight city clusters for centuries is likely to remain a strong force. The forces that encouraged people to move away from city centres during the pandemic will wane significantly, as a new generation of younger workers continues to enjoy the benefits of working and living in relative proximity. Having said this, both working and living patterns will continue to evolve.

In the network era, the previously relentless physical growth of the office economy will slow down. Evolving workstyles have already broken the direct correlation between headcount and square feet/metres of space. As buildings are used in a far more dynamic manner, there will be an overall reduction in previous levels of demand. One consequence is that the office economy will no longer be the mainstay of demand in city centres. It will become concentrated in cities where critical mass creates efficient job markets and choice, and where office businesses benefit from deep, skilled labour pools.

Despite the positive attributes of agglomeration and the sustainability arguments in favour of offices in city centres, there will need to be very significant change. One of the key lessons of recent years has been the fragility of city centres dominated by a monoculture of business. As both the office economy and the retail economy have been undergoing fundamental change, those cities that have proven to be the most resilient are those with a mixed economy of business, housing, leisure, culture and so on. Many cities suffering the severest effects of post-pandemic hybrid working are those with the densest concentrations of 'office monoculture'. In the network era business buildings must become more permeable, more mixed use in character. They will need to be designed so that they are capable of being used for a variety of purposes, rather than for the single purpose of office work.

From castles to condos. In 1996, the author wrote a paper titled, *Less a Castle, More a Condominium*, which argued that offices should be designed, delivered and managed to provide services, amenities and experience to 'guests'.[23] This shift of emphasis implies a change in priorities for the real estate manager from *building* performance to *workplace* performance, from managing desks to welcoming guests. Old style, janitorial facilities management provision will yield to workplace management that provides concierge-type services for employees with a more diverse range of priorities:

- a workplace to support corporate agility and enable change;
- space as a hub and connector to reinforce corporate culture and values;
- a workplace that brings people together to forge a sense of shared objectives;
- design for continuous adaptability and diverse usage patterns;
- creating a memorable experience to attract talent and nurture skills;
- providing amenities and services (food, wellbeing, events etc.), and
- activity-based spaces for collaboration, concentration, creativity and confidentiality.

Weightless and networked firms do not benefit from controlling physical assets; their operational needs are generic, and they attach little symbolic meaning to the exterior of their building. Yet, they are looking out to a property supply process that weighs a great deal indeed. This is most easily seen in a supply process that delivers large buildings – all of a very similar specification – with a view to letting them to single occupiers with strong covenants. Large, impermeable, single-use buildings (castles).

But this is changing. The weightless, networked economy suggests that the asset-driven approach of real estate will become an historic novelty. Where once a building was seen as a castle, in the network era it will be used more like a condominium. It exists to meet a customer need, and everything around its design, delivery and management is focused on that need.

Landscapes for work......more cultural than rational. Office layouts have undergone a number of iterations from almost entirely enclosed to almost entirely open plan. As we have seen various attempts have been made to formalise approaches, not least Bürolandschaft, the Action Office and activity based working, but these were pre-digital and pre-knowledge work. There is a need to explore and find new work landscapes for the network era.

Oseland describes many modern workplaces in terms of 'workplace zoos', drawing a parallel with the poorly designed and poorly managed zoos of the past, with their caged animals in barren surroundings.

> *The Workplace Zoo is the modern interpretation of the open plan office. A high density, overcrowded space with serried rows of desks, few partitions, limited facilities and poor environmental conditions. The focus is foremost on space and cost efficiency rather than designing and planning for human needs to enhance the wellbeing and performance of its occupants.*[24]

Oseland suggests that now is the time to "*remodel the office to address our psychological and physiological needs ... . The design should allow office inhabitants to flourish and thrive rather than simply survive*".[25] He proposes an alternative approach, the Landscaped Office, which builds *"on the original open plan concept Bürolandschaft reinterpreted with a contemporary twist and including elements of the Action Office"*.[26]

The office workplace will continue to have an important role, albeit one that will change significantly. During the past two decades, there has been a slowly growing appreciation that the workplace could contribute to corporate performance, and the role of the office has come to be understood as one of enabling people to interact and collaborate; providing a palette of settings in which individuals and groups can work in more dynamic ways compared with much of the more solitary work of the past. In other words, the office has become less a place to go to work on a set of prescribed tasks, and more somewhere to visit and interact with colleagues.[27]

The emerging priority for tomorrow's workplace will be to energise and motivate people, and to enable them to be productive and effective; it will provide for social interaction and knowledge transfer; it will provide workers with more personal control and it will provide a wide range of support services. It will form part of a more distributed work pattern in which the home, local offices, public transport, client premises, indeed just about anywhere, is used for work.

To respond to this changing role, office workplaces will become less rigid, less formal, less rational. They will bear more resemblance to Bürolandschaft than to the systematised Action Office. There will be a much richer palette of work settings, with traditional desk space taking up a shrinking amount of the typical space budget. The proportion of non-desk space, such as collaborative space and meeting and social space will increase, in order to support a higher level of knowledge sharing and interaction. In parallel, an increase in agile working will result in higher utilisation of desk space.

The supply process: customer focus, demand driven and flexible

Traditionally the property supply process has delivered large, single-use buildings – all of a similar specification – with a view to letting them to single occupiers with strong covenants. For most of the twentieth century, owner occupation and long leaseholds were common, as buildings provided financial meaning in the form of assets and security; operational meaning in the form of bespoke environments and symbolic meaning in the form of architecture. But as the knowledge economy spread, such meanings have largely disappeared. As already noted, many 'weightless' firms do not benefit from controlling physical assets.

In the network era, buildings will need to meet a customer need, there will be greater provision of serviced, or flexible, space, with occupiers signing licences rather than leases, with easy in–easy out terms, supported by service provision, and there might even be serviced apartments for workers to occupy when they stay in town. Growing numbers of companies are opting for this kind of flexibility to allow them to respond to a rapidly changing business environment.

Tenants or customers? The customers of the real estate industry are those organisations that occupy space provided by owners and operators. Their demands on real estate have shifted such that they no longer benefit from controlling physical assets and their operational needs are generic. They are footloose, weightless and agile.

The challenge for the real estate industry, in the face of these contextual changes, is to become an agent of change, not a barrier. And this means that old attitudes and perceptions will need to change; new skills and technologies will be required. Actors in the real estate process will need to balance the traditional skills associated with managing assets with a growing skill base of people trained in service provision and customer satisfaction. And yet, a quarter way through the twenty-first century, they look out to a supply process that continues to use the feudal language of the nineteenth century – *landlord and tenant*.

The changes in the occupational market described here are not forces that the real estate industry can choose to engage with or not. They are a reality. Investors, facility managers, asset managers, brokers, designers and owners – everyone involved in the real estate process – will need to respond and evolve. Nothing less than a shift in culture is required, and that process will be made simpler with a change of language.

For the past 8 decades, the real estate industry has created product, which has been wholesaled either to client-side property teams or intermediaries representing the client body. In both cases, the needs of individual workers were overlooked. Now, as in-house client-side teams shrink and/or disappear, and as property itself is commoditised, the real estate supply process must 'retail' its products and services to the customers who use their workplaces.

The workplace experience and the manner in which it is curated will assume greater importance. Obviously design will continue to provide innovative solutions to emerging problems; but the real focus in the coming years will be a management issue. How workplaces are managed will assume critical importance, not only in an era of heightened health awareness, but also in an era when space is a commodity that customers turn on and off as their businesses require.

Alongside the growth in flexible space, the biggest step that the real estate supply process will need to take in the future is to transition from a culture based on managing assets to one providing services to businesses, or customers. As described above, offices in the network era will resemble condominiums rather than castles; work space is becoming 'commoditised', and its operation and management will more closely resemble that provided in the hotel and leisure industries rather than that of traditional asset management.

Such a shift of mindset might seem radical in today's market; but seen in the context of the 350 years of change described in this book, it is simply a next logical step.

Supply or demand driven process? The changes that are running through the office economy today represent a period of extreme turbulence: yesterday's truths, assumptions and norms are becoming tomorrow's archaic practices. And this means that old attitudes and perceptions must change.

The physical origins of the office economy lie in domestic premises and the Lombard Street tradition with a 'shop' on the ground floor and the clerks living in the upper floors. As merchants became bankers so demand for larger premises grew. In the mid-nineteenth century, developers began to speculate – creating buildings for the open market, for banks to lease discrete floors. But as banks grew in size, matching the rapid growth and development of their colleagues in insurance, so a new building form emerged. Well capitalised joint stock companies began to commission their own buildings, with which they sought to express their sense of their own scale, importance and solidity with increasingly grandiose and ostentatious architecture.

In the mid-twentieth century, the supply process became dominated by banks, insurance companies and pension funds, and in the 1980s, the market internationalised and property funding did likewise, with global investors, sovereign wealth funds and so on. They demanded a form of building designed and managed for long-term occupation by relatively unchanging organisations, and including long-leaseholds and onerous terms. The result was an industry that had two customers – the investment market and the occupier market.

The problem is that the interests of the two parties are different. The occupier customer is looking for flexibility, service, choice, the ability to adapt their occupation to suit changing business needs; while the investor customer is looking for long-term, secure income streams and capital growth. For this, they demand uniformity of product and a dominant grip over the supply process. The two positions are mutually exclusive, but it can be of little surprise that the motives and priorities of the investor customer have until now held sway, while the needs of the occupier customer have taken second place.

The seismic shifts in the economy that have resulted in the changing nature of office occupier businesses, as described in this book, are demanding a new approach – one that is less supply driven and more aligned to the needs of the occupier customer.

The rise of the knowledge economy has exposed the real estate industry to the fact that it is providing space for occupier customers, not faceless investors. And firms' expectations of their workplaces today are a world away from those of even 20 years ago. It is likely that the investment model will have to evolve into one that is based on business-to-business relationships rather than on feudal terms.

One way in which the supply industry can begin to respond is by transitioning from an asset valuation process based on capital growth to one based on quality of income. In this way, owners and providers would be rewarded for service provision and quality of management. Another response will be to embrace the approach taken by the flexible space market.

Flexible and on-demand. As occupier requirements become more fluid and as the SME sector demands a higher quality product and service, so real estate will be procured more like other commodities: by the year, the month, the day. While having been available since the 1990s, the flexible office market expanded rapidly during the early-2000s in direct response to such growing demand. As an alternative to the traditional long lease with onerous responsibilities, landlord and managing agent model, flexible space is offered by an operator on short-term conditions, often accompanied by office services and amenities. The flexible space market provides choice and flexibility both for growing numbers of small businesses seeking to have a presence in the commercial property market on terms that suit their business models, and for larger corporate organisations that wish to optimise organisational agility by occupying real estate on less onerous terms than those offered by traditional leases.

Traditional owners have resisted the flexible space market and will want to continue to do so. But there is an inevitability about economic and technological progress, which suggests they will need to change. One month before the first COVID-19 lockdown, the author wrote that the flexible space market

> *has the potential to become the dominant way in which the property industry conducts its business. Accepting that there will always be demand for long-term space from large corporates looking for core estate security, the rest of the market could evolve to FSM in one form or another.*

Going on to predict that flexible space would cease to be the exception and instead become the new normal, the paper went on to suggest

> *there will need to be much deeper changes to the supply chain – particularly in the surveying and facilities management professions – in order to supply the resources and skills for the expanding market. Traditional asset management skills will need to be balanced with service culture skills.*[28]

The real estate supply process will need to adapt to the new reality with products and services more akin to consumer markets than corporate markets. Small firms are driving demand for flexible space; many are no longer satisfied with secondary space on secondary streets. They aspire to the quality of space that their corporate cousins occupy. And the flexible space market is providing that – along with intensively managed, commodity space which they can turn on and off as their business cycles demand. Eventually, as the flexible market matures from niche to mainstream, the market will provide space that will be as easy to hire as an Airbnb stay: an office will be ordered, created and delivered via a credit card, in an 'Easy Office' version of 'Easy Hotel'.

A final word: from markets to networks

This history of the office began with the market era, with coffee houses and exchanges at the heart of the nascent office economy. This was followed by 350 years of evolution in business, in work and in the workplace: the factory era saw mechanisation of office work, the corporate era saw the emergence of labyrinthine, bureaucratic structures and the digital era enabled the arrival of agile working. Throughout this time, the economy, business, work and the workplace have co-evolved – sometimes smoothly, sometimes less so as lags between innovation and response occurred.

As we stand on the cusp of a new era – posited here as the network era – it is clear that there are strong parallels between the distant past and the immediate future. As corporate business structures and management styles adapt to the demands of the network era, and as the hardware of the digital era becomes either invisible or mobile, so work will more resemble that of the coffee houses and exchanges than of the intervening phases.

The network era will demand a thorough reappraisal of how we engage with the workplace and this will of itself, demand innovative approaches to the design, delivery, occupation and management of offices. As well as providing physical places from where company plans, priorities and purpose can be directed, and places where work 'can get done', the role of offices in the network era will include providing a venue for meeting, collaborating, sharing, learning and socialising.

But adapting the physical workplace to new modes of work is relatively straightforward. The key lesson from this history of the office economy is that the management of business will need to adapt radically in the network era. The COVID-19 pandemic laid to rest the notion that all workers need to attend the office at the same time in a regimented manner to be tightly managed, or controlled.

Social attitudes to work are also changing and the notion of working for a single employer over decades is rapidly disappearing. Growing numbers of workers now 'self-manage' and exercise choice over where, when and how they undertake their work. In addition, the 'corporate islands' of past times are becoming more permeable, with large numbers of contingent workers, consultants and advisors, and complex supply chain relationships that involve countless meetings.

Wright Mill's demiurge will need to yield to a management culture that supports innovation, adaptability and business agility. Management must move from the constraints of efficiency and cost management to one of support and development. The new sense of choice that hybrid working has given people has aligned work more closely with people's broader life experience: while work was seen to be a constraining activity, other aspects of life are liberating. Organisations will need to nurture a sense that the workplace will actively provide support, amenity and service, rather than being perceived as a place of control.

It seems that, for the foreseeable future, we will continue to need office work, but the manner in which it will be undertaken will change.

Once management has gone through the necessary phase of reinventing itself then it will become clearer what kind of physical space is required to support its goals.

Notes

1 Coase R (1937) The Nature of the Firm *Economica* Vol 4 No 16 pp386–405.
2 Donkin R (2001) *Blood, Sweat and Tears: the Evolution of Work* Texere, London and New York p278.
3 Mumford L (1940) *The Culture of Cities* (2nd Edition) Secker and Warburg, London.
4 Terkel S (1972) *Working: People Talk About What they do All Day and How They Feel About What They Do* Pantheon Books, New York pXI.
5 Weber M (1930) *The Protestant Ethic and the Spirit of Capitalism* 11th Impression, 1971 Unwin University Books, London p159.
6 Donkin R (2001) *Blood, Sweat and Tears: the Evolution of Work* Texere, London and New York pXXIV.
7 Donkin R (2001) *Blood, Sweat and Tears: the Evolution of Work* Texere, London and New York p49.
8 BBC (2024) More the a Fifth of UK Adults Not Looking for Work www.bbc.co.uk/news/business-68534537
9 Daniels PW (1975) Office Location: An Urban and Regional Study G Bell & Sons, London pp4–5.
10 Graeber D (2018) *Bullshit Jobs: A Theory* Simon & Schuster, London p15.
11 Cited in: Office of the Future *Financial Times Survey* 23rd September 1998 ppI–IV.
12 Susskind R & Susskind D (2015) *The Future of Professions: How Technology will Transform the Work of Human Experts* Oxford University Press, Oxford p1.
13 Susskind R & Susskind D (2015) *The Future of Professions: How Technology will Transform the Work of Human Experts* Oxford University Press, Oxford p2.
14 Susskind R & Susskind D (2015) *The Future of Professions: How Technology will Transform the Work of Human Experts* Oxford University Press, Oxford p271.
15 Myerson J & Ross P (2022) *Unworking: The Reinvention of the Modern Office* Reaktion Books, London p203.
16 Martin J & Norman AJ (1970) *The Computerised Society: An Appraisal of the Impact of Computers on Society in the Next Fifteen Years* Englewood Cliffs, NJ, Prentice-Hall p32 and p150.
17 Bell D (1973) *The Coming of Post-industrial Society: A Venture in Social Forecasting* Basic Books, New York.
18 Gottmann J (1983) *The Coming of the Transactional City* University of Maryland Institute for Urban Studies Monograph Series No2 College Park.
19 Toffler A (1980) *The Third Wave* William Morrow and Company Inc, New York p210.
20 Cairncross F (1997) *The Death of Distance* Orion Business Books.
21 Glaeser E (2011) *Triumph of the City* Macmillan p248.
22 Haigh G (2020) *The Momentous, Uneventful Day: A Requiem for the Office* Scribe Publications, London p113.
23 Harris R (1996) *Less a Castle, More a Condominium: Taking a Look at the Future of the Office* Gerald Eve Chartered Surveyors Private publication.
24 Oseland N (2022) *Beyond the Workplace Zoo: Humanising the Office* Routledge, Abingdon, Oxon pp7–8.
25 Oseland N (2022) *Beyond the Workplace Zoo: Humanising the Office* Routledge, Abingdon, Oxon p182.
26 Oseland N (2022) *Beyond the Workplace Zoo: Humanising the Office* Routledge, Abingdon, Oxon p184.
27 Ramidus Consulting (2015) *Future Workstyles and Future Workplaces in the City of London* Corporation of London p6.
28 Harris R (2020) Property Ownership in a Flexible World Investment Property Forum www.ipf.org.uk/resourceLibrary/property-ownership-in-a-flexible-world-pdf.html

Index

FSC
www.fsc.org
MIX
Papier | Fördert gute Waldnutzung
FSC® C083411

Kolibri 360 GmbH
Ferdinand-Jühlke-Straße 7
99095 Erfurt, Deutschland
produktsicherheit@kolibri360.de